AF760824

Springer Theses

Recognizing Outstanding Ph.D. Research

Aims and Scope

The series "Springer Theses" brings together a selection of the very best Ph.D. theses from around the world and across the physical sciences. Nominated and endorsed by two recognized specialists, each published volume has been selected for its scientific excellence and the high impact of its contents for the pertinent field of research. For greater accessibility to non-specialists, the published versions include an extended introduction, as well as a foreword by the student's supervisor explaining the special relevance of the work for the field. As a whole, the series will provide a valuable resource both for newcomers to the research fields described, and for other scientists seeking detailed background information on special questions. Finally, it provides an accredited documentation of the valuable contributions made by today's younger generation of scientists.

Theses may be nominated for publication in this series by heads of department at internationally leading universities or institutes and should fulfill all of the following criteria

- They must be written in good English.
- The topic should fall within the confines of Chemistry, Physics, Earth Sciences, Engineering and related interdisciplinary fields such as Materials, Nanoscience, Chemical Engineering, Complex Systems and Biophysics.
- The work reported in the thesis must represent a significant scientific advance.
- If the thesis includes previously published material, permission to reproduce this must be gained from the respective copyright holder (a maximum 30% of the thesis should be a verbatim reproduction from the author's previous publications).
- They must have been examined and passed during the 12 months prior to nomination.
- Each thesis should include a foreword by the supervisor outlining the significance of its content.
- The theses should have a clearly defined structure including an introduction accessible to new PhD students and scientists not expert in the relevant field.

Indexed by zbMATH.

More information about this series at http://www.springer.com/series/8790

Sarah A. M. Loos

Stochastic Systems with Time Delay

Probabilistic and Thermodynamic Descriptions of non-Markovian Processes Far from Equilibrium

Doctoral Thesis accepted by
Technical University Berlin, Berlin, Germany

Springer

Author
Dr. Sarah A. M. Loos
Institute of Theoretical Physics
Leipzig University
Leipzig, Germany

Supervisor
Prof. Sabine H. L. Klapp
Institute of Theoretical Physics
Technical University Berlin
Berlin, Germany

ISSN 2190-5053 ISSN 2190-5061 (electronic)
Springer Theses
ISBN 978-3-030-80770-2 ISBN 978-3-030-80771-9 (eBook)
https://doi.org/10.1007/978-3-030-80771-9

© The Editor(s) (if applicable) and The Author(s), under exclusive license to Springer Nature Switzerland AG 2021
This work is subject to copyright. All rights are solely and exclusively licensed by the Publisher, whether the whole or part of the material is concerned, specifically the rights of translation, reprinting, reuse of illustrations, recitation, broadcasting, reproduction on microfilms or in any other physical way, and transmission or information storage and retrieval, electronic adaptation, computer software, or by similar or dissimilar methodology now known or hereafter developed.
The use of general descriptive names, registered names, trademarks, service marks, etc. in this publication does not imply, even in the absence of a specific statement, that such names are exempt from the relevant protective laws and regulations and therefore free for general use.
The publisher, the authors and the editors are safe to assume that the advice and information in this book are believed to be true and accurate at the date of publication. Neither the publisher nor the authors or the editors give a warranty, expressed or implied, with respect to the material contained herein or for any errors or omissions that may have been made. The publisher remains neutral with regard to jurisdictional claims in published maps and institutional affiliations.

This Springer imprint is published by the registered company Springer Nature Switzerland AG
The registered company address is: Gewerbestrasse 11, 6330 Cham, Switzerland

Für meine Eltern,

die mir gezeigt haben, dass die
Dinge es wert sind, aus mehr als einem
Winkel betrachtet zu werden.

Supervisor's Foreword

In the past few decades, the nonequilibrium behavior and thermodynamics of small systems, which are subject to strong fluctuations due to their coupling to an environment, has become a major focus of scientific interest. In fact, there have been significant experimental advancements in investigating and manipulating nano- and mesoscopic systems, examples reaching from the quantum scale (such as quantum-optical devices) to classical soft matter systems (e.g., colloids in suspension) and complex biological systems. At the same time, much progress in understanding such small, fluctuating systems from a thermodynamic point of view has been achieved, and powerful frameworks such as Stochastic Thermodynamics are now well established. This includes the description of systems manipulated by an external agent, that is a "feedback controller", a famous example being a Maxwell Demon changing the entropy balance of a nonequilibrium system.

New theoretical challenges emerge when the fluctuating system is, in addition, subject to time delay. Indeed, a time lag between the signal detection and the control action is an ubiquitous feature in biological and neural systems. It also plays an important role in many experimental setups involving feedback control, such as a Brownian (colloidal) particle in an optical trap whose center follows the particle's position with some time delay. Thus, the impact of time delay, which was formerly mainly considered in quasi-deterministic systems arising in engineering (e.g., satellite or cruise control) or communication (laser-optical) networks, is now receiving increasing attention also in small, fluctuating systems.

However, even in the classical case, the mathematical description of time-delayed stochastic systems and their interpretation via thermodynamic notions such as entropy production, heat or information exchange is still in its infancy. Indeed, Stochastic Thermodynamics relies on the assumption that the underlying stochastic process is Markovian, that means, it does not have a memory. This assumption is clearly broken in the presence of a discrete time delay or an extended memory kernel affecting the system's dynamics. The methodological challenges in treating such non-Markovian effects become especially severe in presence of nonlinear forces, which are the rule rather than an exception in any realistic system.

The thesis by Sarah A. M. Loos substantially advances this field of research in several directions. The first major contribution concerns the description of time-delayed, stochastic systems by an evolution equation for the space- and time-dependent probability density. For theoretical calculations, this probabilistic (Fokker-Planck) level of description is often a useful complement to a (delayed) Langevin equation. However, already for seemingly simple cases, such as a single Brownian particle with discrete time delay, one is faced with an infinite hierarchy of Fokker-Planck equations which is typically not solvable. The thesis proposes new recipes to handle this problem, including a new, non-perturbative approximation scheme for the one-point density and the derivation of higher order equations via the powerful technique of Markovian embedding. Such probabilistic equations of motion and their approximate solutions are indeed crucial for investigating the dynamics of time-delayed systems via noise-averaged observables, and as an input for calculating thermodynamic quantities.

This brings us to the second main theme of the thesis: the investigation of thermodynamic notions in non-Markovian stochastic systems far from equilibrium. Among the most interesting results concern a paradigmatic nonlinear model system, a Brownian particle in a bistable potential with time-delayed feedback. It is found that time delay alone produces non-zero heat flow, a strong indication for nonequilibrium. Moreover, contrary to the traditional view of an externally driven particle heating up its environment, there can be heat flow from the bath to the particle; an observation, which is, in principle, experimentally verifiable. A further main achievement of the thesis concerns the calculation of the full entropy production and related quantities such as information exchange, which are particularly interesting in the context of feedback control. Due to non-Markovianity the standard approach to calculate entropy production can not be used; the challenge is the intimate relationship between entropy production and time asymmetry at the trajectory level. The thesis elegantly circumvents this problem by expanding the phase space; that is, by introducing auxiliary variables governed by Markovian equations of motion. This not only allows for an explicit inclusion of the "feedback controller" into the theoretical description but also paves the way to formulate entropy production and information exchange using tools from conventional Stochastic Thermodynamics. It turns out that delayed feedback is intimately linked to non-reciprocal coupling between the variables in the extended space, and to constant energy input, the latter being a characteristic property of active systems. Thus, these considerations reveal fundamental connections between memory, feedback and activity. As a matter of fact, the role of non-reciprocity in active systems, as well as connections between activity and feedback, are currently receiving tremendous attention. The thesis of Sarah A. M. Loos significantly advances our understanding of these connections from a thermodynamic point of view.

Berlin, Germany
November 2020

Prof. Sabine H. L. Klapp

Abstract

In addition to the ever-present random fluctuations due to *noise*, many systems in biology, physics and technology involve discrete *time delay*, stemming, e.g., from finite information transmission times. The combination of noise and delay yields *non-Markovian* dynamics describable by time-nonlocal Langevin equations with a delta-peaked memory kernel. Despite their omnipresence, the understanding of stochastic systems with time delay is yet in its infancy. This is especially true in the presence of *nonlinear* forces, which are ubiquitous in realistic situations. In this thesis, we consider a Brownian particle subject to nonlinear forces and time-delayed feedback. This represents an experimentally accessible and analytically treatable prototype of a non-Markovian, intrinsically nonequilibrium system. We consider two important cornerstones of statistical physics: *probabilistic descriptions* and *thermodynamic laws*. Both have proven themselves to be valuable frameworks to describe passive particles with time-local equations. And both are, so far, not well understood for time-delayed systems.

It is already known that the Fokker-Planck description for delayed systems is given by an *infinite hierarchy* of coupled equations. We compare different derivations of this hierarchy, and further present a new, simpler approach by means of a Markovian embedding technique. Further, we consider the first member of the hierarchy, which has been used in earlier literature as starting point for approximation schemes aiming for the one-time probability density of nonlinear systems. One conceptual disadvantage of the existing approximations is that they effectively render Markovian descriptions, thus, do not capture the essence of the process with memory. To tackle this problem, we propose a new scheme, called the force-linearization closure, which yields a non-Markovian description at the cost of negligence of nonlinear effects in higher-order terms. By application to a bistable and a periodic model, we demonstrate that the resulting one-time probability densities are more accurate than earlier approximations from the literature, in particular, the perturbation theory and the small delay expansion. We further show that the insights provided by the one-time probability alone are limited. This is because the dynamics is crucially determined by the memory, and, subsequently, the temporal correlations. In this spirit, we also shed light on the higher members of the hierarchy, and introduce a new approach aiming for the two-time joint probability density, allowing us to finally describe non-Markovian dynamical effects, e.g., delay-induced oscillations. To the best of our knowledge, this is the only approach on the Fokker-Planck level yet that captures this dynamical state.

Turning to thermodynamic quantities, we first focus on the net steady-state heat flow, which is accessible via the framework of stochastic thermodynamics. We show that the feedback inevitably leads to a finite heat flow, even for vanishingly small delay. We link the heat flow at small delay times to "entropy pumping", which is a phenomenon that is already established for Markovian systems subject to velocity-dependent driving. Application to our bistable model further reveals that the feedback induces heating as well as cooling regimes, and leads to a maximum of medium entropy production at coherence resonance conditions. We further discuss the possibility to quantify the irreversibility and the distance from equilibrium. Here, we immediately encounter a fundamental problem, i.e., the notion of total entropy production of non-Markovian systems. As an attempt to address this problem, we additionally include the feedback controller state explicitly in the model. To this end, we step away from the discrete delay and consider systems subject to distributed delay and colored noise. These systems can be represented (via Markovian embedding) by *Markovian networks*, which can be treated by the standard formalism. We demonstrate that a *non-monotonic* memory kernel in the Langevin equation is inextricably linked to *non-reciprocal* coupling. Further, we show that the non-reciprocity alone implies a finite entropy production. We derive a generalized second law involving information flow. Discrete time delay is recovered as the infinite-dimensional limit. Here the entropy production diverges, implying unbounded cost for precisely storing a Brownian trajectory. Our considerations further reveal connections between memory, feedback and activity, and show that these phenomena are all describable by the same unifying concept, i.e., non-reciprocal coupling.

Acknowledgements

This work was funded by the Deutsche Forschungsgemeinschaft (DFG, German Research Foundation)—Projektnummer 163436311—SFB 910.

First of all, I would like to thank Sabine Klapp for giving me the opportunity to be part of your group, allowing me to work very independently and autonomously and for supporting me with advice and guidance.

Warm thanks goes to Henning Reinken for supporting my work, and me personally, through all the years. Thank you for all the discussions and ideas concerning this work. I am very grateful for the incredibly nice experience of sharing the office with such a kind and clever person.

I would like to take the opportunity to thank Édgar Roldán, Roland Netz, Klaus Kroy, Frank Cichos and Viktor Holubec, Christian Kuehn, Hannes Stuke, Isabelle Schneider, Bernold Fiedler, Thomas Ihle and Aljaz Godec, for inviting me and giving me the opportunity to discuss this work with you and your groups. Furthermore, I would like to express my gratitude to Rainer Klages, Alexander Carmele and Gernot Schaller for continuous discussions and valuable advice.

Special thanks also goes to the students Jan Meyer, Simon Hermann and Timo Doerries, who I had the opportunity to co-supervise and mentor. It was a great experience to be continuously challenged by your questions, and to understand myself better through teaching you. I would further like to thank my colleagues for creating a warm work environment. Special thanks goes to Henning Reinken, Nima Siboni, Sascha Gerloff, Christopher Waechtler, Thomas Martynec and Maximilian Seyrich for various interesting discussions. I thank Florian Buchholz, Felix Herrmann, Henning Reinken, Thomas Martynec, Vera and Peter Loos, Yousof Mardoukhi,

Sascha Gerloff, Johanna Strunge and Maximilian Seyrich for giving me valuable feedback on this thesis.

Lastly, I would like to express my gratitude for the constant support by my family and friends. I am especially thankful to Felix Herrmann, who was always there when I needed him, and never let me down.

Publications by Sarah A. M. Loos

[1] Doerries, T. J., Loos, S. A. M. and Klapp, S. H. L. *Correlation functions of non-Markovian systems out of equilibrium: Analytical expressions beyond single-exponential memory*. J. Stat. Mech. 2021, 033202 (2021).

[2] Martynec, T., Klapp, S. H. L. and Loos, S. A. M. *Entropy production at criticality in a non-equilibrium Potts model*. New J. Phys. **22**, 093069 (2020).

[3] Loos, S. A. M. and Klapp, S. H. L. *Irreversibility, heat and information ows induced by non-reciprocal interactions*. New J. Phys. **22**, 123051 (2020).

[4] Loos, S. A. M. and Klapp, S. H. L. *Fokker-Planck equations for time-delayed systems via Markovian Embedding*. J. Stat. Phys. **177**, 95–118 (2019).

[5] Loos, S. A. M. and Klapp, S. H. L. *Heat flow due to time-delayed feedback*. Sci. Rep. **9**, 2491 (2019).

[6] Loos, S. A. M. and Klapp, S. H. L. *Force-linearization closure for non-Markovian Langevin systems with time delay*. Phys. Rev. E **96**, 012106 (2017).

[7] Loos, S. A. M., Claussen, J. C., Schöll, E., Zakharova, A. *Chimera patterns under the impact of noise*. Phys. Rev. E **93** 012209 (2016).

[8] Gernert, R., Loos, S. A. M., Lichtner, K. and Klapp, S. H. L. *Feedback control of colloidal transport*. In Schöll, E., Klapp, S. H. L. and Hövel, P. (eds.) Control of Self-Organizing Nonlinear Systems, 375–392 (Springer, 2016).

[9] Zakharova, A., Loos, S. A. M., Siebert, J., Gjurchinovski, A., Claussen, J. C., Schöll, E. *Controlling chimera patterns in networks: interplay of structure, noise, and delay*. In: Schöll, E., Klapp, S. H. L., Hövel, P. (eds) *Control of self-organizing nonlinear systems*, 375–392. Springer (2016).

[10] Loos, S., Zakharova, A., Claussen, J. C., Schöll, E. *Robustness of chimera states with respect to noise*. Proceedings of the 7th International Conference on Physics and Control, Istanbul (PhysCon 2015) (2015).

[11] Zakharova, A., Loos, S., Siebert, J., Gjurchinovski, A., Schöll, E. *Chimera patterns: influence of time delay and noise*. IFAC-PapersOnLine **48**, 007 (2015).

[12] Schneider, I., Kapeller, M., Loos, S., Zakharova, A., Fiedler, B., Schöll, E. Stable and transient multicluster oscillation death in nonlocally coupled networks. Phys. Rev. E **92**, 052915 (2015).

[13] Loos, S. A. M., Gernert, R. and Klapp, S. H. L. *Delay-induced transport in a rocking ratchet under feedback control.* Phys. Rev. E **89**, 052136 (2014).

Contents

Abbreviations and Symbols

Abbreviations

AOUP	Active Ornstein-Uhlenbeck particles
BD	Brownian dynamics simulation
CR	Coherence resonance
d.o.f.	degree of freedom
e.g.	Latin *exempli gratia*, which means "for example"
EP	Entropy production
EQ	Thermodynamic equilibrium
FDR	Fluctuation-dissipation relation
Fig., Eq., Sec., Ref.	Figure, Equation, Section, Reference
FLC	Force-linearization closure
FP, FPE	Fokker-Planck, Fokker-Planck equation
i.e.	Latin *id est*, which means "that is"
IFT	Integral fluctuation theorem
l.h.s., r.h.s.	left-hand side, right-hand side
LE	Langevin equation
MSD	Mean square displacement
NESS	Nonequilibrium steady state
NR	Non-reciprocal
OU	Ornstein-Uhlenbeck process
PDF	Probability density function
PT	Perturbation theory
(S)DDE	(Stochastic) delay differential equation
small τ	Small delay expansion
ss	steady state (used as subscript)
w.r.t.	with respect to

Symbols

t	Time, s
τ	Delay time, s
τ_{ir}	Intra-well relaxation time, s
τ_K, r_K	Kramers escape time, rate, s, 1/s
J	Probability current, 1/s
τ_B	"Brownian" time $\tau_B = \sigma^2/D_0$, *used as time scale*, s
x	Space coordinate, m
X	Stochastic variable, particle position, m
$\underline{X}$	Stochastic vector $\{X_0, ..., X_n\}$, m^{n+1}
$\mathbf{X}$, $\underline{\mathbf{X}}$	Stochastic path, i.e., random trajectory, m, m^{n+1}
ϕ	History function (initial condition), m
Ω	Domain of x, m
$C(t')$	Position-autocorrelation function, $\langle X(t)X(t+t')\rangle$, m^2
σ	Particle diameter, characteristic length of potential, *used as length scale*, m
ρ	Probability density function (PDF), 1/m
ρ_n	n-point probability density, $1/m^n$
ρ_c	Conditional probability density, 1/m
Z	Normalization constant, m
P	Probability, 1
$\mathcal{P}$	Path probability, 1/m, $1/m^n$
$\langle ... \rangle$	Ensemble averages, --
γ	Friction coefficient, kg/s
D_0	Diffusion coefficient, m^2/s
ξ	Gaussian white noise, $1/\sqrt{s}$
$\mathcal{T}$	Temperature, K
k_B	Boltzmann constant : 1.38×10^{-23} kgm$^2/(s^2K)$, kgm$^2/(s^2K)$
V	Potentials, kgm$^2/s^2$
V_0	Barrier heights, kgm$^2/s^2$
F	Forces, kgm$/s^2$
$k_B\mathcal{T}$	Thermal energy, *used as energy scale*, kgm$^2/s^2$
u, q, w	Fluctuating inner energy, heat, work, kgm$^2/s^2$
S_{Sh}, S_{tot}	Shannon and total entropy, kgm$^2/(s^2K)$
$\mathcal{I}$	Mutual information, 1
$\dot{I}$	Information flow, 1/s
N	Number of realizations (simulations), 1
Δx, Δt	spatial step size, temporal step size (simulations), in units of σ, τ_B

Chapter 1
Introduction

Non-Markov is the rule, Markov is the exception.

—Nicolaas G. van Kampen, 1998 [5]

Many natural and artificial systems are governed by dynamical equations which involve both, noise and time delay [6–8]. Noise due to imperfections or random environmental influences is essentially omnipresent in real-world systems and experimental setups. From a mathematical point of view, it often results from the presence of hidden degrees of freedom that have disappeared in the course of a coarse-graining procedure. Stochastic models that account for the noise are well known to be a powerful tool to effectively describe complex systems by using just relatively few, *mesoscopic* (stochastic) degrees of freedom. Such mesoscale models, for example the *Brownian motion* studied in this thesis, also play a central role in statistical physics. However, many theoretical concepts rely on the Markov assumption, neglecting the fact that realistic systems often have *memory*, i.e., history-dependence, due to, e.g., viscoelasticity [9–12], hydrodynamics [13–15], persistence [16–18], or due to *time delay*.

Time delay can have various sources. A prominent example from electrodynamics are the retarded potentials (e.g., Liénard–Wiechert potentials [19, 20]), where delay arises due to the finite speed of light. In biological systems, maturation times can create delays (as in population growth [21] and prey-predator dynamics [22]). On the biomolecular level, they arise in chemical reaction kinetics (e.g., cell metabolism [23–25]). Neural delays caused by finite refractory times [26–29] can even become apparent on a macroscopic level, as in the human pupil reflex [27], or in stick balancing experiments [1–4]. The latter example, which is sketched in Fig. 1.1, illustrates that delay can have a destabilizing effect (the stick can be stabilized longer in the upright position, if the reaction time delay is shorter). In socio-economic systems, infor-

© The Author(s), under exclusive license to Springer Nature Switzerland AG 2021

S. A. M. Loos, *Stochastic Systems with Time Delay*, Springer Theses,
https://doi.org/10.1007/978-3-030-80771-9_1

Fig. 1.1 A well-known control problem, where time delay matters: *balancing*. As an example, think of the attempt to balance a stick on your finger tip [1–4]. The delay τ is created within the control loop consisting of your eyes, the neural transmission to and back from your brain (or spine), plus the information processing and the muscle reaction time. As everybody knows, it is very difficult to balance the stick, but practice improves the performance, as the delay can be reduced

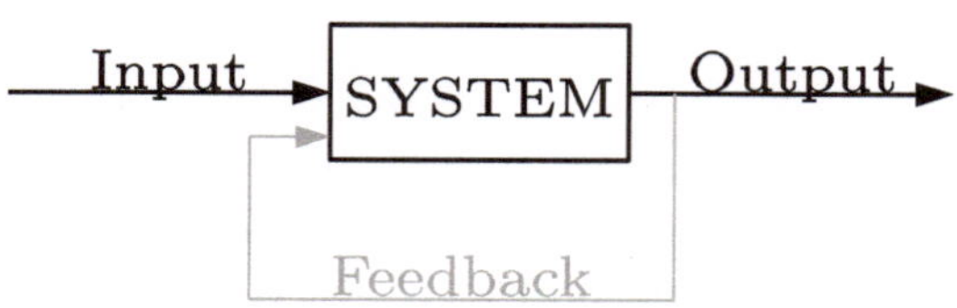

Fig. 1.2 The basic principle of feedback control: feeding back the output signal into a system, via a so-called feedback loop

mation processing and communication times create delays [30] (e.g., financial markets [31–33]).

Beyond these intrinsic delays, time delay is an important topic in control theory, particularly in the context of *feedback*, i.e., a driving that depends on an (earlier) system state [7, 34] (Fig. 1.2). Feedback protocols have an unavoidable time lag between signal detection and action of control [7, 8, 34–38]. An everyday example is the adjustment of the shower temperature (see Fig. 1.3). This example further illustrates that delay may induce oscillatory behavior, which is quite characteristic for delay dynamics. Similarly, they arise in laser systems (due to the traveling time of light in the cavity [38–40]), and quantum-optical systems (coupled to a structured photonic reservoir [41–43]). Because delays induce complex behavior like oscillations, chaos, and can even have stabilizing effects (e.g., on unstable orbits [44–46]), they are also implemented intentionally (as in Pyragas control [7]). Besides numerous technical applications, feedback is a very important mechanism that is naturally encountered in biological systems [47–52], as in gene regulatory networks [53–56]. An important example is the chemotactal motion of bacteria which relies on feedback

Fig. 1.3 An everyday control problem, where we experience the impact of delay: Adjusting the water temperature in your shower. Typically, there is a time lag between adjustment of temperature controller, and actual change of temperature. This can yield unpleasant oscillations between hot and cold (especially when the time lag is long)

which involves delay [47, 57, 58]. In fact, delay often occurs because of feedback, and feedback quasi always includes delay.[1]

In this thesis, we investigate a colloidal particle subject to nonlinear static forces and a position-dependent, time-continuously operating feedback force with a single, discrete delay $\tau \geq 0$. Experimentally, such a feedback can be realized with the help of a co-moving optical trap [59–64]. The control "target", i.e., the particle position, is accessible by an optical microscope due to the typical (micron-scale) size of a colloid [65–67]. The effect of the trap can be well approximated by a linear force towards the delayed position, with a tunable delay time[2] [59, 60]. We model the dynamics of the colloid with time- and space-continuous equations, and consider white noise. Due to the delay, the equation of motion, that is, the (overdamped) Langevin equation (LE), becomes time-nonlocal, or, in a more physical language, non-Markovian. This type of description is also appropriate to model various other systems, including some of the aforementioned examples; and this type of delay LE is subject of numerous recent theoretical works, e.g. [38, 68–87]. The discrete delay can further be considered as a special case of distributed delay, where the memory kernel is a delta-distribution around τ. From a mathematical point of view, the delay LE is a stochastic delay differential equation (SDDE). It is well known that SDDEs are *infinite-dimensional*[3] [7, 88]. The usual concepts and solutions for Markovian LEs then do not apply and the development of a general solution method for SDDEs is yet an unresolved problem. This is especially true in regard to the interplay of noise, delay, and *nonlinear* forces, which are omnipresent in realistic situations. Only for the special case of purely linear equations, closed-form solutions for probability density functions, position-correlations and stability boundaries were obtained in earlier literature [6, 75, 82, 87, 89–91].

[1] Sometimes its rather a matter of preference, which aspect is emphasized.

[2] A small time delay arises naturally, and it can further be artificially enlarged to a desired value within the resolution of the camera.

[3] As reflected by an initial condition with infinitely many degrees of freedom.

As we have seen, despite the fact that stochastic (nonlinear) delay systems are naturally encountered in various contexts, there are still many open problems in their mathematical description. Moreover, the incorporation of time delays, or generally memory, in the theoretical frameworks of statistical physics is still in its infancy.

Statistical Physics Perspective

There is indeed a class of (mesoscale) systems with noise and memory that is already well-established in statistical physics, which comprises, e.g., colloids in fluids with viscoelastic properties [9–12] or hydrodynamic memory [13–15]. This class of systems is typically modeled by generalized Langevin equations [92], which also feature memory, i.e., time non-local terms. Specifically, they involve a memory kernel in the friction force and finite noise correlations (colored noise) resulting from conservative interactions between a colloidal particle (or, more generally, the "slow" variable) and the bath particles. As a consequence, the delayed friction is related to the noise via a fluctuation-dissipation relation [93], which is closely connected to the fact that the system may approach thermal equilibrium (see, e.g., [12, 94–96]). It is important to note that the type of memory treated in this thesis is crucially different. A fundamental difference is that there is no relation between the delay and noise correlations. From the simple comparison with generalized LEs alone, one may already suspect that a system subject to a time-delayed force is *intrinsically out of thermal equilibrium*, even in the absence of additional drivings. This guess turns out to be true, as was previously shown [97], and as we will also demonstrate by various means throughout this thesis. The system is pushed out of equilibrium by a *history-dependent* driving, yielding a crucially different state than time-dependent (or constant) drivings, which are much better understood.

The property of being intrinsically out of equilibrium already on the level of individual constituents is a feature that feedback-controlled systems have in common with *active systems*. This class of systems, which includes robots, active microswimmers, or actively moving motor proteins like kinesin [98, 99], is a prime focus of the latest research in nonequilibrium statistical physics [16, 100–108]. There is another—less obvious—shared feature, that is, the presence of memory. While the discussion of memory, or delay, in feedback loops has a long tradition, this viewpoint is less common in the context of active motion. However, the importance of history-dependence becomes apparent in the form of persistence of active swimmers [16–18], the viscoelastic properties of active gels [15], or the giant Kovacs-like memory effect in the Vicsek model [109]. Despite these apparent similarities, the theoretical description has essentially been worked out separately. Only recently, researchers have started to realize the delicate connections between both [36, 48, 64]. For example, recent experiments [64] and theoretical studies [87, 110] have demonstrated that time-delayed feedback can be used to mimic active particles. We believe that a further cross-fertilization between the fields would be very important not only for the understanding of individual elements, but also for the collective behavior on larger scales; a topic that is already of major interest for active systems [17, 36].

From the viewpoint of statistical physics, one may consider time-delayed stochastic systems as well as active small-scale systems as "novel types of particles". Let us explain this in more detail. The theoretical description of the complex, living or artificial world surrounding us is the main objective of modern statistical physics. The building blocks of this world are subsystems which often operate far from thermal equilibrium themselves, already on the individual level.[4] Experiments and computer simulations show that such nonequilibrium systems display intriguing behavior from the individual to the collective level. For example, recent experiments [36, 64] and numerical studies [108, 111, 112] focus on the collective behavior of time-delayed systems, where delay often arises within the interactions between individual constituents. In [36], large swarms of robots with delayed communications were studied. Similarly, optical feedback was employed to stabilize "colloidal molecules" [64, 87]. These are molecule-like non-equilibrium structures consisting of a few Brownian particles, which exhibit bond length oscillations due to the delay. While these studies offer interesting insights into the statistical physics of nonequilibrium building blocks, the theoretical understanding is still in its infancy. This is because the common assumptions that are the basis of many established theoretical results are violated. Traditionally, the frameworks of statistical physics were designed for passive particles with time-local interactions, like hard disks, harmonic springs, billiard balls or Lennard-Jones particles [113], describable by Hamiltonian equations on the underlying, microscopic level. This implies *reciprocal* (conservative) interactions, which fulfill Newton's third law, *actio* = *reactio*. Such systems can reach thermal equilibrium, if no external driving prevents them.

In contrast to those passive particles, one may consider the aforementioned intrinsically nonequilibrium systems to represent a "novel type of particle". For such intrinsically nonequilibrium building blocks, the established frameworks of statistical mechanics need to be revisited. Theoretical challenges include the absence of an underlying Hamiltonian description, non-Markovianity, non-reciprocal interactions (which we will come back to at the end of this introduction), and the lack of a variational principle with suitable thermodynamic potentials. Colloids subject to (optical) feedback may be viewed as prototypes of a "novel type of particle", particularly useful to study the effect of non-Markovianity as the time delay represents a "tunable memory", which, by itself, already drives them out of equilibrium.

Two particularly relevant open problems in the context of statistical physics of such intrinsically nonequilibrium systems are the *probabilistic description* and the generalization of *thermodynamics*. Both, Fokker–Planck equations and thermodynamic laws have proven themselves of outstanding value for the development of traditional statistical physics of passive particles. Thus, it is natural to also reach out for Fokker–Planck descriptions and thermodynamics notions of time-delayed stochastic systems. This is the objective of the investigations presented in this thesis. In the following, we briefly lay out the main challenges and research questions.

Probabilistic Treatment

It is already known for more than 20 years that for systems with time delay, the Fokker–Planck equation (FPE) of the one-time probability density function (PDF)

[4] For such systems, the traditional picture of a (passive) system subject to external ("open-loop") forces is not appropriate, as the drive is *internal*, or in other words, correlated with the system state.

does not provide a complete, self-contained probabilistic description, because it explicitly involves the two-time PDF. In fact, it represents only the first member of an infinite hierarchy, mirroring the infinite-dimensionality of the underlying LE. Contrary to Markovian systems, the route from the LE to a FPE is more involved [80, 84, 85, 90, 114]. Therefore, writing down explicit expressions for the various members of this hierarchy is already a nontrivial problem. In earlier literature, we could find a general expression solely for the first member. We will consider this mathematical problem from different angles and explore strategies to nevertheless obtain analytical expressions for the FPEs themselves and the PDFs, specifically for *nonlinear* overdamped systems.

As mentioned above, the only class of delayed noisy systems where closed-form solutions are known, are those with linear forces. Here, all the n-time probability densities are given by multivariate Gaussian distributions. Exact solutions can be obtained on the LE [75] and the FPE level [90], or by utilizing both levels of description [82]. In the more general case of nonlinear systems, one has to rely on approximations [6]. The two most established strategies are the "small delay expansion", that is a Taylor expansion of the LE in τ [80, 81], and a perturbation theory on the level of the FPE [84, 85], where the entire delay force is treated as a small deviation from the Markovian dynamics. A conceptual disadvantage of these approaches is that they effectively render Markovian descriptions. This motivates us to ask: Is there a way to approximate the dynamics, but in a non-Markovian way? In this work, we will develop such an approach. The trick is to exploit the closed-form PDF solutions from the linear delay case. By application to two generic nonlinear systems, a colloidal particle in a sinusoidal or in a bistable potential supplemented by a linear delay force, we will compare our new approximation (called "force-linearization closure") against the earlier approaches.

We will further show that some important features of the dynamics are caused by the interplay of non-Markovianity and nonlinearities. For example, the memory of an escape event over a potential barrier can cause a subsequent jump, inducing spontaneous (delay- and noise-induced) particle oscillations between neighboring potential wells. To capture such complex dynamics, one actually also needs to consider higher-order correlations in time. In principle, these are described by the higher members of the Fokker–Planck hierarchy. The fact that explicit equations for the latter are, so far, not known hinders the development of further approximation schemes and solution methods. We will therefore shed light on the higher-order equations and the higher-order PDFs. To this end, we introduce a new derivation of the Fokker–Planck description by means of a *Markovian embedding* technique [95, 115–119] and a subsequent limiting procedure. The Markovian embedding is a generalization of the so-called linear chain trick [6, 120, 121], known in the context of deterministic delay equations. In this way, we find a closed, complete Fokker–Planck equation in an infinite-dimensional space, from which one can derive a hierarchy of equations. From a conceptual point of view, our approach yields an alternative perspective on the mathematical structure of delayed processes. Based on the second member, we will further propose an approximation for the two-time PDF, which is a central quantity in the description of delay systems, as it directly gives the correlation between the present and the delayed state. We will again test this approximation by application to the aforementioned bistable example model, this time focusing on the regime of delay-induced interwell-oscillations.

Thermodynamics

Another central framework in statistical physics is thermodynamics. In the last decades, special interest has been devoted to heat flow and other thermodynamic properties of small (mesoscale) systems, which are dominated by thermal fluctuations [122, 123]. *Stochastic thermodynamics* has emerged as an elegant and consistent framework to generalize thermodynamic notions to the level of noisy trajectories and to systems far from equilibrium [68, 124–130], with numerous applications to soft matter [104], biological [129, 131], and quantum systems [132, 133]. For Markovian systems, concepts like entropy production, fluctuation theorems, stochastic energetics, and information exchange have been successfully formulated (and tested experimentally [61, 134, 135]). From a technical point of view, the FPE and path integrals (and their direct correspondence to the LE) turn out to be very useful. For example, the *total entropy production*, i.e., the *irreversibility* measure, can be directly calculated from path probabilities, or via the PDF and the associated probability current [123]. For non-Markovian systems, where probabilistic descriptions are much more involved, all these concepts have to be revisited. While extensions towards several non-Markovian systems have indeed been carried out in the past [136–145], the application to time-delayed systems with discrete time delay is still in its infancy. But before we dig further into technicalities, let us first think about relevant physical questions and expectations. As is commonly done in stochastic thermodynamics, we will focus again on the long-time behavior, where the system approaches a steady state.

While the incorporation of delay is a major challenge and it is not clear what we shall expect, the control aspect of our system is already well understood. Control is a deep-rooted topic in thermodynamics. In fact, major theoretical advances where initiated in the course of the long-lasting debate about the "Maxwell demon" [134, 146, 147]. This is an imaginary creature that can, by means of a sophisticated control, extract energy from a single heat bath, which means it "defeats the second law of thermodynamics". Later, simpler devices have been proposed, like the Szilárd engine [146, 148]. Interestingly, in the literature one even finds time-delayed, position-dependent feedback protocols suitable for this task, for example the feedback ratchet from Ref. [149]. However, there are two aspects that are fundamentally different from our feedback control. First, in these previous models the control is applied only at solitary instants in time. Between the control action, the system evolves "freely" and is on these time intervals, in fact, Markovian. Second, the control protocols typically break the spatial symmetry of the system externally. For example, in the Szilárd engine and the feedback ratchet, the control is "on" or "off" depending on whether the particle is left or right of a certain position. We wonder: Can the time-continuously operating feedback control considered in this thesis nevertheless also be used for the task of energy extraction from a single heat bath? In our setting, this would correspond to a *reversed*, i.e., negative net heat flow, which we shall thus look for!

Recent research has made clear that the search for a thermodynamic description of systems subject to continuous time-delayed feedback is indeed quite involved, even in the absence of nonlinearities [91, 97, 114, 150]. For the special case of a *linear* delay LE, it has been established that the delay force induces a net heat flow in the steady state [97] (implying nonequilibrium). However, there are three open problems concerning this earlier study. First, the derived heat flow is claimed to be positive definite, with an explanation that is based on an incomplete entropic treatment.[5] Second, the derived heat flow has an apparently unphysical behavior in the limit of vanishing delay time. Third, the approach taken in [97] is limited to linear systems only, which is a severe restriction. We want to clarify these issues, and extend the analysis towards nonlinear systems.

To this end, we will evaluate the heat rate in nonlinear (polynomial) potentials, and particularly focus on the behavior for small delay times. Our calculation explicitly yields a correction term to the results from [97], predicting that the heat rate is discontinuous at $\tau \to 0$, and drops to zero at $\tau = 0$ (where equilibrium is reached). We will discuss the origin of this behavior. We will further explicitly show that the direction of induced heat flow is indeed tunable upon adjusting the delay time and feedback strength (i.e., the laser intensity).

Thus, the "Maxwell demon type"-task can be performed by our feedback. Nowadays it is well known that the underlying reason for the apparent "violation of the second law" in such controlled systems is that additional terms need to be added to the entropy balance, which account for the entropic cost associated with *information*. For example, deleting or storing information by the controller has entropic cost, as stated by Landauer's principle [146, 151]. Thus, in addition to the change of Shannon entropy (which automatically vanishes in steady states) and the medium entropy production (the heat flow divided by the temperature), information-theoretic terms add to the total entropy production. This is why one can have a negative steady-state heat flow, and at the same time a positive total entropy production. It would be very interesting to quantify these terms for the feedback control considered here. To this end, one should evaluate the total entropy production, which can be directly calculated by comparing the path probabilities of the process and its time-reversed image [123]. According to the framework and arguments by [91, 114], one here encounters a severe technical issue: the acausality of the backward process. Notably, the acausal nature of the backward process constitutes a crucial difference to the aforementioned protocols, where the control only acts at singular instants in time.[6]

Considering all of this from a broader perspective, we actually run into fundamental problems of stochastic thermodynamics at this point: the notion of total entropy production of non-Markovian systems, the role of hidden degrees of freedom, the

[5] Information-theoretic terms in the entropy balance were forgotten and based thereon the heat flow is claimed to be positive definite due to the second law. We will come back to this later.

[6] Then one can chop the trajectory in a way that the control actions are at the boundaries only, creating a process whose backward image does not violate causality.

role of auxiliary variables, and coarse-graining. Recent research focuses on different strategies to approach these issues. One attempt is the formulation of "effective thermodynamic" descriptions[7] [91, 103, 105, 107, 114, 150, 152]. In this context, a generalized irreversibility measure for time-delayed systems was suggested in Refs. [91, 114], which fulfills nonequilibrium inequalities, generalizing the second law of thermodynamics and providing bounds to the extractable energy. However, this approach is not applicable for nonlinear systems, which is a major restriction. Furthermore, a general problem of effective thermodynamic descriptions is that the definition of such generalized irreversibility measures is a bit ambiguous.

Here, we will pursue a different strategy. In particular, we will include the controller in the model, via a specific representation of the feedback controller (via a ring network with unidirectional, linear couplings). This representation is closely connected to the aforementioned Markovian embedding.[8] We will explore how much we can learn about the delayed process by studying the thermodynamic properties of the associated Markovian networks. In this way, we can for example access the additional information-theoretical entropy terms, and show that they are directly connected to information flows.

A characteristic feature of the Markovian networks is the involved unidirectional coupling, which is a specific type of a *non-reciprocal* interaction. As we have mentioned above, such non-reciprocal couplings are uncommon in statistical physics. Traditionally, one investigates systems consisting of elements with conservative interactions that are derivable from interaction Hamiltonians, and are thus reciprocal, like the atoms in a polymer, colloids in a dense suspension, or billiard balls. In the absence of driving forces or temperature gradients, such systems can equilibrate and are well described by traditional thermodynamics. However, it has already been realized in earlier literature that the idea of reciprocal couplings (and its thermodynamic implications) breaks down in many living and artificial active systems, where due to high complexity more general interactions between subsystems, or mesoscopic degrees of freedom, can emerge [108, 153]. This resembles the situation encountered here. Indeed, it appears somewhat natural that the feedback loop can be considered as a combination of non-reciprocally coupled entities, mirroring the intrinsic asymmetry of the elements' response to each other. Lately, researchers have started to investigate the consequences of non-reciprocal coupling on statistical properties of collective systems [108, 153, 154]. However, the thermodynamic implications of the non-reciprocity have, to the best of our knowledge, not been considered in earlier literature. This is one focal point of this thesis.

Our investigation will reveal common features of delay dynamics and active matter. To rigorously analyze these connections, we will combine information-theoretical with energetic and entropic considerations. This yields an interesting perspective on the role of memory, activity and feedback for thermodynamic quantities.

---

[7] We will later explain the idea in more detail, Chap. 4.

[8] Which automatically relates time delay to a unidirectionally coupled Markovian network.

1.1 Outline of the Thesis

Here we lay out the structure of this thesis, which is visualized in Fig. 1.4.

Part I is an extended introduction to stochastic systems and provides the methodological background by reviewing important theoretical concepts, specifically, the Langevin equation (Chap. 2), the Fokker–Planck equation (Chap. 3), and stochastic thermodynamics (Chap. 4). At the end of each Chapter, we reconsider the framework with regard to time-delayed feedback, give an overview of the state of the art, and point out apparent open problems and ongoing discussions. These considerations provide the basis and motivate the detailed analysis presented in Parts II and III, which contain the main new results of this thesis.

In Part II, we thoroughly discuss the probabilistic description of systems with time delay. To begin with, we extend two earlier derivations towards higher orders (Chap. 5). We discuss linear delay equations separately, for which we explicitly derive

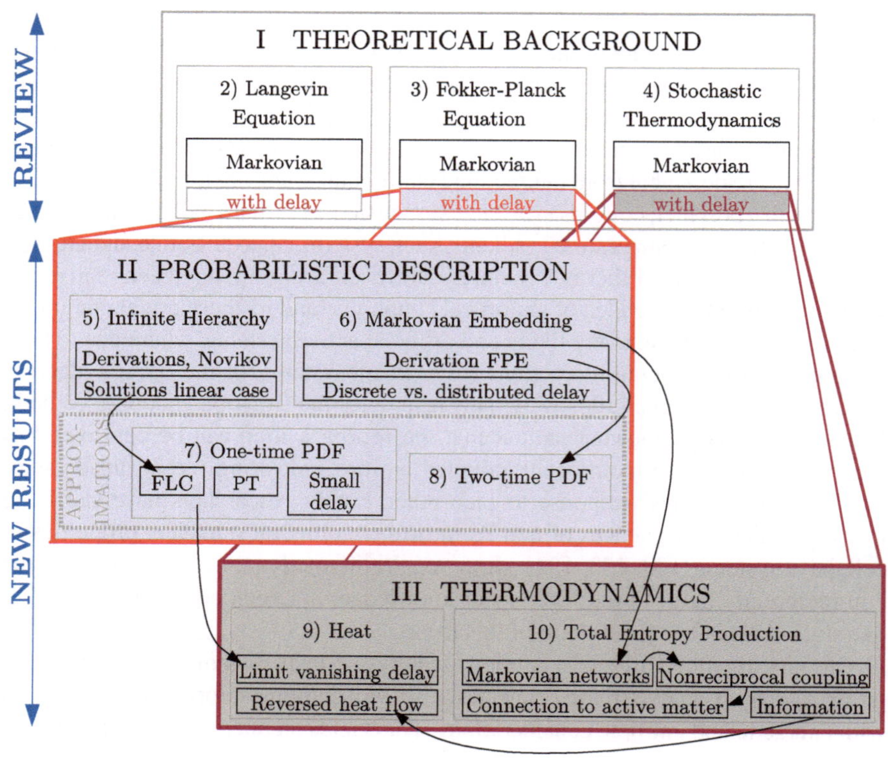

Fig. 1.4 Visualization of the structure of this thesis. The three large boxes represent the superordinate structure: Part I is a long introduction with review character, Parts II and III provide new results concerning probabilistic (II) and thermodynamic (III) descriptions of systems with time delay. The Arabic numbers in the inner boxes represent the respective Chapter numbers. A list of the meaning of the abbreviations can be found on "List of Abbreviations"

higher-order probabilistic solutions (Sect. 5.2). Furthermore, we present an alternative derivation of the Fokker–Planck hierarchy (Chap. 6), based on a Markovian embedding (Sect. 6.1). The embedding makes explicit that the discrete delay can be regarded as the limiting case of distributed delay and further reveals connections between delay and non-reciprocal coupling. We introduce a new approximation scheme for the one-time probability density in Chap. 7 (called force-linearization closure), which exploits the solutions derived before for the linear case. We compare it to earlier approaches and discuss applications to two nonlinear example systems, a periodic and a bistable one. In Chap. 8, we present an approximation scheme aiming at the two-time joint probability density. Again, we apply this approach to the bistable example system, and compare it to the earlier approaches from Chap. 7.

Part III is dedicated to the thermodynamic description of stochastic processes with delay. First, we discuss the mean heat flow induced by the time delay (Chap. 9). Again using the bistable system as an example, we consider on both, the intrawell dynamics and the delay-induced interwell oscillations, and test how well the approximations from Part II can estimate the heat. We reveal connections to "entropy pumping", and investigate the implications of the overdamped limit. In Chap. 10, we turn to the concept of entropy, now considering a specific representation of the feedback controller, which involves non-reciprocal coupling (and is closely related to the Markovian embedding technique introduced in Part II). We investigate the (information-)thermodynamic implications of the encountered non-reciprocity, and discuss common features of delay dynamics and active matter.

Lastly, in Part IV we summarize our main results and provide an outlook for future research on stochastic systems with time delay.

References

1. Milton JG et al (2013) Delayed pursuit-escape as a model for virtual stick balancing. Nonlinear Theory Appl, IEICE 4:129–137
2. Milton JG et al (2009) Balancing with vibration: a prelude for "Drift and Act" balance control. PLoS ONE 4:e7427
3. Ohira T (2010) Delay, noise, and resonance: human balancing and temporal non-locality. Artif Life Robot 15:436–438
4. Ohira T, Milton J (2009) Delayed random walks: investigating the interplay between delay and noise. Delay differential equations. Springer, pp 1–31
5. van Kampen NG (1998) Remarks on non-markov processes. Braz J Phys 28:90–96
6. Longtin A (2010). Complex time-delay systems: theory and applications. In: In: Atay FM. (ed) Springer, Berlin
7. Schöll E, Schuster HG (eds) (2008) Handbook of chaos control. Wiley, New York
8. Schöll E, Klapp SHL, Hövel P (eds) (2016) Control of self-organizing nonlinear systems. Springer, Berlin
9. Zwanzig R (2001) Nonequilibrium statistical mechanics. Oxford University Press, Oxford
10. Goychuk I (2009) Viscoelastic subdiffusion: from anomalous to normal. Phys Rev E 80:046125
11. Rouse PE Jr (1953) A theory of the linear viscoelastic properties of dilute solutions of coiling polymers. J Chem Phys 21:1272–1280

12. Maes C, Safaverdi S, Visco P, Van Wijland F (2013) Fluctuation-response relations for nonequilibrium diffusions with memory. Phys Rev E 87:022125
13. Franosch T et al (2011) Resonances arising from hydrodynamic memory in Brownian motion. Nature 478:85
14. Kheifets S, Simha A, Melin K, Li T, Raizen MG (2014) Observation of Brownian motion in liquids at short times: instantaneous velocity and memory loss. Science 343:1493–1496
15. Marchetti MC et al (2013) Hydrodynamics of soft active matter. Rev Mod Phys 85:1143
16. Narinder N, Bechinger C, Gomez-Solano JR (2018) Memory-induced transition from a persistent random walk to circular motion for achiral microswimmers. Phys Rev Lett 121:078003
17. Nagai KH, Sumino Y, Montagne R, Aranson IS, Chaté H (2015) Collective motion of self-propelled particles with memory. Phys Rev Lett 114:168001
18. Scholz C, Jahanshahi S, Ldov A, Löwen H (2018) Inertial delay of self-propelled particles. Nat Commun 9:5156
19. Driver RD (1969) A "backwards" two-body problem of classical relativistic electrodynamics. Phys Rev 178:2051
20. Will CM (2003) Propagation speed of gravity and the relativistic time delay. Astrophys J 590:683
21. Goel NS, Maitra SC, Montroll EW (1971) On the Volterra and other nonlinear models of interacting populations. Rev Mod Phys 43:231
22. Das K, Srinivas M, Srinivas M, Gazi N (2012) Chaotic dynamics of a three species prey-predator competition model with bionomic harvesting due to delayed environmental noise as external driving force. C R Biol 335:503–513
23. Schiering C et al (2017) Feedback control of AHR signalling regulates intestinal immunity. Nature 542:242
24. Pardee AB (1959) Regulation of cell metabolism. J. and A, Churchill LTD, London
25. Polettini M, Esposito M (2014) Irreversible thermodynamics of open chemical networks. I. Emergent cycles and broken conservation laws. J Chem Phys 141:07B610_1
26. Waibel A, Hanazawa T, Hinton G, Shikano K, Lang KJ (1995) Backpropagation: theory, architectures and applications. Lawrence Erlbaum Associates, Mahwah
27. Longtin A, Milton JG, Bos JE, Mackey MC (1990) Noise and critical behavior of the pupil light reflex at oscillation onset. Phys Rev A 41:6992
28. Cabral JR et al (2014) Exploring mechanisms of spontaneous functional connectivity in MEG: how delayed network interactions lead to structured amplitude envelopes of band-pass filtered oscillations. Neuroimage 90:423–435
29. Karmeshu Gupta V, Kadambari K (2011) Neuronal model with distributed delay: analysis and simulation study for gamma distribution memory kernel. Biol Cybern 104:369–383
30. Voss HU, Kurths J (2002) Modelling and forecasting financial data: techniques of nonlinear dynamics, vol 2. Springer Science & Business Media, Berlin
31. Stoica G (2005) A stochastic delay financial model. Proc Am Math Soc 133:1837–1841
32. Tambue A, Brown EK, Mohammed S (2015) A stochastic delay model for pricing debt and equity: numerical techniques and applications. Commun Nonlinear Sci Numer Simul 20:281–297
33. Callen JL, Khan M, Lu H (2013) Accounting quality, stock price delay, and future stock returns. Contemp Account Res 30:269–295
34. Bechhoefer J (2005) Feedback for physicists: a tutorial essay on control. Rev Mod Phys 77:783
35. Német N, Parkins S (2016) Enhanced optical squeezing from a degenerate parametric amplifier via time-delayed coherent feedback. Phys Rev A 94:023809
36. Mijalkov M, McDaniel A, Wehr J, Volpe G (2016) Engineering sensorial delay to control phototaxis and emergent collective behaviors. Phys Rev X 6:011008
37. Bruot N, Damet L, Kotar J, Cicuta P, Lagomarsino MC (2011) Noise and synchronization of a single active colloid. Phys Rev Lett 107:094101
38. Masoller C (2002) Noise-induced resonance in delayed feedback systems. Phys Rev Lett 88:034102

39. Kane DM, Shore KA (eds) (2005) Unlocking dynamical diversity: optical feedback effects on semiconductor lasers. Wiley, New York
40. Liu Y, Chen H, Liu J, Davis P, Aida T (2001) Synchronization of optical-feedback-induced chaos in semiconductor lasers by optical injection. Phys Rev A 63:031802
41. Hein SM, Schulze F, Carmele A, Knorr A (2015) Entanglement control in quantum networks by quantum-coherent time-delayed feedback. Phys Rev A 91:052321
42. Lu Y et al (2017) Intensified antibunching via feedback-induced quantum interference. Phys Rev A 95:063840
43. Carmele A, Kabuss J, Schulze F, Reitzenstein S, Knorr A (2013) Single photon delayed feedback: a way to stabilize intrinsic quantum cavity electrodynamics. Phys Rev Lett 110:013601
44. Schneider I (2013) Delayed feedback control of three diffusively coupled Stuart-Landau oscillators: a case study in equivariant hopf bifurcation. Philos Trans R Soc A 371:20120472
45. Zakharova A et al (2016) Controlling chimera patterns in networks: interplay of structure, noise, and delay. Springer, Berlin
46. Schneider I, Bosewitz M (2016) Eliminating restrictions of time-delayed feedback control using equivariance. Disc Cont Dyn Syst A 36:451–467
47. Yi T-M, Huang Y, Simon MI, Doyle J (2000) Robust perfect adaptation in bacterial chemotaxis through integral feedback control. PNAS 97:4649–4653
48. Aguilar-Hidalgo D et al (2018) Critical point in self-organized tissue growth. Phys Rev Lett 120:198102
49. Schöll E, Hiller G, Hövel P, Dahlem MA (2009) Time-delayed feedback in neurosystems. Philos Trans Royal Soc A 367:1079–1096
50. Parmar K, Blyuss KB, Kyrychko YN, Hogan SJ (2015) Time-delayed models of gene regulatory networks. Comput Math Meth M 2015:347273
51. Bratsun D, Volfson D, Tsimring LS, Hasty J (2005) Delay-induced stochastic oscillations in gene regulation. PNAS 102:14593–14598
52. Gupta C, López JM, Ott W, Josić K, Bennett MR (2013) Transcriptional delay stabilizes bistable gene networks. Phys Rev Lett 111:058104
53. Lai X, Wolkenhauer O, Vera J (2016) Understanding microrna-mediated gene regulatory networks through mathematical modelling. Nucleic Acids Res 44:6019–6035
54. Josić K, López JM, Ott W, Shiau L, Bennett MR (2011) Stochastic delay accelerates signaling in gene networks. PLoS Comput Biol 7:e1002264
55. Rateitschak K, Wolkenhauer O (2007) Intracellular delay limits cyclic changes in gene expression. Math Biosci 205:163–179
56. Friedman N, Cai L, Xie XS (2006) Linking stochastic dynamics to population distribution: an analytical framework of gene expression. Phys Rev Lett 97:168302
57. Ito S, Sagawa T (2015) Maxwell's demon in biochemical signal transduction with feedback loop. Nat Commun 6:7498
58. Micali G, Endres RG (2016) Bacterial chemotaxis: information processing, thermodynamics, and behavior. Curr Opin Microbiol 30:8–15
59. Wördemann M (2012) Structured light fields: applications in optical trapping, manipulation and organisation. Springer Science & Business Media, Berlin
60. Kotar J, Leoni M, Bassetti B, Lagomarsino MC, Cicuta P (2010) Hydrodynamic synchronization of colloidal oscillators. PNAS 107:7669–7673
61. Kim KH, Qian H (2007) Fluctuation theorems for a molecular refrigerator. Phys Rev E 75:022102
62. Qian B, Montiel D, Bregulla A, Cichos F, Yang H (2013) Harnessing thermal fluctuations for purposeful activities: the manipulation of single micro-swimmers by adaptive photon nudging. Chem Sci 4:1420–1429
63. Balijepalli A, Gorman JJ, Gupta SK, LeBrun TW (2012) Significantly improved trapping lifetime of nanoparticles in an optical trap using feedback control. Nano Lett 12:2347–2351
64. Khadka U, Holubec V, Yang H, Cichos F (2018) Active particles bound by information flows. Nat Commun 9:3864

65. Braun M, Cichos F (2013) Optically controlled thermophoretic trapping of single nano-objects. ACS Nano 7:11200–11208
66. Braun M, Bregulla AP, Günther K, Mertig M, Cichos F (2015) Single molecules trapped by dynamic inhomogeneous temperature fields. Nano Lett 15:5499–5505
67. Haeufle DFB et al (2016) External control strategies for self-propelled particles: optimizing navigational efficiency in the presence of limited resources. Phys Rev E 94:012617
68. Loos SAM, Gernert R, Klapp SHL (2014) Delay-induced transport in a rocking ratchet under feedback control. Phys Rev E 89:052136
69. Gernert R, Emary C, Klapp SHL (2014) Waiting time distribution for continuous stochastic systems. Phys Rev E 90:062115
70. Gernert R, Klapp SHL (2015) Enhancement of mobility in an interacting colloidal system under feedback control. Phys Rev E 92:022132
71. Gernert R, Loos SAM, Lichtner K, Klapp SHL (2016) Feedback control of colloidal transport. In: Schöll E, Klapp SHL, Hövel P (eds) Control of self-organizing nonlinear systems. Springer, pp 375–392
72. Lichtner K, Pototsky A, Klapp SHL (2012) Feedback-induced oscillations in one-dimensional colloidal transport. Phys Rev E 86:051405
73. Lichtner K, Klapp SHL (2010) Feedback-controlled transport in an interacting colloidal system. EPL 92:40007
74. Hennig D (2009) Current control in a tilted washboard potential via time-delayed feedback. Phys Rev E 79:041114
75. Küchler U, Mensch B (1992) Langevins stochastic differential equation extended by a time-delayed term. Stoch Stoch Rep 40:23–42
76. Ohira T, Hosaka T (2005) Control in delayed stochastic systems. Technical, Report
77. Ohira T, Yamane T (2000) Delayed stochastic systems. Phys Rev E 61:1247–1257
78. Wu D, Zhu S (2006) Brownian motor with time-delayed feedback. Phys Rev E 73:051107
79. Milton J, Cabrera J, Ohira T (2008) Unstable dynamical systems: delays, noise and control. EPL 83:48001
80. Guillouzic S, L'Heureux I, Longtin A (1999) Small delay approximation of stochastic delay differential equations. Phys Rev E 59:3970
81. Guillouzic S, L'Heureux I, Longtin A (2000) Rate processes in a delayed, stochastically driven, and overdamped system. Phys Rev E 61:4906
82. Frank TD, Beek PJ (2001) Stationary solutions of linear stochastic delay differential equations: applications to biological systems. Phys Rev E 64:021917
83. Frank TD, Beek PJ, Friedrich R (2003) Fokker-Planck perspective on stochastic delay systems: exact solutions and data analysis of biological systems. Phys Rev E 68:021912
84. Frank TD (2005) Delay Fokker-Planck equations, perturbation theory, and data analysis for nonlinear stochastic systems with time delays. Phys Rev E 71:031106
85. Frank TD (2005) Delay Fokker-Planck equations, Novikov's theorem, and Boltzmann distributions as small delay approximations. Phys Rev E 72:011112
86. Van Vu T, Hasegawa Y (2019) Uncertainty relations for time-delayed Langevin systems. Phys Rev E 100:012134
87. Geiss D, Kroy K, Holubec V (2019) Brownian molecules formed by delayed harmonic interactions. New J Phys 21:093014
88. Mackey MC, Nechaeva IG (1994) Noise and stability in differential delay equations. J Dyn Diff Equ 6:395–426
89. McKetterick TJ, Giuggioli L (2014) Exact dynamics of stochastic linear delayed systems: application to spatiotemporal coordination of comoving agents. Phys Rev E 90:042135
90. Giuggioli L, McKetterick TJ, Kenkre VM, Chase M (2016) Fokker-Planck description for a linear delayed Langevin equation with additive Gaussian noise. J Phys A 49:384002
91. Munakata T, Rosinberg ML (2014) Entropy production and fluctuation theorems for Langevin processes under continuous non-Markovian feedback control. Phys Rev Lett 112:180601
92. Zwanzig R (1973) Nonlinear generalized Langevin equations. J Stat Phys 9:215–220
93. Kubo R (1966) The fluctuation-dissipation theorem. Rep Prog Phys 29:255

94. Krüeger M, Maes C (2016) The modified Langevin description for probes in a nonlinear medium. J Phys: Condens Matter 29:064004
95. Bao J-D, Hänggi P, Zhuo Y-Z (2005) Non-Markovian Brownian dynamics and nonergodicity. Phys Rev E 72:061107
96. Maes C (2014) On the second fluctuation-dissipation theorem for nonequilibrium baths. J Stat Phys 154:705–722
97. Munakata T, Iwama S, Kimizuka M (2009) Linear stochastic system with delay: energy balance and entropy production. Phys Rev E 79:031104
98. Bloom GS, Endow SA (1995) Motor proteins 1: kinesins. Protein Profile 2:1105–1171
99. Kuo SC, Sheetz MP (1993) Force of single kinesin molecules measured with optical tweezers. Science 260:232–234
100. Ramaswamy S (2010) The mechanics and statistics of active matter. Annu Rev Condens Matter Phys 1:323–345
101. Ramaswamy S (2017) Active matter. J Stat Mech Theor Exp 2017:054002
102. Pietzonka P, Seifert U (2017) Entropy production of active particles and for particles in active baths. J Phys A: Math Theor 51:01LT01
103. Shankar S, Marchetti MC (2018) Hidden entropy production and work fluctuations in an ideal active gas. Phys Rev E 98:020604(R)
104. Speck T (2016) Stochastic thermodynamics for active matter. EPL 114:30006
105. Dabelow L, Bo S, Eichhorn R (2019) Irreversibility in active matter systems: Fluctuation theorem and mutual information. Phys Rev X 9:021009
106. Ehrich J, Kahlen M (2019) Approximating microswimmer dynamics by active Brownian motion: Energetics and efficiency. Phys Rev E 99:012118
107. Caprini L, Marconi UMB, Puglisi A, Vulpiani A (2019) The entropy production of Ornstein-Uhlenbeck active particles: a path integral method for correlations. J Stat Mech Theor Exp 2019:053203
108. Durve M, Saha A, Sayeed A (2018) Active particle condensation by non-reciprocal and time-delayed interactions. Eur Phys J E 41:49
109. Kürsten R, Sushkov V, Ihle T (2017) Giant kovacs-like memory effect for active particles. Phys Rev Lett 119:188001
110. Khadem SMJ, Klapp SH (2019) Delayed feedback control of active particles: a controlled journey towards the destination. Phys Chem Chem Phys 21:13776
111. Tarama S, Egelhaaf SU, Löwen H (2019) Traveling band formation in feedback-driven colloids. Phys Rev E 100:022609
112. Piwowarczyk R, Selin M, Ihle T, Volpe G (2019) Influence of sensorial delay on clustering and swarming. Phys Rev E 100:012607
113. Schwabl F (2002) Statistical mechanics. Springer, Berlin
114. Rosinberg ML, Munakata T, Tarjus G (2015) Stochastic thermodynamics of Langevin systems under time-delayed feedback control: second-law-like inequalities. Phys Rev E 91:042114
115. Puglisi A, Villamaina D (2009) Irreversible effects of memory. EPL 88:30004
116. Crisanti A, Puglisi A, Villamaina D (2012) Nonequilibrium and information: the role of cross correlations. Phys Rev E 85:061127
117. Siegle P, Goychuk I, Talkner P, Hänggi P (2010) Markovian embedding of non-Markovian superdiffusion. Phys Rev E 81:011136
118. Siegle P, Goychuk I, Hänggi P (2011) Markovian embedding of fractional superdiffusion. EPL 93:20002
119. Villamaina D, Baldassarri A, Puglisi A, Vulpiani A (2009) The fluctuation-dissipation relation: how does one compare correlation functions and responses? J Stat Mech: Theory Exp 2009:P07024
120. Smith HL (2011) An introduction to delay differential equations with applications to the life sciences, vol 57. Springer, New York
121. Niculescu S-I, Gu K (2012) Advances in time-delay systems, vol 38. Springer Science & Business Media, Berlin
122. Sekimoto K (2010) Stochastic energetics, vol 799. Springer, Berlin

123. Seifert U (2012) Stochastic thermodynamics, fluctuation theorems and molecular machines. Rep Prog Phys 75:126001
124. Esposito M (2012) Stochastic thermodynamics under coarse graining. Phys Rev E 85:041125
125. Sagawa T, Ueda M (2012) Nonequilibrium thermodynamics of feedback control. Phys Rev E 85:021104
126. Abreu D, Seifert U (2012) Thermodynamics of genuine nonequilibrium states under feedback control. Phys Rev Lett 108:030601
127. Barato AC, Seifert U (2014) Unifying three perspectives on information processing in stochastic thermodynamics. Phys Rev Lett 112:090601
128. Mandal D, Jarzynski C (2012) Work and information processing in a solvable model of Maxwell's demon. PNAS 109:11641–11645
129. Parrondo JMR, Horowitz JM, Sagawa T (2015) Thermodynamics of information. Nat Phys 11:131–139
130. Kutvonen A, Sagawa T, Ala-Nissila T (2016) Thermodynamics of information exchange between two coupled quantum dots. Phys Rev E 93:032147
131. Barato AC, Seifert U (2015) Thermodynamic uncertainty relation for biomolecular processes. Phys Rev Lett 114:158101
132. Esposito M, Lindenberg K, Van den Broeck C (2009) Thermoelectric efficiency at maximum power in a quantum dot. EPL 85:60010
133. Strasberg P, Schaller G, Brandes T, Esposito M (2013) Thermodynamics of a physical model implementing a Maxwell demon. Phys Rev Lett 110:040601
134. Koski JV, Maisi VF, Sagawa T, Pekola JP (2014) Experimental observation of the role of mutual information in the nonequilibrium dynamics of a Maxwell demon. Phys Rev Lett 113:030601
135. Imparato A, Peliti L, Pesce G, Rusciano G, Sasso A (2007) Work and heat probability distribution of an optically driven Brownian particle: theory and experiments. Phys Rev E 76:050101
136. Speck T, Seifert U (2007) The Jarzynski relation, fluctuation theorems, and stochastic thermodynamics for non-Markovian processes. J Stat Mech: Theory Exp 2007:L09002
137. Vaikuntanathan S, Jarzynski C (2009) Dissipation and lag in irreversible processes. EPL 87:60005
138. Kutvonen A, Ala-Nissila T, Pekola J (2015) Entropy production in a non-Markovian environment. Phys Rev E 92:012107
139. Hasegawa H (2011) Classical open systems with nonlinear nonlocal dissipation and state-dependent diffusion: dynamical responses and the Jarzynski equality. Phys Rev E 84:051124
140. García-García R (2012) Nonadiabatic entropy production for non-Markov dynamics. Phys Rev E 86:031117
141. Roche B et al (2015) Harvesting dissipated energy with a mesoscopic ratchet. Nat Commun 6:6738
142. Whitney RS (2018) Non-Markovian quantum thermodynamics: laws and fluctuation theorems. Phys Rev B 98:085415
143. Mai T, Dhar A (2007) Nonequilibrium work fluctuations for oscillators in non-Markovian baths. Phys Rev E 75:061101
144. Schmidt R, Carusela MF, Pekola JP, Suomela S, Ankerhold J (2015) Work and heat for two-level systems in dissipative environments: strong driving and non-Markovian dynamics. Phys Rev B 91:224303
145. Cui B, Zaccone A (2018) Generalized Langevin equation and fluctuation-dissipation theorem for particle-bath systems in external oscillating fields. Phys Rev E 97:060102
146. Sagawa T (2012) Thermodynamics of information processing in small systems. Progress Theoret Phys 127:1–56
147. Maxwell JC (1871) Theory of heat. Dover Publications Inc, Mineola
148. Roldán É, Martínez IA, Parrondo JMR, Petrov D (2014) Universal features in the energetics of symmetry breaking. Nat Phys 10:457
149. Cao FJ, Feito M (2009) Thermodynamics of feedback controlled systems. Phys Rev E 79:041118

150. Rosinberg ML, Tarjus G, Munakata T (2017) Stochastic thermodynamics of Langevin systems under time-delayed feedback control. II. Nonequilibrium steady-state fluctuations. Phys Rev E 95:022123
151. Landauer R (1961) Irreversibility and heat generation in the computing process. IBM J Res Dev 5:183–191
152. Polettini M, Esposito M (2017) Effective thermodynamics for a marginal observer. Phys Rev Lett 119:240601
153. Agudo-Canalejo J, Golestanian R (2019) Active phase separation in mixtures of chemically interacting particles. Phys Rev Lett 123:018101
154. Lavergne FA, Wendehenne H, Bäuerle T, Bechinger C (2019) Group formation and cohesion of active particles with visual perception-dependent motility. Science 364:70–74

Part I
Theoretical Background and State of the Art

Mathematicians may atter themselves that they possess new ideas which mere human language is as yet unable to express

—James Clerk Maxwell, 1865 [155]

In this Part I, we review some important theoretical concepts of the description of stochastic systems, in particular, the Langevin equation, the Fokker-Planck equation, and stochastic thermodynamics. We will always start with introducing established frameworks that are already subject of standard literature. As we will see, these frameworks are often limited to Markovian cases, while non-Markovian systems are by far less understood. At the end of each Chapter, we will reconsider the frameworks with regard to *time-delayed* dynamics. We will give an overview of the state-of-the-art research on stochastic delay equations and hint to ongoing debates.

Besides providing the methodological background, the purpose of these considerations is to make the Reader aware of apparent technical and conceptual problems in the description of non-Markovian, nonequilibrium or active systems, and to formulate relevant research questions. Another goal is to differentiate our findings from previous research. In this way, Part I builds up the basis for our new results presented mainly in Parts II and III of this thesis.

Chapter 2
The Langevin Equation

2.1 The Stochastic Way of Describing Things

2.1.1 Brownian Motion

While you are reading this thesis, at every instant of time countless particles of the surrounding air hit your skin due to their irregular thermal motion. There are of the order of $\sim 10^{23}$ molecules in every liter of air [1, 2].

We cannot feel this bombardment, since the particles are tiny. But now imagine you would be much smaller, lets say, of the size of a single cell,[1] yet larger than the molecules of your surrounding. Then, you would feel the bombardment very dramatically, in fact, these collisions would make you move randomly around yourself, even when you would try to stand still. You would perform a *Brownian motion*, like every small particle, called *Brownian particle*, suspended in a fluid at a finite temperature does [3]. Mathematically, this type of erratic, random motion is described by a *Langevin equation*.

Langevin equations describe systems subject to random perturbation. It is indeed typical that the system itself is "intermediate-sized", and studied on a *mesoscopic* length and timescale. This means, it is small enough to expose the collisions in the form of thermal fluctuations (by performing the random Brownian motion). This randomness would average out, if the system was much larger, or if it was considered on a *macroscopic* scale, i.e., a much bigger time- and length scale. Yet, the system is larger than the subsystems responsible for the perturbations, whose impact is described only implicitly by a single random term and a friction force. Typically, this is appropriate if multiple collisions take place at the same time, see Fig. 2.1 for an illustration. In contrast, a *microscopic* description would mean that the positions and velocities of all particles are precisely described. To this end, one would need to

[1] Like in the famous TV Series "Il était une fois...la vie" [*English title*: "Once upon a time...life"], where this effect is completely neglected, drawing a physically incorrect picture of the processes in our body.

© The Author(s), under exclusive license to Springer Nature Switzerland AG 2021

S. A. M. Loos, *Stochastic Systems with Time Delay*, Springer Theses,
https://doi.org/10.1007/978-3-030-80771-9_2

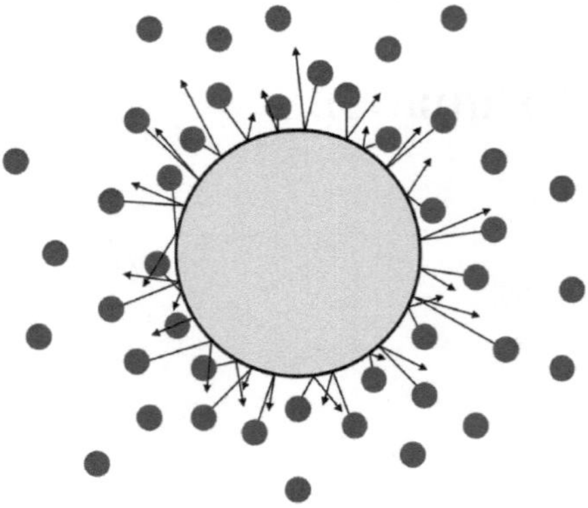

Fig. 2.1 Illustration of a colloid (large, light gray sphere) immersed in a fluid of much smaller particles (dark spheres). Multiple collisions take place at the same time, creating a Brownian motion of the colloidal particle. In the mesoscopic, i.e., coarse-grained description, this bombardment is modeled by a random force and a friction force

solve a vast amount of equations[2] at every instant of time (about $\sim 10^{23}$, in the case of a colloidal particle immersed in one mole of fluid).

The stochastic way of modeling (for, e.g., colloidal suspensions) is the best description we have to date to describe complex phenomena on the nano- to micrometer scale. In the modern framework of statistical physics, stochastic modeling by the Langevin equation, and in particular Brownian motion, plays a central role. It provides a key step on the route from microscopic first principles to macroscopic laws, and enables a fully probabilistic formulation of statistical mechanics. Furthermore, it forms the backbone of a dynamical interpretation of statistical effects [3].

A key feature of Langevin equations is that they are *not* limited to systems in thermal equilibrium,[3] which is of crucial importance in many applications. They can thus also be used to model various *active* particles, i.e., living or artificial matter with an internal energy source [5], like bacteria or robots (see Sect. 4.7). This will be important for our considerations, specifically the second Part of this thesis.

2.1.2 *Colloidal Suspensions*

Colloids are a common example of Brownian particles, useful to experimentally investigate Brownian motion. The term colloidal particle denotes a few nanometer- to micrometer-sized, i.e., $\sim 10^{-9}$m -10^{-6}m, objects immersed in a fluid of atomically scaled particles, i.e., several Ångström, i.e., $\sim 10^{-10}$m. Colloids can be single

[2] An example for a numerical approach is multi-particle collision dynamics [4]. Analytically, however, there is no hope.

[3] Except for an equilibrium assumption for the bath, which we will explain in Sect. 2.3.1.

macromolecules, lipid drops in water, artificial spheres, or pollen grains, as in the case of Robert Brown's name-giving, original observation of this motion [6]. Various further examples of colloidal suspensions can be found in [7]. From an experimental perspective, a pleasant property of colloidal particles is that they are comparably easy to access in experiments, as they are usually visible in optical microscopes. They can further be manufactured in a controlled manner, making it possible to design colloids with desired properties (size, shape, charge) within a quite wide range [7]. Also the interactions between colloidal particles are often controllable, e.g., by external electrical fields [8, 9] or optical feedback [10]. Furthermore, experimental techniques enable precise dynamical control of the position of a colloidal particle, for example by optical traps (this tool is particularly relevant in the context of this thesis, and will be reviewed in Sect. 2.4.1). On the collective level, colloids are used to study friction, phase transitions, glassy dynamics, or critical phenomena [8]. Further, by tracking the movement of individual colloidal particles, one can study hydrodynamic effects, viscoelastic properties of the surrounding fluid, or diffusion properties. Colloidal suspensions can also be found in nature. For instance, macro-molecules, viruses, individual and small clusters of cells are on this very (colloidal) length scale, which is why the stochastic way of modeling introduced in the following, is particularly useful in (cell-)biology.

2.1.3 Side Note: A More General View

Besides the example of Brownian motion, there are endlessly more processes in physics and other disciplines, where a detailed description at all length scales is impossible, because there are too many components, due to uncertainties in the model resulting from the fact that not all relevant processes are known precisely, or because of external random influences. Thus, it is no surprise that stochastic modeling is very widely spread, and Langevin, or Langevin-like, stochastic differential equations have proven themselves extremely useful to understand complex dynamical systems. Examples include gene expression [11], population growth dynamics [12], climate prediction [13], the stochastic motion of stock market prizes [14], the human pupil reflex [15], neural networks [16], just to name a few.[4] Other common examples are the fluctuating electric voltage in RC circuits [17, 18], and the concentration of a chemical component in a mixture [19, 20]. We will mostly consider colloidal systems, but many results presented in this thesis also apply to other stochastic processes.

For pedagogical and comprehensive introductions to stochastic systems in physics, we refer the interested Reader to the books by Gardiner [2] and Risken [21], and to [22] by Karatzsas and Shreve for a more mathematical perspective. A great book on stochastic thermodynamics, which also contains an educational introduction to stochastic modeling in physics, is the book by Sekimoto [23].

[4] By the way, in all these examples, time delays are in fact also known to play a role, but let us first focus only on the stochastic part here and take it step by step.

2.2 The Markovian Langevin Equation

The motion of a Brownian particle is mathematically described by the Markovian *Langevin equation* (LE)

$$m\frac{\mathrm{d}^2X(t)}{\mathrm{d}t^2} = -\gamma\frac{\mathrm{d}X(t)}{\mathrm{d}t} + F + \sqrt{2k_\mathrm{B}\mathcal{T}\gamma}\,\xi(t), \tag{2.1}$$

which is a stochastic differential equation with the time $t \in [0, \infty[$, and $X \in \Omega$. $\Omega \subseteq \mathbb{R}$ denotes the spatial domain. X is a stochastic variable and represents the dynamical quantity of interest. Depending on the system at hand, this can be the position of a Brownian particle, the angle of a stochastically rotating system, the concentration of a chemical component, or any other *degree of freedom* (d.o.f.) whose stochastic dynamics is describable by (2.1). We will nevertheless call X *position* throughout this thesis for the sake of simplicity. The term *Markovian* refers to the fact that the LE (2.1) is *local* in time, which means that one only needs to know the system state at time t to infer the probabilities of its future evolution, while additional knowledge of past states does not provide further useful information for this task (we will give a more rigorous definition based on transition probabilities in Sect. 3.1.2).

The LE is a force balance, with friction and noise.Let us comment on all terms of (2.1) from left to right:

- $m\ddot{X}(t)$ is the inertia term with the mass of the particle m and the second derivative of the position, i.e., the particle acceleration. We will later consider the overdamped limit which amounts to neglecting this term. Note that we use both notations for time derivatives: $\dot{X}(t') = \frac{\mathrm{d}X(t)}{\mathrm{d}t}\big|_{t=t'}$.
- $-\gamma\dot{X}(t)$ is a Stokes friction force[5] that is proportional to the velocity with the friction coefficient γ. In Sect. 2.3.1, we will sketch a possible derivation of this term for a toy model with harmonic oscillators.
- F denotes the sum of all *external forces* acting on the particle. They may originate from a static potential $V(x)$, then $F(x) = -V'(x) = -\partial V(x)/\partial x$ is conservative, but they may also stem from other external sources. In particular, F can also be time-dependent and nonconservative, or be a combination of both, conservative and nonconservative parts. We will later introduce time-delayed (feedback) forces, making it *history*-dependent. We will denote the forces in F *deterministic*, to distinguish them from the random Langevin force (next bullet point).
- The last term is the Langevin force, with the noise strength $\sqrt{2k_\mathrm{B}\mathcal{T}\gamma}$, and the *noise* ξ. It involves the temperature $\mathcal{T}$ of the surrounding fluid, the Boltzmann constant k_B, and the friction constant γ. As the Langevin force considered here is not X-dependent, this noise is referred to as *additive noise* (as opposed to multiplicative noise "$g[X(t)]\xi(t)$" with some function g).

---

[5] The simple form of the friction is justified because we consider a fluid at low Reynolds numbers with laminar flow. According to Stokes law, the friction coefficient is related to the diameter of the colloid σ, and the viscosity of the fluid ν, via $\gamma = 3\pi\sigma\nu$.

An overview over all coefficients and variables, including the respective SI units, is given at the beginning of this thesis on "List of Symbols". Further, a list of all abbreviations is given on "List of Abbreviations".

In the following, we will further specify and discuss the individual components of the Langevin equation, introducing the suitable mathematical framework and provide physical explanations. Special attention needs to be drawn to the noise term. In particular, we need to define the noise and the stochastic calculus (i.e., Itô, Stratonovich) below, to give (2.1) mathematical meaning. While we will use a notation indicating one-dimensionality of the considered system, many of the definitions and considerations can indeed readily be applied to multidimensional systems as well, for example by simply replacing X by a vector $\underline{X} = \{X_0, X_1, \ldots, X_n\}$, and the noise ξ by a vector $\underline{\xi} = \{\xi_0, \xi_1, \ldots, \xi_n\}$. We will come back to this in later parts of this thesis (e.g., Chaps. 6 and 10).

2.2.1 Gaussian White Noise

The noise ξ is assumed to be Gaussian distributed, hence, it is fully determined by the first two moments. Further, we will assume *white noise*,[6] that means

$$\langle \xi(t) \rangle = 0 \tag{2.2}$$
$$\langle \xi(t)\xi(t') \rangle = \delta(t - t'), \tag{2.3}$$

involving the Dirac delta distribution, and the *ensemble average* $\langle \cdots \rangle$ (which will be defined in Sect. 2.2.2). The temporal correlations $\langle \xi(t)\xi(t') \rangle$ decay infinitely fast. This amounts to the assumption that the random collisions of the colloid and bath particles are so numerous and fast that they are independent from each other at different instants of time. In contrast, noise terms with account for finite correlations in time (not a delta-distributed correlation) are denoted *colored noise*. We will consider a particular case of colored noise in Chaps. 6 and 10. The white noise (2.2) has infinitely many "jumps" on each time interval, no matter how short it is. Indeed, this is physically not reasonable, but yields a good approximation for many situations, and is much easier to deal with from a technical point of view, as opposed to correlated, i.e., colored noise. Figure 2.2 shows exemplary, numerical realizations of the white noise process.

[6] "White" refers to the power spectrum of the noise, which is constant, resembling the spectrum of white light. The power spectrum is given by the Fourier-transformation of the correlation function according to the Wiener–Khintchine theorem [21].

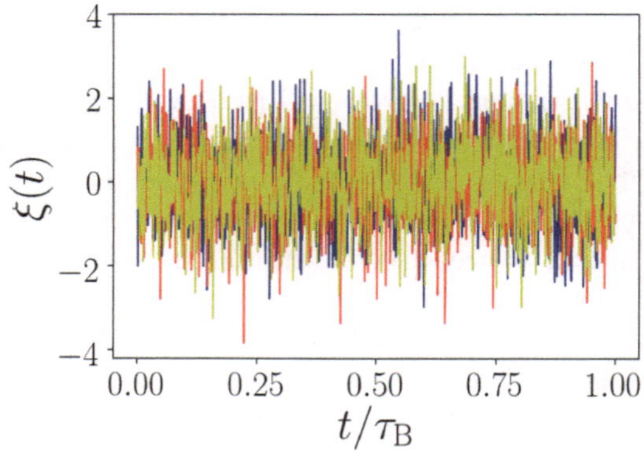

Fig. 2.2 Numerically generated white noise with a temporal discretization of $dt = 10^{-3}\tau_B$, with the here arbitrary timescale τ_B (we will later give physical meaning to τ_B). Three different random number seeds where used (indicated by different colors)

The *path probability* to observe a specific realization $\boldsymbol{\xi} = \{\xi(t)\}_{t_1}^{t_2}$, of the white noise on a time interval $t \in [t_s, t_f]$ is given by the path integral [24] (also see Sect. 2.2.6 below)

$$\mathcal{P}[\boldsymbol{\xi}] \propto e^{-\int_{t_s}^{t_f} \frac{\xi(t')^2}{2} dt'}. \tag{2.4}$$

A mathematically rigorous treatment can be found in [24].

We will come back to the mathematical description of white noise below, when we introduce the Wiener process and the stochastic calculus.

Gaussianity—Central limit theorem Besides making the mathematical treatment much easier, the assumption of *Gaussianity* of the noise term is, in various cases, indeed very reasonable from a physical point of view. The underlying reason is the *central limit theorem*. It states, loosely speaking, that the sum of a great number of independent random events[7] will asymptotically approach a Gaussian distribution [1, 2, 25]. Thus, Gaussian noise in a sense "automatically comes up" whenever random processes are modeled that are a result of many (approximately) uncorrelated, random processes. This is for example the case for all the collisions between the Brownian particle and the surrounding bath particles, and among the latter.

The heat bath and the temperature The friction force and the Langevin force both stem from the interaction between the Brownian particle and the surrounding fluid particles. The collection of these fluid particles is called the *reservoir*, the *heat bath*, or simply the *bath*. Friction slows the particle down, while the Langevin force prevents the particle from standing still, creating the erratic motion.

As the surrounding particles are much smaller and thus lighter than the Brownian particle, the timescale of their motion is also much shorter (lighter particles typically move much faster[8]). Thus, both timescales are well-separated, making it plausible to assume that the bath particles can exchange their momenta and distribute their

[7] Under fairly general conditions on their individual statistics, as long as they have finite variances.

[8] The reason is the equipartition theorem, see below.

positions "infinitely fast" from the perspective of the "slow" Brownian motion one is mainly interested in. Further, the bath is assumed to be *infinitely large* and at a constant temperature $\mathcal{T}$. This amounts to the important assumption that *the bath is in thermal equilibrium at any time*. The temperature is connected to the mean quadratic velocity (in fact, the kinetic energy) of the bath particles via the *equipartition theorem*. This theorem states that in equilibrium, the energy is (on average) equally distributed among the degrees of freedom [2], with

$$\frac{1}{2} m_{\text{bath}} \langle v^2 \rangle = \frac{d}{2} k_B \mathcal{T}, \tag{2.5}$$

for d-dimensional systems, with m_{bath} being the mass of the individual bath particles. The associated thermal energy of a particle in the bath is $k_B \mathcal{T}$. One should stress that the temperature is defined on the basis of the bath only, not taking into account the state of the Brownian particle (which does not have to be in equilibrium, and can thus also move faster or slower than the equipartition theorem would state). In Sect. 2.3.1, we will sketch how friction and noise emerge from interactions with the bath particles, and how the temperature is introduced in the stochastic description. The infinite size of the bath, which corresponds to a thermodynamic limit, makes it plausible that the temperature can stay constant, even if external driving induces a steady net heat flow from the particle to the bath, which we will discuss in Chap. 9.

2.2.2 *Ensemble Averages and Probability Density*

An ensemble is an imaginary (infinite) set of identically prepared systems (with equal temperature, same initial conditions,...), each yielding one possible realization of the random process. The ensemble average [e.g., in Eq. (2.2)] is the average over multiple, in principle *all* possible, realizations of the random process, i.e.,

$$\langle A(X) \rangle = \frac{1}{N} \sum_{i=1}^{N} A(X_i) \tag{2.6}$$

for an arbitrary function A of the random system state X. Here X_i denotes an individual realization, and N is the total number of all realizations ($N \to \infty$). For *ergodic* processes, the ensemble average is equivalent to the time-average of the process (over infinitely long trajectories) [2]. All processes considered in this thesis are assumed to be ergodic. As opposed to the random process X, ensemble averages are deterministic quantities (i.e., noise-free), and hence, contain "reliable" information about the process. In some contexts, they are hence of main interest.

Probability density A particularly important ensemble-averaged quantity is the *probability density function* (PDF)

$$\rho_1(x,t) = \langle \delta[X(t) - x] \rangle, \tag{2.7}$$

with $x \in \Omega$ (note the distinction between X which is the stochastic system state, i.e., the colloidal position, and x which is the spatial variable). The probability of finding the particle at time t within a given interval $[x, x + \Delta x]$ is $\int_x^{x+\Delta x} \rho_1(x', t)\mathrm{d}x'$. As the overall probability to find the particle somewhere in the spatial domain Ω has to be one, the normalization is $\int_\Omega \rho_1(x', t)\mathrm{d}x' = 1$. The subscript emphasizes that this is a PDF for a single stochastic event $\{x, t\}$, as opposed to joint PDFs which will be consider in Sect. 3.2. Finding an analytical expression for the PDFs of systems with time delay will be the main focus of Part II of this thesis. Like (2.6), the PDF can as well be used to calculate ensemble averages of a function $A(X)$, through

$$\langle A(X) \rangle = \int_\Omega A(x)\rho_1(x,t)\mathrm{d}x. \tag{2.8}$$

Steady states In the *long time limit* many systems approach a stationary PDF

$$\text{Steady states:} \quad \partial_t \rho_{1,\mathrm{ss}}(x,t) = 0, \tag{2.9}$$

which is the defining criterion of *steady states*. The latter can be distinguished as thermal *equilibria* (EQ), or *non-equilibrium steady states* (NESS), which we will precisely define in Chap. 4. Due to their time-independence, steady states generally allow for more analytical treatment than transients. They play a central role in statistical physics, in general, and stochastic thermodynamics, in particular (see Chap. 4), comparable with the central role of EQ in traditional macroscopic thermodynamics. Hence, they are a major focus in our investigations.

Throughout the thesis, we will occasionally use the subscript "ss" to emphasize that an ensemble average or PDF refers to the steady state (however, we will not always indicate steady states in that way).

Correlation functions Another common type of ensemble-averaged quantities are *autocorrelation functions*, for example the (temporal) *position-autocorrelation function*

$$C(\Delta t) = \langle X(t)X(t + \Delta t) \rangle_{\mathrm{ss}}. \tag{2.10}$$

As already indicated by the argument here, in steady states the correlation functions only depend on the time-difference Δt, which is the typical scenario that we will consider in this thesis. When considering correlation functions of other quantities than X, we will denote them with corresponding subscript, for instance, $C_\nu = \langle \nu(t)\nu(t + \Delta t) \rangle_{\mathrm{ss}}$ is the autocorrelation of ν. Correlation functions contain a lot of information about a (random) process, and, in particular, tell how different instants are connected with each other. Thus, they are on the one hand important to describe dynamical behavior, and on the other hand, give us insight into the memory of a process.

In fact, for processes with Gaussian-distributed probability densities, the complete probabilistic information is encoded in the position-autocorrelation function plus the mean value $\langle X \rangle$, because Gaussian distributions are fully determined by their first two moments. As we will later see, the steady-state solutions of *linear* LEs are typically Gaussian PDFs.

2.2.3 *Solutions of the Langevin Equation and the Overdamped Limit*

Ballistic timescale Let us consider (2.1) in the absence of external forces, i.e.,

$$m\frac{\mathrm{d}^2X(t)}{\mathrm{d}t^2} = -\gamma\frac{\mathrm{d}X(t)}{\mathrm{d}t} + \sqrt{2k_\mathrm{B}\mathcal{T}\gamma}\,\xi(t), \tag{2.11}$$

describing a "free" particle. Setting the velocity v_0 as initial condition, i.e., $\frac{\mathrm{d}X(t)}{\mathrm{d}t}\big|_{t=0} = v_0$, this equation has the formal solution

$$\dot{X}(t) = v_0 e^{-(\gamma/m)t} + \int_0^t e^{-(\gamma/m)|t-t'|}\sqrt{2k_\mathrm{B}\mathcal{T}\gamma/m^2}\,\xi(t')\mathrm{d}t'. \tag{2.12}$$

The term "formal" refers to the fact that the solution still contains the random force, owing to the stochastic character of the Langevin equation. Such formal solutions are very valuable for analytical calculations, and will hence play an important role in the remainder of this thesis. We stress that for nonlinear LEs, formal solutions are most often not available (see, e.g., Ref. [26] for a discussion of exact solutions of nonlinear differential equations), making their treatment much more involved.

Equation (2.12) shows that the initially set velocity v_0 only appears in an exponentially decaying term, and that the velocity relaxes on a timescale m/γ, called the *ballistic* or *velocity-relaxation timescale*. This is typically the shortest characteristic timescale of the process.

Using the properties of white noise (2.2), one finds from (2.12) the velocity-auto correlation function

$$\left\langle \dot{X}(t_1)\dot{X}(t_2)\right\rangle = v_0^2\, e^{-(\gamma/m)(t_1+t_2)} + \frac{k_\mathrm{B}\mathcal{T}}{m}\left[e^{-(\gamma/m)|t_1-t_2|} - e^{-(\gamma/m)|t_1+t_2|}\right], \tag{2.13}$$

which in the long time limit $t_1 \gg 1$, $t_2 \gg 1$ relaxes to the stationary value

$$\left\langle \dot{X}(t_1)\dot{X}(t_2)\right\rangle_\mathrm{ss} = \frac{k_\mathrm{B}\mathcal{T}}{m}\, e^{-(\gamma/m)|t_1-t_2|}, \tag{2.14}$$

(see [21] for further details). Thus, the velocity correlations decay exponentially on the ballistic timescale.

The value of the stationary velocity-autocorrelation at $t_1 = t_2$ further implies that the mean kinetic energy of the free Brownian particle in the long-time limit is

$$\langle E_{\text{kin}} \rangle = \frac{m}{2} \langle \dot{X}^2 \rangle_{\text{ss}} = \frac{k_B \mathcal{T}}{2}. \tag{2.15}$$

Thus, as one should physically expect, the free Brownian particle approaches thermal equilibrium, where its mean velocity is adjusted by the bath's temperature and the equipartition theorem holds.

Overdamped Langevin equation In many situations of interest, the velocity relaxation (which takes place on the ballistic timescale m/γ), is very fast as compared to timescales one is mainly interested in, such as the diffusive timescale (which we will discuss below). Thus, one can often assume the velocity relaxation to be infinitely fast. This gives rise to the *overdamped* LE

$$\gamma \frac{\mathrm{d}X(t)}{\mathrm{d}t} = F + \sqrt{2k_B \mathcal{T} \gamma}\, \xi(t), \tag{2.16}$$

representing the overdamped limit, i.e., $m/\gamma \to 0$, of (2.1). We will mainly consider the overdamped LE in this thesis. We will refer to (2.1) as the *underdamped* LE.

Diffusive timescale Now, we again consider a free Brownian particle (i.e., $F = 0$), this time described by the overdamped Eq. (2.16). Figure 2.3 displays some example trajectories. Starting at position $X(0)$, the formal solution reads

$$X(t) = X(0) + \int_0^t \sqrt{2k_B \mathcal{T}/\gamma}\, \xi(t')\mathrm{d}t'. \tag{2.17}$$

This formal solution further implies a linear increase in time of the so-called *mean square displacement* (MSD) $\langle [X(t) - X(0)]^2 \rangle$, specifically,

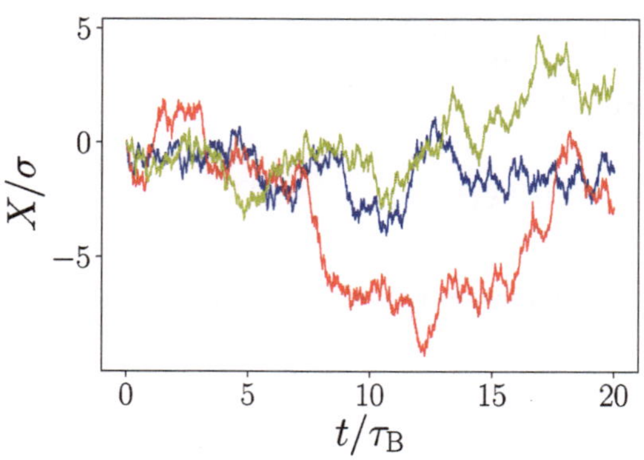

Fig. 2.3 Three example trajectories of a free Brownian particle in the overdamped limit with $X(0) = 0$. The data stem from Brownian dynamics simulations (see Appendix A.1 for details). At the same time, these trajectories represent a Wiener process (2.26) with $W(t) \hat{=} X(t)$, see below

$$\langle [X(t) - X(0)]^2 \rangle = \frac{2k_B\mathcal{T}}{\gamma} \int_0^t \int_0^t \delta(t' - t'')\mathrm{d}t'\mathrm{d}t'' = 2\frac{k_B\mathcal{T}}{\gamma} t = 2D_0 t. \quad (2.18)$$

The MSD is a measure for the diffusive spreading of the Brownian particle due to the noise. Its linear increase is characteristic for Brownian motion. Importantly, the underdamped equation has the *same* behavior for times larger than the ballistic timescale.[9] In the last step of (2.18), we have brought the MSD in a more common form by introducing the *diffusion coefficient* D_0 and by employing the *Einstein relation* [27]

$$k_B\mathcal{T} = \gamma D_0. \quad (2.19)$$

This renowned relation connects the diffusion coefficient with the temperature and represents a *fluctuation-dissipation relation*. It further implies another way of expressing the thermal energy γD_0.

Considering a typical length scale σ of the system, the average time a free particle would need to "explore", or travel, this distance by diffusion is according to (2.18) given by[10]

$$\tau_B = \sigma^2/D_0, \quad (2.20)$$

which is called the *Brownian time*. This is a *diffusive timescale*. From now on, we will measure time in units of τ_B throughout this thesis. For a free particle, σ could be, for example, the particle's diameter. Later, σ will be a typical length scale of an external potential V_s in our example systems.

In the last decades, there was a lot of interest in the search for models that produce (for long times) nonlinearly growing mean square displacements, in particular, $\langle [X(t) - X(0)]^2 \rangle = t^\alpha$ with $\alpha \neq 1$, called *anomalous diffusion* processes [28, 29]. (Specifically, $\alpha < 1$ denotes sub-diffusion, $\alpha > 1$ super-diffusion.) To find concrete examples, we refer the interested Reader to [30] and references therein. Later, we will also briefly consider the long-time diffusion behavior of a system with delay from this perspective, see Sect. 7.5.5.

2.2.4 Ornstein–Uhlenbeck Process

Another prominent example system is an overdamped particle in a quadratic or "harmonic" potential $(a/2)x^2$. This potential confines the particle's motion around its minimum, leaving no way to escape, and is hence sometimes denoted a "*trap*". The motion of the particle is described by a fully linear LE like the one of a free particle, but this time with an additional linear restoring force $F = -aX$ in (2.16). Then, the LE generates a so called *Ornstein–Uhlenbeck process* (OU) in the position.

---

[9] If one does not neglect inertia, the MSD increases approximately quadratically $\sim t^2$ on very short times up to m/γ, and then passes over to the diffusive regime with $2D_0t$.

[10] In other words, according to (2.18), the MSD equals $2\sigma^2$ at time τ_B.

We have in fact already discussed another OU above, that is Eq. (2.11) for the free particle with inertia, defining an OU in the velocity. Accordingly, the solution of the overdamped particle in a trap is analogously to (2.12) given by

$$X(t) = x_0 e^{-(a/\gamma)t} + \sqrt{2k_B\mathcal{T}/\gamma} \int_0^t e^{-(a/\gamma)|t-t'|}\xi(t')\mathrm{d}t' \tag{2.21}$$

with $X(0) = x_0$. In the long-time limit, this system approaches a stable steady state (if $a > 0$), with a Gaussian-distributed PDF with zero mean. The Gaussianity, which is "preserved" by the *linear* LE, stems from the noise. The position-autocorrelation function reads

$$\big\langle X(t_1)X(t_2)\big\rangle = x_0^2\, e^{-(a/\gamma)(t_1+t_2)} + \frac{k_B\mathcal{T}}{a}\left[e^{-(a/\gamma)|t_1-t_2|} - e^{-(a/\gamma)|t_1+t_2|}\right], \tag{2.22}$$

and in the steady state one thus finds

$$C(\Delta t) = \langle X(t)X(t+\Delta t)\rangle_{\mathrm{ss}} = \frac{k_B\mathcal{T}}{a}\left[e^{-(a/\gamma)\Delta t}\right] = \frac{\gamma D_0}{a}\left[e^{-(a/\gamma)\Delta t}\right]. \tag{2.23}$$

In contrast to the free particle, the MSD reaches a constant value $\gamma D_0/a$, due to the confinement by the quadratic potential. For example, when starting at $X(0) = 0$, the MSD reads

$$\begin{aligned}\langle [X(t) - X(0)]^2\rangle &= \langle X(t)^2\rangle - 2\langle X(t)X(0)\rangle + \langle X(0)^2\rangle \\ &\overset{(2.22)}{=} \frac{\gamma D_0}{a}\left[1 - e^{-2(a/\gamma)t}\right]\end{aligned} \tag{2.24}$$

Thus, the MSD saturates at the timescale where the autocorrelation function decays. As can be seen from Eqs. (2.23), (2.24), the corresponding relaxation time is of the order of γ/a.

Figure 2.4 displays some exemplary trajectories, as well as the corresponding autocorrelation function (2.23) of a Brownian particle in an harmonic trap in the steady state.

2.2.5 White Noise—Wiener Process—Stochastic Calculus

Although very common in physics literature, the treatment of stochastic equations with white noise (2.2) is less popular among mathematicians, who typically write down the LE in the form

$$\gamma \mathrm{d}X(t) = F\mathrm{d}t + \sqrt{2k_B\mathcal{T}\gamma}\,\mathrm{d}W(t), \tag{2.25}$$

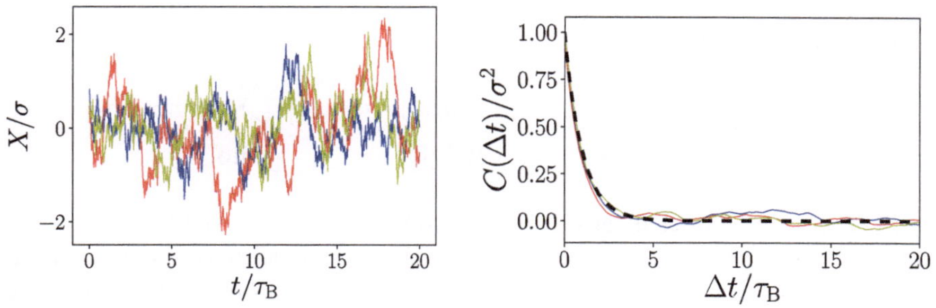

Fig. 2.4 *Left panel:* Three example trajectories of an overdamped Brownian particle in a harmonic trap, whose position X is performing an Ornstein–Uhlenbeck process. *Right panel:* Corresponding position-autocorrelation function, calculated from the three single trajectories (in the respective colors), and from the analytical results (2.23) shown by a dashed black line. We have cut off about 10^5 time steps to ensure steady state conditions. The only free parameter a is set to $a = \gamma/\tau_B$. It may seem strange to calculate the correlation function from a single trajectory. However, the system is ergodic, i.e., in the steady state the ensemble average is equivalent to the time-average of the process

with the *Wiener process* $W(t)$. The latter is a random process defined by independent, Gaussian-distributed increments $\mathrm{d}W(t) = W(t + \mathrm{d}t) - W(t)$, with

$$\begin{aligned} W(0) &= 0, \\ \langle W(t + \mathrm{d}t) - W(t)\rangle &= 0, \\ \langle W(t)W(t + \mathrm{d}t)\rangle &= t, \ \forall\, \mathrm{d}t > 0. \end{aligned} \tag{2.26}$$

As opposed to ξ, the Wiener process $W(t)$ is *continuous* in time. It is not differentiable.

The reason why mathematicians rather favor the description with the Wiener process is self-evident. While $\mathrm{d}X$ in (2.25) is well-defined, the velocity $\dot{X}$ in the LE (2.16) is not well-defined and its absolute value is almost surely infinity at each instant of time [2]. We will nevertheless consider white noise equations in this thesis, consistent with physics literature. After all, both representations have to be treated with equal care, and require the definition of the stochastic calculus.

Itô versus Stratonovich The LEs in the form (2.25), (2.16) [or (2.1) for underdamped motion] are in fact not sufficient to fully characterize the noisy dynamics, but one additionally has to specify the *stochastic calculus*. This is not only true for multiplicative noise, but also for additive noise (despite occasional misconception in the literature), as we explain in the following. The stochastic calculus decides on the question of how to interpret integrals of the type $\int_0^t X(t)\mathrm{d}W(t)$, and roots in the issue of the zero-infinite nature of the *white noise*, i.e., its delta-autocorrelation. Two common choices are Itô, i.e., the starting-point rule, and Stratonovich, i.e., the mid-point rule, which amount to the integration of some arbitrary functional $\mathcal{A}$

$$\int_0^T \mathcal{A}[X(t)]\,\mathrm{d}W(t) = \begin{cases} \sum_{i=0}^{k-1} \mathcal{A}\left[\frac{X(t_{i+1})+X(t_i)}{2}, \frac{t_{i+1}+t_i}{2}\right]\left[W(t_{i+1}) - W(t_i)\right], & \text{Stratonovich} \\ \sum_{i=0}^{k-1} A\left[X(t_i), t_i\right]\left[W(t_{i+1}) - W(t_i)\right], & \text{Itô} \end{cases}. \quad (2.27)$$

See [21, 31] for further details. More generally, both are special types of α-type calculus,[11] where Itô's corresponds to $\alpha = 0$ and Stratonovich to $\alpha = 0.5$. Since they are very uncommon in physics literature, we do not discuss other α-type calculuses here.

As one can see from (2.27), only in the Itô calculus X is *non-anticipating* (or an adapted process), as it does not depend on future X. An important difference is that Stratonovich preserves the normal rules of calculus known from deterministic equations, like the chain rule, while Itô yields different rules. Another distinction is given by the connection between LE and PDF. When Itô calculus is applied, the associated Fokker–Planck equation (see Sect. 3.2) has an additional term, called the *spurious* or *noise-induced drift*, as opposed to the corresponding Stratonovich FPE, see e.g., [21, 31]. This additional drift term, which is only present in Itô, yields a different solution, i.e, a different PDF.

However, the noise-induced drift vanishes for additive noise (as it involves the spatial derivative of the noise strength in the Langevin-force, which is zero for additive noise), thus, both calculus yield the exact same PDF, making the distinction on that level impossible. For this reason, some authors argue that the distinction of the calculus is not necessary at all in this case. This is indeed not true. For additive noise, *other* quantities may still differ, for example the cross-correlations $\langle X(t)W(t')\rangle$. For example, Itô calculus implies $\langle X(t)W(t)\rangle = 0$, whereas in Stratonovich $\langle X(t)W(t)\rangle = t/2$, [for initial condition $X(0) = 0$]. This underlines that the one-time PDF, after all, does not contain the full information about the noisy process.

However, one should emphasize, that *both* calculuses can be used in any case, and the corresponding equations can be *transformed into each other in a unique way* by means of Itô's formula [2]. This is for example very useful for path integrals, which are easier to handle in Itô calculus (where the Jacobian of the transformation from noise to space coordinates, i.e, $\mathcal{J}$ in (2.28), is unity, see [32]). Consistency is important, whenever using a formula one needs to make sure which calculus it was defined in (an example being the stochastic heat, see Chap. 4).

In the 1970's and 1980's there was a debate about the interpretation (and "correct" calculus in physical systems), see, e.g., [33] for a historical review. The main outcomes of this debate may roughly be summarized as follows: Stratonovich typically shall be used to model physical systems that are time-continuous. On the contrary, intrinsically discontinuous systems shall be treated with Itô's rule (e.g., financial markets, population growth dynamics). While there seems to be general consent on this point, there is in fact still an ongoing controversy about the more appropriate calculus in physical systems [33]. We aim to add to this debate the following argu-

---

[11] Note that this α has nothing to do with the exponent defining anomalous diffusion.

ment: when one thinks of a physical system on the molecular scale, the question of continuity versus discreteness may actually be fuzzy (the number of Adenosine triphosphate, short ATP, molecules, fueling a process is discrete, while the fueled molecular motor [34] itself is moving continuously through space). Due to improved experimental techniques there is even renewed interest in the question of what the "correct" calculus is, see, e.g., [35–38] (also arguing that other calculus than the α-type might be more appropriate [39]). However, as pointed out in [33], some of the arguments and questions should be viewed critically.

We will employ *Stratonovich calculus* throughout the thesis, because this is, in general, the more common choice in physics literature, and, in particular, consistent with the formulation of the stochastic energetics [23] employed in Part III, see Chap. 4.

2.2.6 Path Integral Representation

We have already seen how the probability density function $\rho_1(x, t)$ can be defined on the basis of the stochastic process X [see Eq. (2.7)]. It gives the probability to observe a certain value at a given time t. Besides the PDF, another way of expressing the process's probabilities are *path probabilities*, which tell how likely it is to observe an entire trajectory. In particular, one defines *paths* as $\mathbf{X} = \{X(t)\}_{t_s}^{t_f}$, i.e., $X(t)$ for all $t \in [t_s, t_f]$ giving one specific realization of the random process. For a Markovian process with white noise (2.4), the path probability $\mathcal{P}$ conditioned on the starting point $X(t_s) = x_s$, is given by the exponential of the *Onsager–Machlup action* [40, 41]. The latter readily follows from the Markovian LE, yielding (for the overdamped case)

$$\mathcal{P}[\mathbf{X}|x_s] = \mathcal{J}\, e^{-\int_{t_s}^{t_f} \left(\dot{X}(t') - \gamma^{-1} F[X(t')]\right)^2 / (4 k_B \mathcal{T} \gamma) \mathrm{d}t'}. \tag{2.28}$$

The $\mathcal{J}$ factor is a Jacobi determinant stemming from a coordinate transformation from noise to spatial coordinates, $\xi(t_j) \to X(t_j)$, see Ref. [32] for details. In Itô calculus $\mathcal{J}$ is unity, in Stratonovich not. As we will mostly be interested in path probability ratios in this thesis, these factors anyway cancel out. We will thus not further discuss them.

In the framework of stochastic thermodynamics, which we will introduce in Chap. 4, the path probabilities play a central role, in particular to measure irreversibility. More generally, they can be used to calculate the ensemble average of a given functional $\mathcal{A}$ along an entire trajectory, via

$$\langle \mathcal{A}[\mathbf{X}] \rangle = \int_{t_s}^{t_f} \mathcal{A}[\mathbf{X}] \mathcal{P}[\mathbf{X}]\, \mathfrak{D}\mathbf{X} \tag{2.29}$$

with $\mathcal{P}[\mathbf{X}] = \mathcal{P}[\mathbf{X}|x_s] \rho_1(x_s, t_s)$.

The concept of path probabilities can be generalized to non-Markovian processes. For example, Ref. [42] discusses path integrals for processes with colored noise, and Ref. [43, 44] employs path probabilities for system with time delay. A crucial issue in stochastic thermodynamics of non-Markovian processes is the path probability of the backward process. We will come back to this issue in Chap. 4.

In the following, we will make a little detour, exploring the aforementioned connection between mechanical laws and stochastic dynamics in more detail. We will proceed with the properties of the process generated by the LE in Sect. 2.4, where we will finally introduce the time delay.

2.3 Generalised Langevin Equations—How Stochastic Motion Emerges from First Principles

The introduction of time-reversal asymmetry

2.3.1 *Infinite Harmonic Oscillators Bath—An Example of a Mori–Zwanzig Projection*

There is an established and well-understood link between the Langevin equation (2.1), (2.16) and *first principles*, via Mori–Zwanzig projection method. To be more specific, in the picture of the Mori–Zwanzig derivation, the "microscopic world" consists of point particles governed by Newton's laws (i.e., fulfill *actio* = *reactio*), obeying momentum and energy conservation. These laws are considered the first principles here.[12] In the following, we give a rough overview over this derivation. More details can be found, e.g, in Ref. [20, 23, 47, 48].

One starts with Hamiltonian dynamics for *all* particles with various, presumably identical, "bath particles", say $Y_{j\in\mathbb{N}}$, and one special particle, say X. The latter could be the position of a colloid. One particularly simple possibility is to consider for the Y_j *harmonic oscillators* with frequencies ω_i, which are all linearly coupled to X, but not among each other (see, e.g., [20] p. 22). X is reciprocally (i.e, the interaction fulfills *actio* = *reactio*) coupled back to each harmonic oscillator, and may further "feel" external forces F. This situation is described by the following coupled equations

$$\begin{aligned} m\ddot{X} &= F + \sum_j (b_j Y_j - (b_j/\omega_j)^2 X) \\ \ddot{Y}_j &= -\omega_j^2 Y_j + b_j X. \end{aligned} \tag{2.30}$$

[12] As it is also the case in kinetic gas theory [45, 46].

These are the "microscopic" equations for the present case. Due to the *linearity* of the Y_j-equations, this set of equations can be handled quite well analytically.

In particular, one can *project* all equations onto the dynamical equation of the colloid, by formally solving and substituting them into each other. To this end, we perform the Laplace-transformation $\hat{Y}(s) = \int_0^\infty Y(t)e^{-st}\mathrm{d}t$ of the harmonic oscillator equations, i.e., the second line of (2.30). This yields for each oscillator j

$$
\begin{aligned}
s^2\hat{Y}_j - sY_j(0) - \dot{Y}_j(0) &= -\omega_j^2\hat{Y}_j + b_j\hat{X} \\
\Rightarrow \hat{Y}_j &= (s^2+\omega_j^2)^{-1}[b_j\hat{X} + sY_j(0) + \dot{Y}_j(0)]. \qquad (2.31)
\end{aligned}
$$

Transforming back to real space, this yields

$$
\dot{Y}_j(t) = b_j \int_0^t \frac{\sin(\omega_j t')}{\omega_j} X(t-t')\,\mathrm{d}t' + \cos(\omega_j t)Y_j(0) + \frac{\sin(\omega_j t)}{\omega_j}\dot{Y}_j(0). \qquad (2.32)
$$

In the last step, we have utilizes the convolution theorem to transform the first term. Using partial integration, one can further rewrite this as

$$
\begin{aligned}
\dot{Y}_j(t) = &-\int_0^t \frac{b_j}{\omega_j^2}\cos(\omega_j t')\dot{X}(t-t')\,\mathrm{d}t' - \frac{b_j}{\omega_j^2}\cos(\omega_j t)X(0) + \frac{b_j}{\omega_j^2}X(t) \\
&+ \cos(\omega_j t)Y_j(0) + {\omega_j}^{-1}\sin(\omega_j t)\dot{Y}_j(0). \qquad (2.33)
\end{aligned}
$$

Now, we consider again the first line of (2.30), and substitute the $\dot{Y}_j(t)$-terms by the corresponding expressions (2.33), yielding the following equation for X [20]

$$
m\ddot{X}(t) = F - \int_0^t \underbrace{\sum_j (b_j/\omega_j)^2 \cos(\omega_j t')}_{=\gamma(t')}\,\dot{X}(t-t')\mathrm{d}t' + R(t). \qquad (2.34)
$$

This expression describes the exact same X dynamics as Eq. (2.30). In Eq. (2.34), we have already grouped the interaction with the bath into two terms, that is the term $R(t)$ given below in (2.35), and the integral which contains the "friction" kernel $\gamma(t')$ defining a form of memory (as it depends on past times). The integral depends on the past and instantaneous velocity $\dot{X}$ (and the constants b_j, ω_j) and gives the "friction". This term does not depend on the initial conditions of the bath oscillators. Interestingly, by choosing the b_j and ω_j accordingly, one can in this way generate arbitrary friction kernels, because in the limit of infinitely many oscillators the sum converges to a Fourier-integral .

Further, the R term reads

$$
R(t) = \sum_j \dot{Y}_j(0)\frac{b_j}{\omega_j}\sin(\omega_j t) + \sum_j b_j\left[b_j Y_j(0) - (b_j/\omega_j)^2 X(0)\right]\cos(\omega_j t). \qquad (2.35)
$$

It involves the initial position $X(0)$ [but not the initial velocity $\dot{X}(0)$], and all initial positions and velocities of the bath degrees of freedom, $Y_j(0)$ and $\dot{Y}_j(0)$, respectively. For a given set of initial conditions, this a *deterministic* term, describing the impact of Y_j on the dynamics of X. In contrast to the friction, it does not depend on $\dot{X}$ at any time t [nor on $X(t > 0)$].

In the here demonstrated simple form, the projection method is only applicable for heat baths with linear microscopic equations. However, there is also a way to project out more general microscopic equations, e.g., nonlinear baths, which involves transformation to the Hilbert-space and application of projection operators [20]. However, the main results [a generalized LE (2.37) with FDR (2.36), as given below] and the key arguments are, in principle, always the same. We further note that we will later in Sect. 6.1 perform a projection very similar to the one shown here (however, there we start from coupled stochastic equations).

Let us return to Eq. (2.34). Yet, we have not obtained a mesoscopic description. Now comes an important, final step, which transforms equation (2.34) into a stochastic one.

2.3.2 Coarse-Graining—Forgetting Some Details

It is unpractical and indeed not necessary to carry around the cumbersome terms $R(t)$ and $\gamma(t)$. Further, for more realistic systems, which have typically much more complicated underlying equations (where some details could be unknown), it may even be impossible to work with deterministic, i.e., precise equations such as (2.34). To overcome these complications, one typically performs a *coarse-graining* procedure by throwing away some information about the process. In this way, one can create a simpler, stochastic (approximate) description for X, by a mesoscopic equation.

In the present context, one *assumes* to this end that

- there are infinitely many bath d.o.f.
- the bath d.o.f. initial conditions $Y(0)$ are uncorrelated among each other [which they are not in naturally evolving dynamics from (2.30) at $t > 0$].

These assumptions bring one in the position to apply the *central limit theorem*. Further, one assumes that

- the equipartition theorem holds, i.e., the mean kinetic energy is $\langle \dot{Y}_j(0)^2 \rangle = k_B \mathcal{T}$ and the potential energy in the springs is $\langle \omega_j^2 [Y_j(0) - (b_j/\omega_j^2) X(0)]^2 \rangle = k_B \mathcal{T}$.
- alternatively to the previous items, one can assume that the initial conditions are independently drawn from a Boltzmann distribution $\sim e^{-H_B/(k_B \mathcal{T})}$, with the total bath energy H_B. The result is equivalent.

In other words, the bath is assumed to be infinitely large and in thermal equilibrium. This is a conceptually very important step, as it introduces stochasticity and, at the same time, an element of time-asymmetry (which we elucidate further below). By application of the central limit theorem, these assumptions allow to replace the entire

R term by a Gaussian, zero-mean *noise* ν, whose correlations can be proven [20] to obey the famous *second fluctuation-dissipation relation* (FDR)

$$\text{FDR:}\quad \langle \nu(t)\nu(t')\rangle = k_{\mathrm{B}}T\ \gamma(|t-s|), \tag{2.36}$$

also called Kubo relation of second kind [49]. Due to the construction of the LE, in particular, the assumptions on the bath which yield the FDR, the LE describes systems that relax to thermal equilibrium in the absence of external non-conservative forces F. More precisely, the FDR is connected to a vanishing entropy production. While this cannot immediately be seen from Eq. (2.36) alone, it will become clear later in Part III (see Chap. 10). Thus, for a wide class of systems, the existence of an FDR implies the possibility to reach thermal equilibrium [49, 50].

However, one should be aware that corresponding statements are not trivial in the presence of different driving mechanism. For example, in the presence of velocity-dependent forces, the system can reach a state with vanishing heat dissipation that fulfills detailed balance[13] [40], where the FDR is broken. The fundamental connections between FDR, heat dissipation and EQ, are particularly well-understood for Markovian systems, see, e.g., [51]. For systems subject to feedback, the connections seem to be less understood. We could not find rigorous statements in the literature.

In total, starting from mechanical laws and plugging in, by hand, uncorrelated initial conditions with equally distributed energy among them, one obtains a *non-Markovian* LE, called *generalized Langevin equation*

$$m\ddot{X}(t) = F - \int_0^t \gamma(t'-t)\dot{X}(t')\mathrm{d}t' + \nu(t) \tag{2.37}$$

with *friction kernel* $\gamma(t-t')$ and colored noise ν, which fulfills the FDR (2.36). The generalized LE is and important equation for the description of processes like transport in viscoelastic media or diffusion in the presence of hydrodynamic flows.

The term *non-Markovian* refers to the fact that the future of the process X generated by this LE, at each instant of time t, cannot only be predicted based on $X(t)$ alone, but one further needs information about its *past* $X(t'<t)$. Both friction and noise have history-dependence, i.e., *memory*. Despite its non-Markovianity, there are a lot of results known for generalized LEs. The FDR is an important theoretical result in regard to the physical understanding and furthermore represents a practical tool in the mathematical treatment of GLEs. An example for the latter is the unique mapping from the correlation functions to a generalized LE provided by Volterra-like memory equations, for which we give a concrete example below (the uniqueness of such a mapping only exists when FDR holds).

Vanishing correlations Another implication of the construction of the bath, particularly of the assumed independence of the initial velocities, is that

[13] Which is a clear indicator of thermal equilibrium.

$$\begin{aligned} &\langle R(t)\dot{X}(0)\rangle = 0 \\ \Rightarrow\ &\langle \nu(t)\dot{X}(0)\rangle = 0. \end{aligned} \tag{2.38}$$

This can readily be seen from Eq. (2.35). Thus, the future noise is not correlated with the colloid's velocity at time $t = 0$. We will come back to this point in Sect. 2.3.6. One should stress that (2.38) is not true for other times $t' > 0$, i.e., $\langle \nu(t)\dot{X}(t')\rangle$ not necessarily vanishes.

Memory equation Here we briefly derive a Volterra-like *memory equation*, which uniquely maps the velocity correlation function $C_v(t) = \langle \dot{X}(0)\dot{X}(t)\rangle$ to a specific generalized LE. To this end, we will make three important assumptions: First, the force F shall be linear in X, i.e., $F = aX$ with some constant a. Second, (2.38) holds. We further assume that the ensemble shall be in equilibrium already at the beginning $t = 0$, such that $C_v(t) = \langle \dot{X}(t')\dot{X}(t'+t)\rangle_{\mathrm{ss}}$, although the initial point was indeed a special instant of time [e.g., fulfilling (2.38), which would normally not hold in a naturally evolving system (2.30)]. Formally, the last assumption is not strictly necessary, but makes the formula much more general and applicable, and is hence usually made. This is a delicate aspect, as this assumption contradicts (2.38).

Now we start with multiplying both sides of (2.37) with $\dot{X}(0)$ and taking the ensemble average of the entire expression. This yields

$$m\langle \dot{X}(0)\ddot{X}(t)\rangle = \langle \dot{X}(0)\,[-aX(t)]\rangle - \int_0^t \gamma(t'-t)\langle \dot{X}(0)\dot{X}(t')\rangle \mathrm{d}t' + \langle \dot{X}(0)\nu(t)\rangle. \tag{2.39}$$

Utilizing (2.38) and the identities

$$\begin{aligned} \frac{\mathrm{d}}{\mathrm{d}t}C_v(t) &= \frac{\mathrm{d}}{\mathrm{d}t}\langle \dot{X}(0)\dot{X}(t)\rangle, \\ \int_0^t C_v(t')\mathrm{d}t' &= \langle \dot{X}(0)X(t)\rangle, \end{aligned} \tag{2.40}$$

one readily finds

$$\frac{\mathrm{d}}{\mathrm{d}t}C_v(t) = -\int_0^t m^{-1}[a + \gamma(t'-t)]C_v(t')\mathrm{d}t'. \tag{2.41}$$

This gives a unique mapping from C_v to γ, which, in combination with (2.36) gives a unique generalized LE. Such a relation is important to find suitable models for experimental data, as the involved correlation function is often a measurable quantity. However, one shall bear in mind the necessary assumptions, e.g., linearity of the LE, that the system is in thermal equilibrium (already at the initial time), and the causality condition (2.38).

2.3.3 Side Node: Taking this Simplified Model Serious

Although the picture of infinitely many harmonic oscillators is clearly rather a metaphor for a real fluid, than an actual, realistic description, it is still a very wide spread model for a heat bath of undeniable importance from the viewpoint of theory development. Its obvious advantage is the simple, linear form. For this reason it is worth to take a closer look onto the "noise" R [from Eq. (2.35)].

To get an impression of how well the approximation by a Gaussian noise is, for this particular construction of the heat bath, we consider possible realizations of R from (2.35). Figure 2.5 shows the term $R(t)$ itself supplemented with its autocorrelation function. We have chosen a uniform (equidistant) distribution of ω_j and Gaussian-distributed random velocities $\dot{Y}_j(0)$, and we have depicted two different cases for the positions: once with randomly distributed initial positions $Y_j(0)$, and once with equidistantly distributed, that means *correlated*, $Y_j(0)$. Only in the former case, a truly irregularly looking temporal evolution emerges, while the latter case is rather oscillatory, with irregular amplitudes. Indeed, a random positional distribution, but equidistant velocity distribution has a similar effect (not shown here). Hence, fully uncorrelated initial conditions are essential. We conclude that the initial disorder is enough to create an irregularly looking $R(t)$ (despite the linearity of the

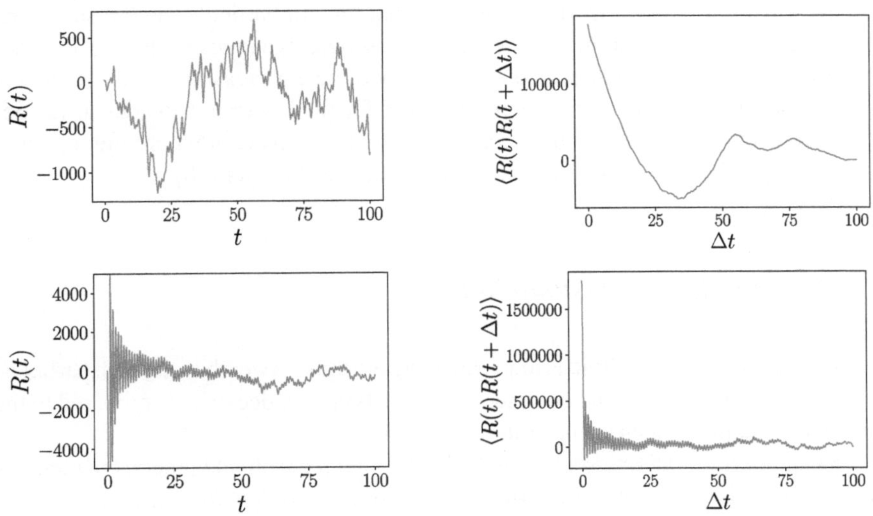

Fig. 2.5 *Left panel:* R from Eq. (2.35) and *right panel:* its autocorrelation function, for a system with $N = 10^5$ harmonic bath oscillators with equidistantly distributed frequencies ω_j among the oscillators within $[0, 2\pi]$, i. e,. $\omega_j = j\, 2\pi/N$ with $j = \{1, \ldots, N\}$. The initial velocities $\dot{Y}_j(0)$ are randomly drawn from a Gaussian distribution (with variance 1), while the initial positions differ between upper and lower plots. *Upper panels:* Gaussian-distributed random initial positions $Y_j(0)$, with variance 0.27. *Lower panels:* Equidistantly distributed positions $Y_j(0)$ within $[-0.2, 0.2]$. The parameters are in both cases chosen such that the equipartition theorem is fulfilled (see text). Other parameters: $X(0) = 0$, $b_j = 1\ \forall j$

model), as is confirmed by testing different variations[14] among the three distributions of ω_j, $Y_j(0)$, $\dot{Y}_j(0)$, and other system sizes. Remarkably, the system size seems to play a minor role. We have checked different numbers of harmonic oscillators N between $N = 10^4$ and $N = 10^6$, and found surprisingly similar results (almost not distinguishable).[15] In no case, we found an evolution comparable to the white noise (plotted in Fig. 2.2). At most, R realizations resemble a Wiener process (compare with Fig. 2.3).

It is noteworthy that in our tests, a system with randomly distributed ω_j seems to produce an irregular $R(t)$, even if $Y_j(0)$, $\dot{Y}_j(0)$ are correlated (results not shown here). This suggests that a (frozen) disorder in the system parameters has a somewhat stronger effect than "initial disorder". However, we have not analyzed this in detail.

These considerations show that the noise approximation is justified, but only for uncorrelated initial conditions. In a sense, the randomness must hence be introduced in this model by hand. This is indeed not surprising given the linearity of all forces. A nonlinear model (which would anyway be more realistic), is of course expected to yield more disorder, and can even be truly *chaotic* for finite system sizes [53]. It is interesting to note that there is some debate about under which conditions a linear differential equation can create chaotic trajectories. This appears to be only possible for infinitely large systems, see, e.g., [54].

At this point one may further hint to the long-standing debate in statistical physics concerning the true *origin of the randomness* of Brownian motion and the diffusive behavior from microscopic, deterministic equations. In particular, is it a sufficient, or is it a necessary requirement that the micropscopic dynamics is chaotic, or is the extreme high dimensionality of the phase space the true reason that by reduction causes the randomly appearing motion? We refer the interested Reader to [3, 55] and references therein. Famous theoretical models in this context are the (chaotic) Lorentz-Gas [56, 57] an the (non-chaotic) "wind-tree" model [58].

2.3.4 The Markov Assumption

In many situations, the colloid is much heavier, and thus slower than the bath particles. Then the memory kernel and colored noise both typically decay fast compared to the characteristic timescales of the colloidal motion.

When the colloidal timescales can be well-separated from the others, the memory kernel and noise-correlations are often assumed to decay *infinitely fast*, i.e.,

$$\langle \nu(t)\nu(t')\rangle \to 2\,k_\mathrm{B}T\;\delta(t-t'),$$

[14] We have not studied the effect of random b_j distribution (but simply set them all to unity here).

[15] Except for the recurrence time, e.g., how long it takes until the initial configuration is recovered, which we do not discuss here. This time strongly depends on N [52], as can indeed easily be checked from (2.35).

and at the same time, due to FDR,

$$\gamma(t - t') \to 2\,\delta(t - t').$$

This is the *Markov assumption*, yielding a Markovian Langevin equation with *instantaneous* friction and *white* noise (2.1), free of any memory effects.

Returning to the "actual" system, i.e., a colloid in a surrounding fluid, the Markov assumption amounts to assuming the collisions between the bath particles and the colloid to be infinitely fast, and, in fact, neglecting any temporal correlations among the bath particles.

2.3.5 Real-World Complications

In the real world, especially when one considers complex systems which contain various different length and timescales, the idealized, strict separation of timescales is not always possible. This can be due to the fact that the heat bath particles themselves have internal degrees of freedom. For example, a heat bath containing polymers or worm-like micelles [29], which have internal relaxation times on timescales comparable to the colloidal motion, results in viscoelastic memory [20, 28, 50, 60, 61]. Another example of colloidal systems, where memory effects do play a non-negligible role, are suspensions with significant hydrodynamic flows [59, 62–64]. The hydrodynamic timescale of a colloidal sphere in acetone was measured in Ref. [59], see Fig. 2.6 for a schematic overview. Further, for an even much more complicated example system, think of diffusion in a cell, where the different cellular components, liquids and lipids form a highly nontrivial environment. Here, the picture of one big colloidal particle in

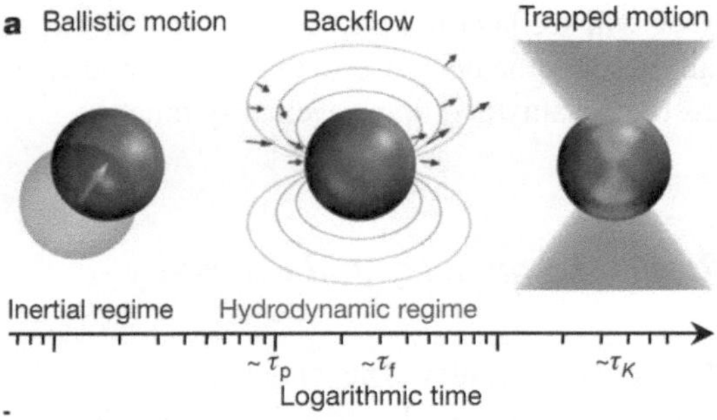

Fig. 2.6 This Figure stems from Ref. [59]. On very short timescales ($t \ll \tau_p \approx m/\gamma$), the particle undergoes ballistic motion (*left*). On the timescale τ_f, hydrodynamic backflow develops (*center*); the gray lines indicate the emerging fluid velocity field; the arrows stem from computer simulations, see [59]. For $t > \tau_K$, the harmonic potential from the optical trap (see Sect. 2.4.1) confines particle diffusion (*right*). The relation between τ_f versus the diffusive timescale τ_B decides whether the Markov assumption is appropriate

a bath of identical, tiny particles, might loose meaning. In all these counter example, the Markov assumption is not appropriate.

Moreover, the equilibrium assumptions on the bath particles, which yield the FDR between friction and noise, are as well not always justified. An example are active systems, which we briefly discuss in Sect. 4.7. Again, the aforementioned living cell may serve as an example. Due to the fact that various components, like molecular motors, transform chemical energy into motion [34], this system is intrinsically out of equilibrium, such that the FDR is not expected to hold.

Recent research aiming for generalizations An interesting research direction are the various recently derived generalization of the fluctuation-dissipation relation to non-equilibrium systems, see, e.g., [50, 65–68]. Immediate applications are systems in active baths, where the bath particles themselves are self-propelled. We will later introduce active systems and explain the terminology (see Sect. 4.7). We are not aware of related research in regard to time-delayed processes, e.g., a bath or colloid under time-delayed feedback. In this thesis, we have not considered such non-equilibrium fluctuation-dissipation relations. However, it would be very interesting to see whether rigorous results can be found, which might be possible for purely linear systems.

Time delay In most parts of this thesis, we will consider another kind of non-Markovian LE than the generalized Langevin equations discussed here, namely the delay LE. It describes a very different type of system, where the memory comes from "outside" (not from interactions and correlations within the bath). Consequentially, there is no relation between the delay and noise correlations. Hence, we can already expect the treatment to be even more involved. Compared to the situation discussed here, the delay LE is so to say the "Markovian limit" of the generalized LE (i.e., the memory effects due to the noise are neglected), with a time delay in the feedback force F, which again renders the motion non-Markovian. Thus, for delay LEs, the non-Markovianity is "tunable", and the assumption that the system can find an equilibrium is not *per se* justified and needs to be revisited (in fact, the system is intrinsically out of equilibrium, as we will see later in Part III).

Before we finally discuss the delay LE, let us make another brief excursion discussing the notions of causality and time-reversal symmetry.

2.3.6 Time-Reversal Symmetry and Causality

The mechanical time-reversibility The "microscopic level" in the derivation sketched above is given by point particles governed by Newton's laws. Such mechanical, Hamiltonian systems are *time-reversal symmetric*. For the sake of illustration, imagine a set of billiard balls bumping into each other and then spreading apart, conserving total momentum and energy (we neglect friction, which slows them down). Now imagine two observers, who see a short sequence, e.g., around the moment where the billiards hit, lets say of length $\{t_0, t_1\}$ with $t_0 < t_1$. One observer should make a prediction of where each of the balls will be at a short time *later* (at $t_1 + \Delta t$), while the other observer shall make a guess where all the balls have been an (equally long) short time *before* the observed sequence (at $t_0 - \Delta t$). *Who has the better chances to be correct?* Naively one might think that the first observer may have the easier task, as he has observed the "cause" and should guess the "effect". However, in fact, both observers have the exact same chances (which are, for observers who can measure and compute perfectly, 100%).

The "microscopic", mechanical world (recall that we neglect friction) is completely reversible in time. Each mechanical trajectory seems equally realistic as its time-reversed picture,[16] and one can just as well predict the "cause" by observing the "effect", as the other way round. This property of mechanical laws is called the *mechanical time-reversal symmetry* [23, 69].

The stochastic (ir)reversibility and causality Things change on the "mesoscopic", stochastic level, where a *causality* principle becomes important that distinguishes between future and past.

For example, the LE with white noise describes a process where the correlations between the system state X and the past noise, $\langle X(t_2)\xi(t_1)\rangle$ and $\langle \dot{X}(t_2)\xi(t_1)\rangle$, with $t_1 < t_2$, can be *non*-zero, but where in sharp contrast [70]

$$\langle X(t_1)\xi(t_2)\rangle \equiv 0 \tag{2.42a}$$

$$\langle \dot{X}(t_1)\xi(t_2)\rangle \equiv 0 \tag{2.42b}$$

strictly hold. The stochastic variables cannot be correlated with the future noise, but with its past noise, which is often denoted causality principle. Remarkably, this does indeed not necessarily disagree with our physical intuition, and almost seems self-evident. Yet, it implies a sharp distinction between future and past, contradicting the above described mechanical time-reversal symmetry [23]. For the physical understanding of this symmetry breaking it is important that it concerns ensemble average, i.e, averages over all possible realizations. One should further emphasize that (2.42) even holds in thermal equilibrium, when the system obeys *thermodynamic reversibility*. The latter is another concept, which we will explain below.

Further, the non-Markovian generalized LE with colored noise considered above, yields the asymmetric correlations

$$\langle \dot{X}(t_1)\nu(t_2)\rangle \neq \langle \dot{X}(t_2)\nu(t_1)\rangle. \tag{2.42c}$$

(The strict statement $\langle \dot{X}(t_1)\nu(t_2)\rangle = 0$ is only true for white noise). Again, this implies breaking time-reversal symmetry, even in equilibrium.

The Eqs. (2.42) demonstrate that the LE does not fulfill time-reversibility in the way the mechanical laws do, i.e., the LE already distinguishes between future and past. The underlying reason is an element of asymmetry introduced within the derivation, which is often done by a specific choice of initial conditions. In the aforementioned derivation by Mori and Zwanzig that is the assumption of *un*correlated initial

[16] This concepts will be further elucidated in Chap. 4 in the context of entropy production.

conditions of the bath degrees of freedom. Specifically, at the end of the projection methods, one must assume that all bath particles have independent initial positions and momenta. Although not directly formally related, this assumption resembles the Stosszahlansatz or molecular chaos hypothesis[17] from kinetic gas theory [45, 46]. This step in the derivation is essential to obtain a LE with Gaussian white or colored noise, which fulfills the second FDR. Indeed, such a step is also crucial in the famous derivation of the fluctuation-dissipation relation by Kubo [49], as is explicitly discussed in Ref. [73]. The derivation by Kubo is somewhat more general (not relying on a specific model for the bath), but in fact is build upon the assumptions $\langle \nu(t)\dot{X}(0)\rangle = 0$, and, more trivially, $\langle \nu \rangle = 0$ (which is consistent with the expected translational symmetry of the overall system[18] [74].). The first assumption, which is identical to (2.38), appears to be essential. All these considerations show that a causality argument, or another way of time-reversal symmetry breaking which may be introduced by the initial conditions, is crucial to find a stochastic description. Such an assumption is also denoted *axiom of causality* [75].

Considering the following course of this thesis, this topic is relevant for three reasons. First, we will employ (2.42) multiple times in Parts II and III. Second, we will encounter a mathematical and physical problem in the formulation of stochastic thermodynamics for non-Markovian processes, which is related to the acausality of the backward process. This will be discussed in Chap. 4. As a primal step towards this problem, it is necessary to be aware of the role of causality in the considered stochastic models. Third, it is interesting and important to understand the *different notions of reversibility*.

We have already explained mechanical reversibility, and the stochastic irreversibility above. Further, in thermodynamics, (ir)reversibility is a central concept, which we will introduce in Chap. 4. *Thermodynamic reversibility* signals equilibrium situations, where the system is stationary and energy (and probability) flows vanish on average (we will elucidate this concepts later in Chap. 4).Under suitable conditions, Langevin equations can indeed describe such thermodynamical reversible equilibrium trajectories, despite the inherent mechanical time-reversal symmetry loss. Furthermore, Langevin equations are also capable of describing nonequilibrium processes, which attempt to maximize their entropy (more precisely, their free energy) possibly hindered by external driving mechanisms. However, Langevin equations *cannot* generate "unphysical" trajectories that violate the second law of thermodynamics (i.e., reduce the total entropy, see Chap. 4), although purely mechanical systems *can*.[19]

[17] Here one should also mention Lochschmidt's paradox [71, 72], which concerns the impossibility to derive time-asymmetric laws (referring to the second law of thermodynamics, see Chap. 4) from time-symmetric dynamics (mechanical microdynamics).

[18] Or equivalently the overall momentum conservation of all particles including bath's, as given by Noether's theorem

[19] There are, in fact, equally many possible trajectories of a mechanical system that minimize and maximize entropy. For every trajectory which increases entropy, there is one which decreases entropy, obtained by reversing all velocities of the end point configuration and use this as new initial condition.

We will come back to this point and extend the discussion about causality in LEs in Sect. 4.4.

Macroscopic reversibility We briefly aim to mention that there is indeed yet another notion of reversibility relevant in this context, describing macroscopic processes. It is an important concept in traditional macroscopic thermodynamics, for example for the realization of a Carnot cycle [1]. It refers to the possibility of retracing the history of macrostates in the opposite order in time, which often requires external driving of the system in a *quasi-static* manner, i.e., so slow that the system has the chance to "reach equilibrium" at each step. To this end, the macroscopic states shall be infinitesimally close to each other [76]. In this thesis, we are not concerned with such (often cyclic) processes, and will thus not consider this type of macroscopic reversibility. However, it is indeed build upon the aforementioned thermodynamic reversibility of equilibrium systems, which is, in contrast, concerned with individual trajectories.

Experimental confirmation of the time-reversal asymmetry In real systems, the white noise assumption never strictly holds, such that one shall expect $\langle \dot{X}(t_1)\nu(t_2)\rangle \neq 0$ in the corresponding stochastic model. As we have mentioned before, the causality principle then amounts to the asymmetry of $\langle \dot{X}(t_1)\nu(t_2)\rangle \neq \langle \dot{X}(t_2)\nu(t_1)\rangle$. In Ref. [62], this asymmetry was experimentally confirmed for a colloid in a trap (realized by "optical tweezers", see Sect. 2.4.1). Due to hydrodynamic coupling between colloid and bath, the colloid's velocity is also correlated with the future noise, as one indeed would expect considering the underlying mechanical motion of all individual particles. The measured asymmetry of the correlations underlines the time reversal symmetry-breaking (in this equilibrium system).

2.4 Introduction to the Langevin Equation with Time Delay

Now we turn to the main topic of this thesis, that is, stochastic systems with time delay. In the last Section, we have seen that stochastic motion emerges by coarse-graining out of a microscopic world which is governed by Newton's laws. We have further seen that this stochastic motion is describable by a generalized Langevin equation involving memory in the form of a friction kernel and noise correlations, which are connected by the FDR, as follows from the underlying first principles. The FDR guarantees that the equation describes a system which, in the absence of external driving, approaches thermal equilibrium. In the Markovian limit, which is appropriate when the timescales between coarse-grained environment and particle dynamics are well-separated, this gives rise to a Langevin equation with white noise and instantaneous friction.

In realistic systems, however, memory may also arise from other sources (not only from the bath itself), such that this type of description is not always the end of the story. For example, in various biological, physical or socio-economic systems, memory in the form of *time delay* naturally arises. Such a delay can have various origins, including finite information processing or decision making times [77–80], signal transmission can create delays in the communication between different agents [81], or finite refractory times [15, 82–84]. Another very important origin of delay are feedback loops [85, 86].

In the following, we make a little detour and briefly discuss an important experimental technique that can be used to realize time-delayed feedback on colloidal particles, creating a process describable by a Langevin equation with time delay. After this, we will return to the theoretical considerations and review some important features of the LE with delay.

2.4.1 Optical Traps—An Experimental Tool to Control

"The possibility of using light [...] to move physical objects may make one think of Star Trek and tractor beams, and sound like pure science fiction."

—Anders Irbäck, from the Nobel prize ceremony speech, 2018 [87]

Arthur Ashkin received in 2018 the Nobel prize in physics for his development of a device which enables trapping without damage of various small sized objects, called *optical tweezers* [88], "and for their applications to biology" (official prize motivation [89]). Enabling targeted trapping and moving of individual colloids, molecules, or atoms, i.e., objects from Ånström to micrometer scale, this device introduced a whole new way of investigating small scale systems and of probing theoretical ideas. Today, this tool is very widely used [90–97], and experiments based on it represent the backbone of single-molecule biophysics, as well as stochastic thermodynamics.

The optical tweezers or, more generally, *optical traps* consist of a Gaussian-profiled laser beam. An example for a colloid which can be controlled in this way is a transparent latex sphere, whose refraction index is higher than the surrounding fluid. Due to refraction of the laser beam at the surfaces of the latex sphere, its non-uniform light field exerts a force onto the sphere (stemming from radiation pressure). This force is directed towards the center of the beam, where the light intensity is highest. Thus, the sphere gets trapped in the center, and by moving the laser beam one can move the colloid around in a very precise manner. When the object is never too far away from the center of the laser beam, the force can often be well approximated by a *linear* force [93, 98]. This technique can also be used to control other objects, like (bio)molecules, or single atoms [99]. By clever choices of laser frequencies, beam geometry and light intensity, it is even possible to trap individual cells or organisms, like blood cells [100], algae, or bacteria [101], as well as individual components within a living cell [102] while keeping the heating effects from the laser moderate, such that living organisms can survive this procedure (see [103] for an estimation of the heating effect due to the laser light).

Experimental applications of optical traps Here we briefly mention some experimental techniques based on optical traps, which have proven themselves particularly important in the field of stochastic thermodynamics in the last years.

First, an optical trap can be used like a spring scale.[20] This enables measurement of forces, e.g., involved in biomolecular processes. For instance, by trapping the cargo of a Kinesin enzyme walking along a microtubule, its strength can be measured [104]. Kinesin is a type of molecular motor, fueled by ATP, which is essential for transport of, e.g., protein and membrane components within cells [105]. Similarly, when attaching a protein, e.g., DNA, to two trap- and movable latex spheres, the forces and the (fluctuating) work necessary to fold and unfold the protein can be tested [106]. (From such nonequilibrium experiments, one can even infer the free energy by means of the Jarzynski relation. The latter relation will be given later in Chap. 4.)

Second, precisely tunable optical trap can be used to build potential landscapes [97, 107]. The latter can further be used for "compression" or "expansion", similar to a movable piston on a macroscopic gas. In particular, a stiffer or looser trap yields more or less confinement of the thermal fluctuations, which is essential for the mesoscopic realization of a Carnot engine [108].

Finally, by using optical traps within *feedback control* loops, offers an even more sophisticated way to experimentally realize desired forces on collective or single particle systems. It can, for example, be used to generate arbitrary *interactions* between colloidal particle [10, 109] (however, inevitably involving some sort of delay). Further, it can also be used to experimentally realize a motion of the colloid as described by the delay Langevin equation studied in this thesis. In the following, we will briefly outline such an experiment.

2.4.2 Time-Delayed Feedback

Time-delayed feedback can be used to introduce delay into the stochastic motion of nano- to microscopic systems. Such a feedback can be implemented in experiments via optical tweezers [110, 111] as schematically depicted in Fig. 2.7. To this end, the particle position is measured via a camera, and the tweezers are then focused onto the previously measured position. As measurement and information processing take time, this feedback loop inevitably involves a time delay τ, such that the optical tweezers are actually focused onto the *delayed* position. Of course, the delay can in principle further be intentionally enlarged to an arbitrary value by storing the past measured positions and focusing the laser onto the position a specific time τ ago. It should be noted that there are also other, similar techniques to realize feedback on colloidal systems or single molecules, for example by temperature gradients, see, e.g., [10, 112–115].

[20] The force can be measured through measurement of the distance to the trap center.

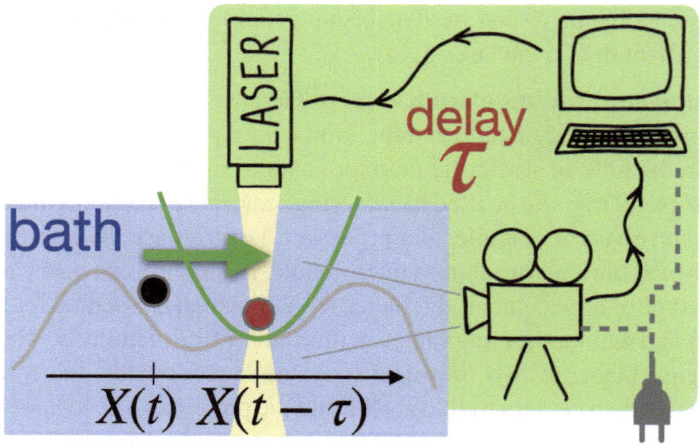

Fig. 2.7 Sketch of a possible experimental realization of a feedback control loop acting on a colloidal particle (black disk). The particle at position $X(t)$ feels a linear force (green arrow and green parabola) towards the delayed particle position $X(t-\tau)$ (red disk). This force stems from the feedback loop (green area) consisting of a camera, a computer and a laser (optical tweezers) and involves a time delay τ. The colloidal particle is immersed in a fluid which represents a heat bath (blue area), and is further subject to an external potential (gray line)

In a real experiment, the measurement is typically not continuous, but happens at discrete time, due to a finite temporal resolution of the camera. However, in this thesis, we will assume that the camera precision is high compared to the other timescales of interest, giving rise to consider the limit of infinitely fast measurements, i.e., a time-continuously operating feedback.

While recent experiments even realize delayed feedback control on multiple colloids at the same time [10], we will here only consider single Brownian particles, as there are still a lot of open questions for this case. The interplay of delay and noise makes the analytical investigation of such systems particularly challenging, but is, at the same time, found to play a crucial role for its behavior. Colloidal particles under optical tweezers feedback represent an ideal system to study this interplay.

The theoretical study of the effects of such a time delay on the particle position fluctuations is the main focus of this thesis. While the corresponding Fokker–Planck equation is not trivially given and requires special attention (this is the goal of Part II), the equation of motion of the colloid, that is, the Langevin equation can readily be written down.

2.4.3 The Langevin Equation with Time Delay

Returning to our theoretical considerations, we now investigate the overdamped delay Langevin equation

$$\dot{X}(t) = \gamma^{-1} F[X(t), X(t-\tau)] + \sqrt{2D_0}\,\xi(t), \tag{2.43}$$

where F denotes the deterministic force which is now *nonlocal in time* and is given by some (generally nonlinear) function depending on the instantaneous and on the delayed system state, $X(t)$ and $X(t-\tau)$, with $\tau \geq 0$ being the single *discrete* time delay. From a mathematical point of view, (2.43) represents a *stochastic delay differential equation* (SDDE).

In the main examples considered in this thesis, we typically separate the force into a conservative part $F_s[X(t)] = -\partial_x V_s(x)|_{x=X(t)}$ stemming from a static potential $V_s(x)$, and a "delay force" F_d, which explicitly depends on the delayed particles position $X(t-\tau)$. This force introduces *feedback* into the system, which is *a drive that depends on an (earlier) system state.* (An overview over different aspects of feedback control can be found in [85].) This feedback may stem from an external controller as discussed in the last Section. The total deterministic force can hence typically be represented as

$$F[X(t), X(t-\tau)] = F_s[X(t)] + F_d[X(t), X(t-\tau)]. \tag{2.44}$$

Due to its simple functional form, a common choice for the delay force, or "feedback force", is a linear force of type

$$F_d[X(t), X(t-\tau)] = -k[X(t) - X(t-\tau)]. \tag{2.45}$$

For instance, Eq. (2.45) is a common choice in control theory, defining a Pyragas control (see Sect. 2.4.4). Further, at $k > 0$, such a force can be used to approximate the force of optical traps. Particularly, one can consider (2.45) to be stemming from a *history-dependent* quadratic potential

$$V_d(x, x_\tau) = (k/2)[x - x_\tau]^2, \tag{2.46}$$

which can be used to model an optical trap centered around $X(t-\tau)$ [93, 98], with $F_d[X(t), X(t-\tau)] = -\partial_x V_d(x, x_\tau)|_{x=X(t),\, x_\tau=X(t-\tau)}$.

Positive/negative feedback The parameter k in (2.45) denotes the *feedback strength*, or delay force strength. In the case of the optical tweezers it corresponds to the laser intensity (which can only be positive). More generally, its sign decides whether the particle is pulled *towards* the delayed position for $k > 0$, which is usually called *negative feedback*, or if its pushed *away* from the delayed position for $k < 0$, denoted *positive feedback*. Note that negative k corresponds to positive feedback, and *vice versa*, which can be confusing! For the sake of illustration, one may think of the former type as attractive feedback, and to the later as repulsive feedback (in a metaphorical sense). In a steady state, where $X(t) \approx X(t-\tau)$, negative (i.e., attractive) feedback has the tendency to counteract perturbations, while positive (repulsive) feedback rather has a destabilizing effect. In regard to these tendencies, the terminology used here is in accordance with general control theory [85]. We will see, however, that both types can also have more versatile effects than that.

We consider linear delay forces in most parts of this thesis, while the other deterministic forces will, in general, be *nonlinear*. Please note that while we will consider position-dependent feedback in this thesis, feedback forces may, in general, also involve other dynamical quantities, like velocities. Examples can be found in [94, 116, 117].

As can immediately be seen from the definition (2.45), the force loses its history-dependence in the trivial cases $k = 0$ or $\tau = 0$. Thus, the dynamics becomes Markovian in these limits. If we do not want to distinguish between both cases, we will sometimes refer to the (general) Markovian limit with $k\tau = 0$ for the sake of a simpler notation.

Initial conditions Due to the non-locality in time, the initial condition is a *history function* $\phi(-\tau \leq t \leq 0)$, i.e., $X(t) = \phi(t)$, for $t \in [-\tau, 0]$. It can be fixed for the entire ensemble or be drawn from a distribution $\mathcal{P}(\phi)$. As setting $\phi(t)$ amounts to specifying infinitely many X values, i.e., infinitely many degrees of freedom, the initial condition indicates that the delayed process is *infinite dimensional*. We will encounter mathematical and physical properties connected to this infinite-dimensionality at various times throughout the thesis.

Memory kernels and distributed delay In general, the control could also involve *distributed delay* [84, 118–128], where the delay force

$$F_{\mathrm{d}}\left[\int_{-\infty}^{t} K(t-t')X(t')\mathrm{d}t'\right] \tag{2.47}$$

involves a convolution with a *memory kernel* K specifying the distribution of delay, and may depend on the entire past trajectory. The discrete delay introduced above corresponds to the special case of a delta-distributed memory kernel $K(t-t') = \delta(|t-t'|-\tau)$. In the case of distributed delay, the initial conditions in general require to specify a function on $]-\infty, 0]$.

We will in this following mostly focus on the discrete delay case, and later give more insight into the distributed delay cases in Parts II and III, (e.g., see Sect. 6.1, and Chap. 10). In Sect. 6.4 we will compare trajectories of systems with discrete and distributed delay.

2.4.4 Side Note: Delay Differential Equations

Before we begin with discussing dynamical properties of Langevin equations with time delay, it is instructive to first consider the deterministic case, given by a *delay differential equations* (DDE). To this end, let us for a moment forget about the noise term in the Langevin equation (2.43), i.e., $\mathcal{T} \to 0$. Alternatively considering the ensemble average of (2.43)

$$\langle \dot{X}(t) \rangle = \gamma^{-1} \langle F[X(t), X(t-\tau)] \rangle + \underbrace{\langle \sqrt{2D_0}\, \xi(t) \rangle}_{=0}. \tag{2.48}$$

renders the same DDE.

The reason to start with the noise-free case is that DDEs have already been quite well studied in mathematics, see [121, 129–134] for comprehensive introductions, and Ref. [121] for applications to life sciences. The discussion of DDEs (which are a special type of functional differential equations) is far beyond the scope of this thesis, and we will here only discuss a few selected aspects relevant for our work.

First, we mention that delay is known to have a smoothing effect on the solutions. For example, when the initial condition has a discontinuity, this is typically washed out in the course of time rendering smooth solutions for $t \gg \tau$.

Furthermore, it is generally known that delay can induce *oscillatory behavior*. For example, a first-order, scalar (one-variable) differential equations with time delay, i.e., a scalar DDE, can have oscillatory solutions, while an ordinary scalar differential equation cannot. We illustrate this using the simplest possible example

$$\dot{X}(t) = -X(t-\tau). \tag{2.49}$$

It is quite easy to see that at $\tau = \pi/2$, this equation has an oscillatory solution $X = \sin(\omega t + \psi)$, with ψ depending on the initial condition.

A very interesting, and often overlooked property of DDEs is that their *backward solutions are not unique*. When instead of the initial condition, one fixes the last interval (of minimal length τ), and tries to infer the corresponding solution on the preceding time intervals, which means solving the equation backwards in time, this problem has no unique solution. In fact, even when fixing the solution on $[\tau, \infty)$, it is not guaranteed that the solution on $[0, \tau]$ is uniquely defined. See Ref. [130] p. 248 for a concrete example. This is in sharp contrast to ordinary differential equations (ODEs), which can be solved equally in both directions in time.

Now we turn to the stability properties. Delay can have both, destabilizing and stabilizing effects. Indeed, the linear delay force in Eq. (2.45), which is in mathematical literature denoted Pyragas control [135–137], can be used for stabilization. It has been extensively studied in the context of chaos control [86], and is still subject of ongoing research, for example, aiming for generalizations and proofs of controllability [138, 139]. For deterministic systems, Pyragas is a *non-invasive* control, that means, it vanishes when the system is in a steady state, or on a periodic orbit with period τ. This is because in both of these cases $X(t) \equiv X(t-\tau)$. Although being zero on a targeted periodic orbit or fixed point, the control term can nevertheless change its stability and, thus, provides a practical tool for stabilization.

On the contrary, delay can also destabilize a system. Again, we look at our simple exemplary DDE (2.49), which is in the absence of delay, $\tau = 0$, clearly a stable system which will always eventually approach $X = 0$, no matter how the initial condition is chosen. It turns out that in the presence of delay it is also stable, as long as $0 \leq \tau < \pi/2$. But it becomes unstable at $\tau \geq \pi/2$. In particular, the "fixed

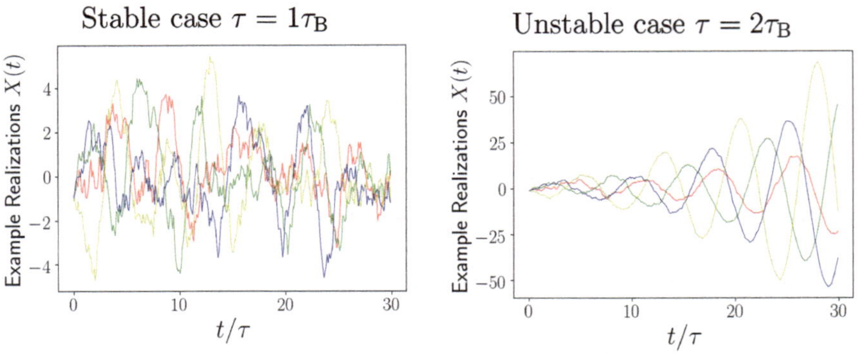

Fig. 2.8 Change of stability upon the increase of delay time. Both panels depict transient exemplary trajectories (in units of σ) of the system with LE (6.46), with history function $X(t') = -\sigma, \forall\, t' \in [-\tau, 0]$. This system is stable at short delay times, e.g., at $\tau = 1\tau_B$ (*left panel*), while it is unstable if $\tau > (\pi/2)\tau_B$, e.g., at $\tau = 2\tau_B$ (*right panel*). The stability boundaries have been calculated using the deterministic limit $\mathcal{T} \to 0$ (see text). Figure 6.3 shows corresponding steady-state trajectories and the autocorrelation function of this system in the stable regime

point" at $X = 0$ changes at $\tau = \pi/2$ its stability. This can be shown by a linear stability analysis. To this end, one makes an exponential ansatz $X = Ce^{\lambda t}$ (with some constants C, λ). If one finds $\lambda < 0$, this indicates that the system is stable, while $\lambda > 0$ signals instability. In the present example, the exponential ansatz readily yields $\lambda X = -Xe^{-\lambda\tau}$, thus, implying the so called characteristic polynomial

$$\Delta(\lambda) = \lambda + e^{-\lambda\tau} \stackrel{!}{=} 0, \tag{2.50}$$

whose roots mark the point where the stability changes. Due to the delay, the characteristic polynomial is a "transcendental" algebraic equation with infinitely many complex[21] solutions λ, given by the "branches" of the Lambert W function $\lambda = W_k(-\tau)$ (k marking the branch). The solution λ with the highest real part changes its sign at $\tau = \pi/2$, where the system gets unstable. For $\tau < \pi/2$, (2.49) may exhibit decaying oscillations, e.g., when $X(t < 0) = \text{const} \neq 0$, while the oscillations increase for $\tau > \pi/2$. Thus, stable oscillations only occur right at very bifurcation point, i.e., the parameter where the system's stability changes. Figure 2.8 shows exemplary realizations of the transient behavior of (2.49), at a delay time before and after the stability boundary $\tau = \pi/2$. Here, we have already added noise to the system.

A quite generally applicable strategy to find explicit solutions of DDEs is by solving them in an iterative manner, using the so-called *method of steps* [140]. This amounts to solving the equation from the Nth interval $[(N-1)\tau, N\tau]$, to the next, $N+1$st one, each time utilizing the previous solution to replace the delay term, and starting with the history function for $N = 0$. By construction, when applied directly to the equation of motion, this technique is not suitable to analytically access the steady-

[21] The imaginary part signals oscillatory solutions.

state dynamics, as one would have to repeat the procedure in principle infinitely many times (while numerically, this can be done). In Sect. 5.2.1, we will explicitly demonstrate how this technique can be used to access steady-state properties, by applying it to the autocorrelation function.

Stochastic delay differential equations Now we return to the Langevin equation with noise and delay. This is a stochastic delay differential equations, which is a specific the type of DDE. Some results for the corresponding DDEs carry over, an example being the stability boundaries (discussed above). However, due to the combination of noise and delay, the analytical treatment of SDDEs is very involved, and, in general, much fewer rigorous results are known. For example, no general solution method has been found till date. An exception is the case of linear SDDEs, which we will discuss next (in Sect. 2.4.5). In this special case, formal solutions can be found for example by employing the aforementioned method of steps.

SDDEs are subject of ongoing research. A good introduction is provided in [140], numerical aspects are discussed in [141], and examples for further recent studies in mathematical literature include [142–144], in physics [145], and nonlinear dynamics [146]. Stability of systems with distributed delay was recently considered in [118, 122]. However, for many nonlinear cases, which we will extensively consider in this thesis, the method of steps and many other techniques are not applicable. We remind the Reader that for nonlinear differential equations often no explicit solutions are known, even in the absence of noise and delay (see, e.g., Ref. [26]).

2.4.5 Linear Systems with Time Delay

To begin with, we consider a very important example system, that is the purely *linear system with time delay*

$$\gamma\dot{X}(t) = -\alpha X(t) - \beta X(t-\tau) + \sqrt{2D_0}\,\xi(t), \tag{2.51}$$

which was already discussed by Kuechler and Mensch in the important 1992 paper [147]. The LE (2.51) can be used to describe a Brownian particle in a static harmonic trap $V_s = \frac{a}{2}x^2$ subject to linear feedback $F_d = -k[X(t) - X(t-\tau)]$, from optical tweezers (see Sect. 2.4.1), then $\alpha = a + k$, $\beta = -k$. Figure 2.9 shows an illustration of this system.

We will consider (2.51) at multiple times throughout this thesis. We will see that for these purely linear models, many mathematical problems due to the delay are, in fact, not that dramatic after all. For example, the steady-state position-autocorrelation function can be computed analytically [147].

Autocorrelation function of linear delay equations The temporal position-autocorrelation function (2.10) between the systems state at time t and time $t+z$, i.e., $C(z) := \langle X(t)X(t+z)\rangle_{ss}$ on $z \in [0, \tau]$, is given by

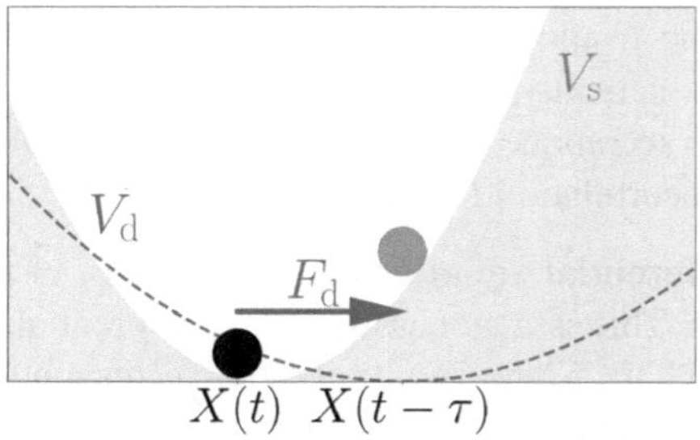

Fig. 2.9 Visualization of the linear system with time delay from Eq. (2.51). The particle moves in a static quadratic potential V_s and is subject to a feedback force $F_d = -k[X(t) - X(t-\tau)]$ stemming from a quadratic trap around the delayed position

$$C(0) = D_0 \frac{1 + (\beta/\omega)\sinh(\omega\tau)}{\alpha + \beta\cosh(\omega\tau)}, \tag{2.52}$$

$$C(z) = C(0)\cosh(\omega z) - (D_0/\omega)\sinh(\omega|z|), \tag{2.53}$$

with $\omega = \sqrt{\alpha^2 - \beta^2} \in \mathbb{C}$. Note that ω becomes imaginary if $|\alpha| < \beta$, such that the hyperbolic functions in Eq. (2.52) convert to trigonometric ones. This explicit closed-form expression for the correlation function on the time difference interval $[0, \tau]$ was derived in earlier literature [70, 148]. The first derivation (based on the LE) stems from Kuechler and Mensch [147]. Later in Sect. 5.2.2, we will extend this result by presenting a derivation for $C(z)$ up to 2τ.

In Sect. 2.4.4 we have briefly discussed a way of finding stability boundaries based on a linear stability analysis. For the system with delay and noise, another indicator for stability is the finiteness of the correlation function: A diverging correlation function implies that there exists *no stable* steady state. Based on the explicit expressions for $C(0 \leq z \leq \tau)$, stability conditions for the linear delay LE were rigorously proven in [147]. In particular, if $|\alpha| < \beta$, there are critical values of the delay time $\tau_c\,\omega = \arccos(-\alpha/\beta) + 2\pi\kappa,\ \forall\kappa \in \mathbb{Z}$, for which C diverges. Furthermore, no stable state exists, if $-\beta \leq \alpha$, or $\tau\alpha \leq -1$. Only if $-\alpha < \beta < \alpha$, the system is stable for all delay times τ. In Fig. 2.8 we have shown examples for stable and unstable transient dynamics of the system (2.51) with $\alpha < \beta$ (in particular, $\alpha = 0$, $\beta = \gamma/\tau_B$).

Now we focus on regions where the system is stable, such that we can consider steady states of (2.51). Let us first have a look at (numerically generated) trajectories and see what we can learn from them. Figure 2.10 (left panels) displays exemplary steady-state trajectories for different signs of β. First we notice that the irregular trajectories look, in fact, rather similar for positive and negative β, although these two cases are quite different ($\beta > 0$ corresponds to positive, repulsive feedback, i.e., a force *away* from the delayed particle position, while the feedback pulls the particle back if $\beta < 0$).

To further get an impression for the impact of delay, we compare them to the trajectories of the (non-delayed) Ornstein–Uhlenbeck process depicted in Fig. 2.4 (see Sect. 2.2.4), which corresponds to the linear delay model (2.51) at $\tau = 0$ (with

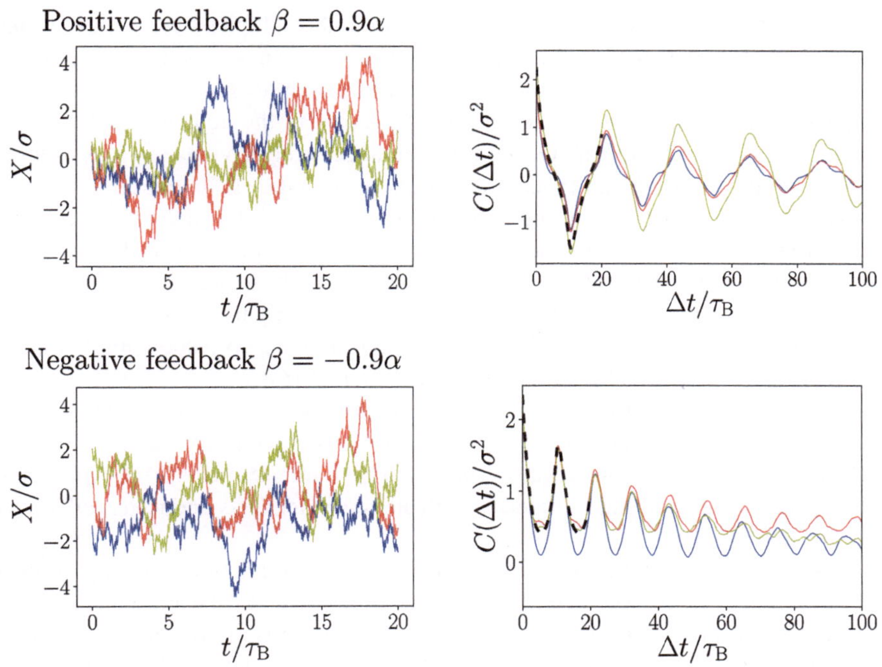

Fig. 2.10 *Left panel:* Three numerically generated example steady-state trajectories of an Brownian particle in a harmonic trap $V_s = [(a-k)/2]x^2$, subject to a linear time-delayed feedback $-k[X(t) - X(t-\tau)]$. The LE reads (2.51), with $\alpha = a, \beta = -k$. The delay time is rather *large*, $\tau = 10\tau_B$. *Right panel:* Corresponding position-autocorrelation function from the three single numerical trajectories (recall that the process is ergodic), and as shown by a dashed black line from the analytical results (2.53), (5.23) for $t \in [0, 2\tau]$. The derivation of the explicit expression (5.23) for $t \in [\tau, 2\tau]$ will be presented later in Chap. 5. To ensure steady state conditions, we have cut off about 10^5 time steps in the numerical simulations (see the Appendix A.1 for further details). Here, $\alpha = 1(\gamma/\tau_B)$, $\beta = \pm 0.9(\gamma/\tau_B)$. The upper panel case is positive ("repulsive") feedback, where the particle is pushed *away* from the delayed position, while it is pulled towards it in the lower panels, corresponding to so-called negative ("attractive") feedback

$a = \alpha + \beta$). We observe that the trajectories[22] of (2.51) are not clearly distinguishable from those of a colloid subject to a static restoring force. Thus, visually inspecting the erratic trajectories alone, does not provide interesting insights.

Delay-induced oscillations in the linear system While the erratic trajectories did not yield an interesting view on the impact of delay in general, or positive versus negative feedback, the position-autocorrelation functions, which are depicted in the right panels of Figs. 2.4 and 2.10, are found to have very distinct characteristics. For example, only the non-delayed case has a monotonically decaying correlation function (Fig. 2.4, right panel). As a quite generic feature of the cases with delayed

[22] Please note that we purposefully used the same random number seeds in all example trajectory plots shown in this Chapter, in order to facilitate the comparison.

feedback, we here detect a pronounced peak around (multiples of) τ. These peaks indicate oscillatory behavior of the particle position, which occurs for both, positive and negative feedback. These position oscillations have a stochastic character, as can be seen from the fact that they are not really visible in the trajectories. This is in sharp contrast to oscillations induced by external periodic drivings (e.g., in systems without no delay), where all trajectories would be periodic and in phase.

More closely inspecting the autocorrelation functions, we find that the *mean oscillation period* is about 2τ for the case of positive (i.e., repulsive) feedback, i.e., a force away from the delayed particle position, and roughly τ for negative (attractive) feedback, which pulls the particle back. Furthermore, only the positive feedback is found to yield negative values of C, i.e., *anti*-correlations, corresponding to a high probability to have different signs at times $X(t)$ and $X(t+t')$.

As we have already mentioned in Sect. 2.4.4, the occurrence of oscillations is a quite ubiquitous phenomenon encountered in various dynamical systems with time delay. In the absence of noise, they typically have a transient character, while they here appear in the steady state. Later, we will also consider delay-induced oscillations in nonlinear systems.

Physically, one might expect that the delay-induced oscillations are only observable when τ is large compared to the other timescales. In the present case, the relevant timescale is the diffusive relaxation within the static quadratic potential. Assuming that this timescale was independent of the time-delayed force, the relaxation timescale can be estimated to be about $\gamma/a = \gamma/\alpha = \tau_B$ here (see Sect. 2.2.3). Indeed, the oscillations are very pronounced if $\tau = 10\tau_B$ (Fig. 2.10), but they are not visible for $\tau = \tau_B$. The latter case is depicted in Fig. 2.11, where again positive and negative feedback are shown.

2.5 Nonlinear Example Systems with Time Delay

In more realistic scenarios, the purely linear case is often of limited use. An important step towards applications is thus the incorporation of *nonlinearities* in the description, which are essentially omnipresent in real-world systems, and are, at the same time, very poorly understood in the context of time-delayed, stochastic systems. For this purpose, we will consider two nonlinear toy models, which are already well established in the context of diffusion and transport of Markovian Brownian particles. In this Section, we will introduce these models; specifically, the involved bistable and periodic static potentials.

Due to their prototypical character, the here introduced exemplary systems are suitable to investigate the effect on time delay in quite generic nonlinear equations. In Parts II and III, they will serve as toy models for the theoretical considerations, and be used as applications for the proposed approaches.

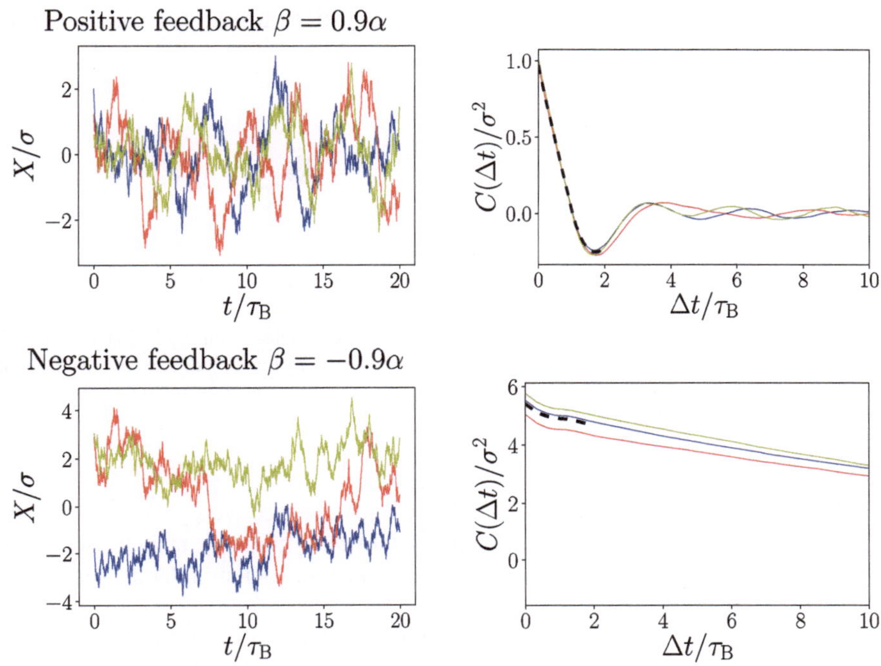

Fig. 2.11 Example steady-state trajectories and corresponding correlation function as in Fig. 2.10, but here in the case of a rather *short* delay time, $\tau = \tau_B$. In both Figures, only the positive (i.e., "repulsive") feedback, i.e., a force *away* from the delayed particle position is found to yield negative values of C, i.e., anticorrelations

2.5.1 Bistable System: The Doublewell Potential

As our prime example system, we will study a particle in a *doublewell potential*

$$V_s(x) = V_0\,[(x/\sigma)^4 - 2\,(x/\sigma)^2]. \tag{2.54}$$

This potential has two minima at $x = \pm\sigma$, separated by a barrier of height V_0. By the quartic potential walls, the particle is confined, which means the chances of finding the particle decay to zero for $x \to \pm\infty$. Typically, the probability will be highest in the vicinity of one of the two minima, i.e., within the center of one well or valley. However, due to the thermal motion, the particle can also leave its valley and jump to the other one. The distance between barrier and minima introduces a length scale, σ. We will thus measure lengths in units of σ.

A Brownian particle in such a doublewell potential is indeed a very well known, generic model for bistable, noisy dynamics. For example, it is commonly used to study escape problems [2, 21, 149, 150]. This is further used to model reaction kinetics, the particle representing a reaction coordinate [150]. Also individual memory cells of memory devices may be modeled in this way [151] (e.g., the particle being

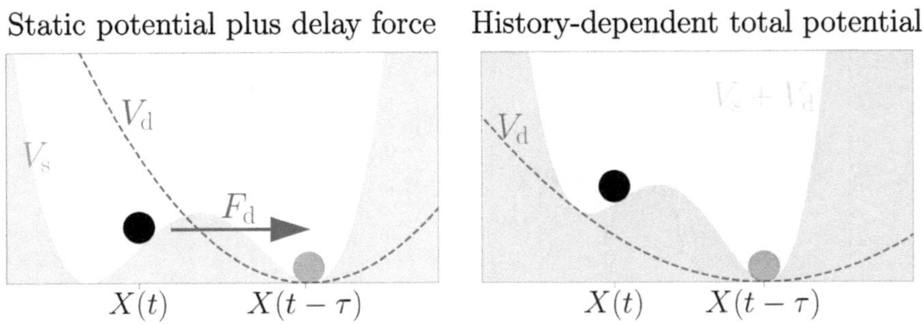

Fig. 2.12 Two ways to visualize the bistable system with time delay from Eq. (2.55). *Left panel:* Static doublewell potential V_s (2.54) supplemented by a linear delay force F_d defined in (2.45). *Right panel:* Total potential $V_s(x) + V_d(x, x_\tau)$ (2.56). Due to the dependency on the delayed position, the total potential is history-dependent

on the left or right side may correspond to 0 and 1, respectively). More generally, this model may describe any continuous system that randomly switches between two characteristic states. Experimentally, such a bistable potential landscape can be generated by optical traps [97].

We will consider the combined impact of this nonlinear potential and linear delay force stemming from a quadratic trap (2.45), giving rise to the LE

$$\gamma \dot{X}(t) = (4V_0/\sigma)[-(X(t)/\sigma)^3 + (X(t)/\sigma)] - k[X(t) - X(t-\tau)] + \sqrt{2D_0}\xi(t). \tag{2.55}$$

One can equivalently consider the forces to stem from a *total potential*

$$V_s(x) + V_d(x, x_\tau), \tag{2.56}$$

with $V_d = (k/2)[X(t) - X(t-\tau)]^2$ from Eq. (2.46). Due to the delay, the total potential is a *history-dependent* energy landscape. Figure 2.12 shows a schematic illustration of the system (2.55) (from both viewpoints).

In recent years, bistable, noisy systems have indeed already been considered under the impact of a linear delay force by different authors [152–157]. Experimental realizations have been presented in Refs. [155, 156], where X represents the polarization dynamics of lasers with optical feedback (and τ is associated with the external cavity length). However, these studies are rather concerned with dynamical phenomena, and thus have a different focus then the work presented in this thesis, which mainly aims at a probabilistic description and thermodynamic notions of delayed systems. We will refer to earlier results in the respective sections at various occasions in Parts II and III.

2.5.2 Periodic System: The Washboard Potential

Second, we will consider the periodic *washboard potential*

$$V_s(x) = -(V_0/2)\cos(x/\sigma) \tag{2.57}$$

with infinitely many barriers of height V_0 and minima at $x = 2\pi\sigma i$ and $i \in \mathbb{Z}$. Again, σ represents a typical length scale and will thus used as a unit for distances.

Particles in sinusoidal potentials (in the absence of delay forces) represent a well-studied paradigm to model transport on rough surfaces [107, 158, 159], and can be realized experimentally, e.g., by optical traps [107]. More generally, this gives a simple model for spatially extended (not confined) systems with a structured energy landscape. Analogously to the bistable example, we will consider the motion of a particle subject to delayed linear feedback in the static potential, described by

$$\gamma \dot{X}(t) = -(V_0/2)\sin[X(t)/\sigma] - k[X(t) - X(t-\tau)] + \sqrt{2D_0}\,\xi(t), \tag{2.58}$$

A schematic plot of the system is shown in Fig. 2.13.

Exact steady-state solution, normalization of periodic system In contrast to the linear trap and the doublewell, this potential is not confining, thus, the probability to find the particle does not quickly vanish for $|x| \gg 0$. Therefore, we have to be a bit more careful in regard to the normalization of the probability densities, on the one hand, and the numerical simulations, on the other hand.

In the absence of external driving (which implies $k\tau = 0$), this Markovian system naturally approaches a steady state in the long time limit, which, in this case, is the thermal equilibrium [40], (see Chap. 4). The PDF is then simply the Boltzmann-distribution

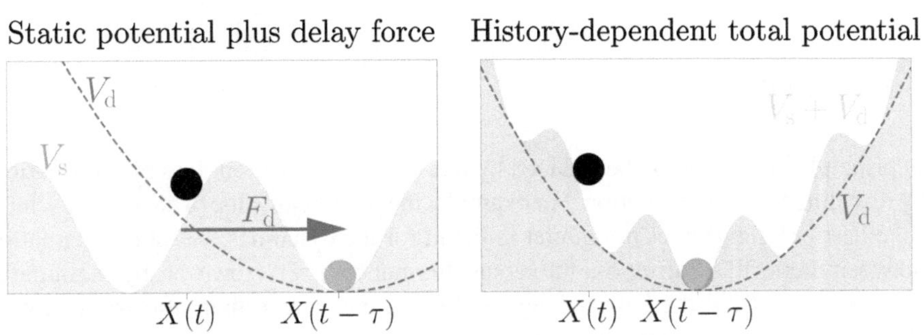

Fig. 2.13 Two ways to visualize the periodic system with time delay from Eq. (2.58). *Left panel:* Static doublewell potential V_s (2.57) supplemented by a linear delay force F_d defined in (2.45). *Right panel:* Total potential $V_s(x) + V_d(x, x_\tau)$ (2.56). Due to the dependency on the delayed position, the total potential is history-dependent

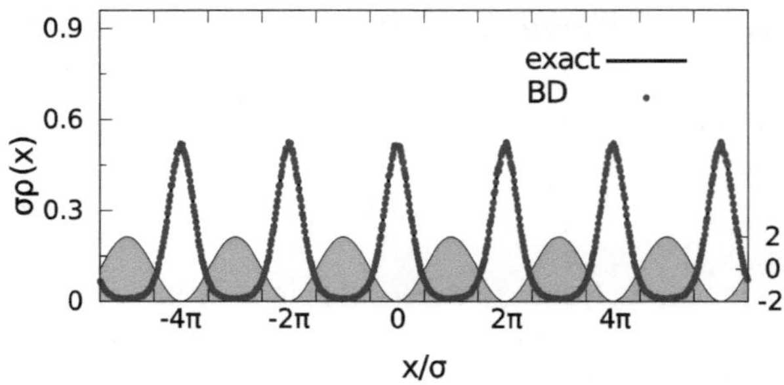

Fig. 2.14 The steady-state PDF in the washboard potential (2.59) with barrier height $V_0 = 2k_BT/\sigma^2$. Black line: exact results (2.59), blue disks: numerical results from BD simulations. Parameters: $N = 9\times 10^5$, $t = 150\tau_B$

$$\rho_{ss}(x) = \frac{1}{Z}\exp\left[-\frac{V_s(x)}{\gamma D_0}\right] = \frac{1}{Z}\exp\left[\frac{V_0\cos(x/\sigma)}{\gamma D_0}\right]. \tag{2.59}$$

However, due to the system's infinite extent, the normalization constant $Z = \int_\Omega \exp[-V_s(x)/(\gamma D_0)]$ is somewhat problematic. Contrary to the other examples, we will hence consider a PDF[23] on a reduced interval $[-\pi, \pi]$ with normalization $\int_{-\pi}^{\pi} \hat{\rho}(x)\mathrm{d}x = 1$. Please note that this does not imply periodic boundary conditions at $\pm\pi$. The latter would not be suitable once we add the delay force. While this reduced viewpoint on the system is appropriate for the steady state and one-time PDFs, it is less useful to consider transient dynamics, or higher order correlation functions.

The PDF (2.59) is shown in Fig. 2.14, which also displays the results of a Brownian dynamics simulation (see Appendix A.1). In the simulations, uniformly distributed initial conditions were used. The agreement between simulation results and exact solution confirms our numerical method.

To simulate the system with washboard potential, we have implemented a co-moving simulation box, and whenever moving the box, we have made sure that the delayed particle position is transformed accordingly. This is particularly important to simulate the long-time diffusive behavior, which we later briefly discuss in Sect. 7.5.5.

2.5.3 *Scaling*

In principle, the LE could be reduced by one parameter, without loss of information, by rescaling the entire equation. For example, the particular value of each energy-term (like barrier height) does not matter in detail for the dynamics, but only the relation between them (like barrier height versus thermal energy). However, by scaling one parameter out (like the barrier height), this parameter can then afterwards not be changed anymore (without changing all other parameters). As we will be interested in the varying different parameters in different Chapters of this thesis, we would have to use alternative rescalings. For the sake of more consistency, we therefore

[23] This reduced PDF already contains the full information due to the spatial 2π-periodicity of the system.

choose not to rescale the equation, and instead consider the equation with all physical coefficients, like friction and diffusion coefficients, γ and D_0, and length scale σ. This might further make comparison with experimental data easier.

We are not going to choose specific values for the parameters in SI units, like $V_0 = \ldots$ kg m^2/s^2, but rather express the values by their ratios (for example, we will say the barrier height is three times larger then the thermal energy, i.e., $V_0 = 3k_B\mathcal{T}$), making use of this scalability. We nevertheless give the corresponding SI units of all coefficients on "List of Symbols". To find suitable real values, we refer the interested Reader to the cited experimental works, like [59, 160], and to the book [7].

Moreover, we emphasize that the theoretical frameworks discussed in this thesis do not only apply to Brownian particles, but also to other natural and artificial systems. Having this in mind, the friction and diffusion coefficient, the Brownian timescale and the particle diameter introduced before just provide appropriate energy, time- and length scales, which one can adjust to the specific system under consideration.

Further, if one has a more mathematical viewpoint on the problem and is less interested in the physical context, it might be more convenient to consider corresponding dimensionless LEs, which can be obtained by rescaling[24] position $x^* = x/\sigma$ and time $t^* = t/\tau_B$, $\tau^* = \tau/\tau_B$, and by rescaling $V_0^* = V_0/(\gamma D_0)$ and $k^* = k(\tau_B/\gamma)$.

2.6 Timescales

The stochastic motion is characterized by various characteristic timescales. We have already introduced some of them in earlier Sections of this Chap. 2. Let us still, at the cost of redundancy, give here a brief overview over all main timescales, from the shortest to the longest. This provides an interesting view on the investigated motion. First, we will consider timescales known for Markovian systems. We recall that lists of all abbreviations and symbols can be found on "List of Abbreviations" and "List of Symbols", respectively.

Ballistic motion On very short times $0 \le t \le m/\gamma$, the motion is *ballistic*, i.e., dominated by inertia, resulting in a quadratic increase of the MSD $\sim t^2$. However, this aspect of the dynamics is not resolved by the overdamped equation considered in this thesis (which neglects the second temporal derivative in the LE, see Sect. 2.2.3). The overdamped limit is appropriate when the particle "mass m is small compared to the friction γ", such that the timescale m/γ is negligible as compared to, e.g., τ_B.

Hydrodynamic memory Further, in our colloidal example system, hydrodynamic effects may be important above the ballistic timescale. In particular, the motion of the colloid affects the flows in the surrounding fluid, which, at a later time then, in turn, impacts the particle position. This *hydrodynamic backflow* gives rise to negative position-autocorrelation functions at intermediate times [59]. However, this effect is

[24] A quite popular alternative is to simply give all parameters dimensionless numbers. This could be done at any point in the following calculations and results.

typically neglected in the description of single colloidal particles, giving rise to a Markovian LE with white noise and instantaneous friction (also see Sect. 2.3.1). We do not further discuss the impact of hydrodynamics within the scope of the thesis.

Diffusion At yet larger times, the dynamics is dominated by *diffusion*. Here, a free particle has a linear MSD $\sim t$. A characteristic scale is the *Brownian time* $\tau_B = \sigma^2/D_0$ describing the typical time it takes for a free (in the absence of deterministic forces) particle to diffuse over a distance σ. Except for the purely linear system, which has no typical length scale (which is the reason why it is convenient to have the somewhat abstract length scale σ), the minima and maxima of the considered potentials have a distance of order σ.

Furthermore, any external potential will influence the dynamics, yielding additional timescales.

Intrawell relaxation For example, a (quadratic) trapping potential $V_s = (a/2)x^2$ confines the motion. This yields a saturation of the mean square displacement with a characteristic saturation time, called *intrawell relaxation time* τ_{ir}, which can be estimated by γ/a (see the description of the Ornstein–Uhlenbeck process in Sect. 2.2.3, and see Sect. 7.5.5 later). More generally, a potential well does not necessarily fully confine the particle, but can also allow for escapes out of the valley (as we discuss next). Still, the MSD will show a saturation on the intrawell relaxation time (before the escapes occur). During this time, the particle position has "relaxed" within the well, meaning, initial condition effects have decayed (and the PDF is given by the Boltzmann distribution). Employing the estimation explained and derived in Ref. [2], p. 348, one can use $\tau_{ir}/\tau_B \approx \gamma/V_s''(x_{min})$, yielding $\tau_{ir}/\tau_B \approx 1/8\,(k_B\mathcal{T}/V_0)$ for doublewell potential (2.54) at $k = 0$.

When the considered static potential has multiple (local) minima, *escape events* can occur, i.e., the particle spontaneously leaves its current valley (due to random kicks from the heat baths). These escapes are characterized by further typical timescales.

Jump duration time First, the time that the particle actually on average needs to cross the separating barrier, that is, after it passed one minimum for the last time, until it reaches the next minimum for the first time. The average *jump duration time* depends on the shape of the potential, and, of course, on the barrier height as compared to the thermal energy of the particle. In this thesis, we will not study this timescale in detail, but refer the interested Reader to Ref. [161], where it was thoroughly discussed for Markovian systems. Instead, we will here only use numerical estimations.[25] This timescale is found to be important for a delay-induced oscillatory state, as we will describe later.

Kramers escape time Furthermore, the typical time that passes between two jump events occur is the famous *Kramers escape time* τ_K. For Markovian processes, this

[25] We will employ the minimum-based technique of Ref. [161]: Basically, one numerically detects the times it takes for the particle to sequentially pass the three positions: one minimum—the maximum—the adjacent minimum, and then average over the results.

timescale is described by the Kramers theory and the Arrhenius formula. Due to the importance of the Kramers escape time, we will dedicate a little section to it in the following. (An explicit example for the impact of the escape time on the MSD will be given later in Sect. 7.5.5.)

Time delay Of course, the time delay itself introduces an additional timescale into the system, namely τ. Only by thinking about the well-known timescales mentioned above, it is already clear that we shall expect different delay-induced dynamical phenomena, depending on the value of τ as compared to all these timescales. A prominent example are the delay-induced oscillations in the linear system, see Sect. 2.4.5. We will come back to this thought multiple times in the remainder of this thesis.

Moreover, the presence of a delay force may also shift the other timescales. However, we will often still use "Markovian estimates" for the timescales, which are, indeed, nevertheless found to be appropriate in many cases.

2.6.1 Kramers Escape Times

For the Markovian case ($k\tau = 0$), the total waiting time between two jumps Δt_{jump} is described by the Kramers theory [21, 149, 150]. This framework predicts exponentially decaying waiting time probabilities, with an average value, called *mean escape time* given by Arrhenius formula [given by Eq. (2.60)], and relies on several assumptions and simplifications, which we briefly lay out in the following.

The Kramers theory, which was developed for *Markovian* systems characterized by a static potential landscape U. It gives an approximation for the mean escape time τ_K over a barrier of height ΔU induced, or as one sometimes says "activated", by white noise. Importantly, the potential barriers ΔU shall be large compared to the thermal energy, i.e., $\gamma D_0 \ll \Delta U$. This justifies a quasistatic assumption of the dynamics within the well. Further, around both extrema (the well and the barrier), the potential shall be well-approximated by a quadratic potential.[26] The Kramers theory provides an estimation for the escape rate by the Arrhenius formula, which reads [2, 149, 150, 152, 161, 162]

$$r_K = \sqrt{U''(x_{\min})|U''(x_{\max})|}/(2\pi\gamma)\, e^{-\Delta U/\gamma D_0}, \tag{2.60}$$

with the corresponding potential minimum $x_{\min}$ and maximum, $x_{\max}$. There are many further results known for this and similar problems, see, e.g., [150] for a detailed overview.

The estimated escape time from the rate, depends on the number of ways to exit the potential well. In our one-dimensional models, this can either be one, as in the

[26] Generalizations of the Kramers estimate have been worked out, which take higher order terms of a expansion around the extrema into account and can be used if this assumption is not appropriate for the potential at hand, see e.g. [21].

case of the bistable potential, or two, as for the period system, yielding $\tau_K = 1/r_K$ or $\tau_K = 1/(2r_K)$, respectively.

What happens in the presence of time delay? We stress that the estimate (2.60) relies on a quasistatic approximation for ρ_1 and does not involve ρ_2, or any higher order correlation between different times and spaces. Applying it to the non-Markovian system, as we will do in Parts II and III, must be considered a crude approximation. There are actually generalizations towards non-Markovian systems, for example [150, 163]. An interesting comparison between the Kramers predictions and alternative, more sophisticated approximations to the actual escape times for a system with time delay were reported in [164]. Indeed, this is a research topic on its own and not the focus of the present thesis. We will here instead only employ Eq. (2.60). The main goal of the approximation presented in Part II, is rather to check how far we can get with the approach for the PDF presented there.

2.7 Delay-Induced Oscillations and Coherence Resonance

In this Section, we will discuss an important dynamical state present in our nonlinear example applications, that are spontaneous, delay-induced oscillations between different potential wells. As it has already been realized in earlier literature (for similar models) [152, 153, 160, 165, 166], these oscillations are characterized by resonance between noise and delay. We introduce this state here, because it will serve in Parts II and III as a particular interesting state to test our theoretical approaches.

2.7.1 Delay-Induced Oscillations

The delay-induced stochastic oscillations of the particle position, which we have encountered in the linear system with time-delayed feedback (see Sect. 2.4.5) , can also be observed in the nonlinear systems. Despite many similarities, the oscillatory behavior is be more rich, due to the additional timescales and the more complex environments. A particularly important difference is that in the nonlinear systems, the oscillations can occur within one potential well, which we will denote "intrawell oscillations", but they can also be between different potential valleys, denoted "interwell oscillations" (e.g., depending on the barrier heights). This is the dynamical state we will be most interested in throughout the thesis.

Let us first discuss universal features. The aforementioned association between *positive/negative* feedback and the *mean period* $\sim 2\tau / \sim \tau$ of the induced oscillations, is also found to be characteristic in the nonlinear example system [152, 153, 162, 165, 167, 168] for both intrawell and interwell oscillations. Further, the oscillations are as well not transient (as typical for DDEs), but manifest in the autocorrelation

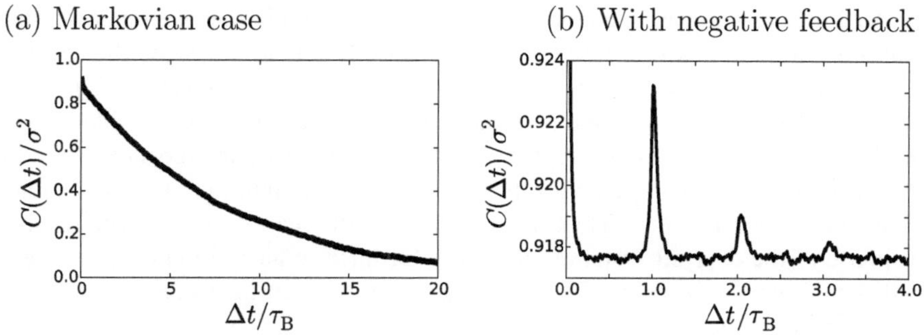

Fig. 2.15 The delay-induced oscillations manifest in the position-autocorrelation function $C(\Delta t) = \langle X(t)X(t + \Delta t)\rangle$. Numerically calculated C in the doublewell with barrier height $V_0 = 4k_B\mathcal{T}$. (a) No delay force, (b) under negative feedback with delay force strength $k = 12k_B\mathcal{T}/\sigma^2$ and delay time $\tau = \tau_B$

function of the steady state. They have again stochastic character. The interwell oscillations are easier to see in the trajectories. They randomly set in and pause, and have varying, random period.

Figure 2.15 shows as an example the position-autocorrelation function in the doublewell potential without and with negative feedback. In the latter case, the correlation function has pronounced peaks at multiples of τ. Here, the particle position oscillated between the two valleys, and are thus much clearer visible as in the linear system. The delay-induced oscillations are subsequent escape events with a waiting time between two escape events around $\tau/2$ (or τ in the case of positive feedback). We further observe that the oscillations randomly set in and pause. The oscillations also manifest in the waiting time distributions, as we will later show in Sect. 7.5.4.

We recall that the purely linear, time-delayed (overdamped) system is characterized by two relevant timescales, the relaxation within the well, and the time delay. The oscillations are pronounced, if the delay is larger than the relaxation time. This is also true for intrawell oscillations in the nonlinear systems. Similarly, particle oscillations between neighboring wells set in, if the delay is large compared to the jump duration times (see above Sect. 2.6). These are delay-induced interwell oscillations. For the latter, yet another timescale is important, that is, the Kramers escape time. This additional timescale is indeed connected to an intricate dynamical phenomenon, called *coherence resonance*.

2.7.2 *Coherence Resonance*

If τ is large compared to the jump duration time, delay-induced interwell oscillations of the particle position are visible (e.g., in the position-autocorrelation function). We consider parameter regimes, where in the absence of noise, no (transient) intrawell escape events could be induced by the delay force (i.e., k is somewhat small compared to V_0). For a very low temperature of the system, the oscillations are random events that often stop after two or three oscillation periods. Upon increasing the temperature, the Kramers escape time is drastically reduced, as can be seen from Eq. (2.60). When the Kramers escape time is in the regime of τ, the interwell oscillations are particularly pronounced and regular. For even larger temperature, the motion is dominated by the noise. The particle's thermal energy is so high that it does not really "see" the barrier anymore. Then, the regularity of the jumps decays. The resonant response of the delayed system to the noise for intermediate, finite temperatures is called *coherence resonance* (CR). Importantly, in contrast to *stochastic resonance*, there is no periodic external driving.

These oscillations and CR have already been studied previously. Experimentally confirmations of CR in delayed, bistable models were reported in Refs. [153, 160, 165]. A particularly important theoretical contribution is [152], where an approximate expression for the corresponding position-autocorrelation function was derived, based on a discretization of the dynamics. We will elucidate this 2-state reduction in more detail and employ the result from [152] in Chap. 9. However, some aspects of the dynamics which are relevant in the context of our investigations were not studied in the earlier literature. This motivates us to further elucidate this dynamical state more elaborately in Parts II and III of this thesis, particularly in Chap. 9. To analyze these phenomena, we will consider the position-autocorrelation function (e.g., see Fig. 9.7), as well as waiting time distributions (e.g., Figs. 7.10 and 9.6). Furthermore, we will study how thermodynamic quantities behave around CR (see Chap. 9).

Throughout this thesis, our analysis of CR is focused on the bistable model. However, our numerical investigations indeed suggest that the periodic model as well exhibits CR.

2.7.3 Bifurcation Theoretical Perspective on Delay-Induced Oscillations

Let us briefly revisit the delay-induced interwell oscillations from the viewpoint of *bifurcation theory* [53]. In general, bifurcation scenarios are best understood for deterministic systems. Thus, we here start with the corresponding deterministic, i.e, noise-free LE ($\mathcal{T} = 0$). Furthermore, for the sake of simplicity, we here focus on the bistable model, and $k > 0$. The idea can be generalized to the washboard potential, and to $k < 0$. We recall that one can define a total potential (2.56) consisting of the static doublewell potential plus the harmonic trap around the delayed position, which is then a history-dependent potential.

Using the terminology of nonlinear dynamics (see, e.g., [53]), the bistable system with linear delay force (and not too small τ) has a bifurcation upon the increase of k. In particular, for any k, the system has two solutions $\{X(t) = \pm\sigma, X(t-\tau) = X(t)\}$, corresponding to $X(t)$ and $X(t-\tau)$ being in the same of the two minima of V_s. These two solutions are always stable. Further, if k reaches a certain critical value (which

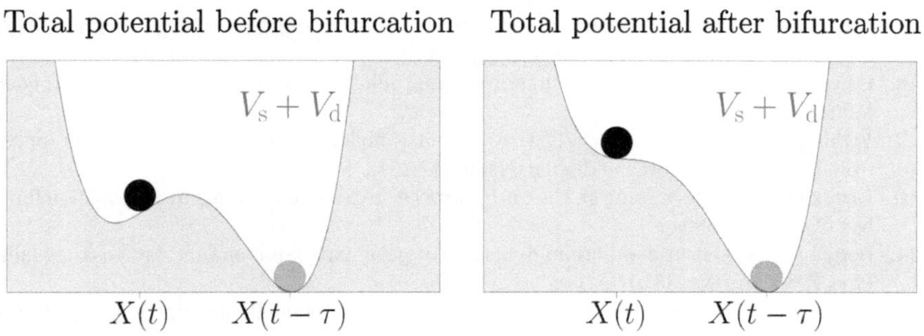

Fig. 2.16 Schematic plots of the history-dependent total potential $V_s[X(t)] + V_d[X(t), X(t-\tau)]$ for two different values of $k\sigma^2/V_0$. Here $X(t-\tau) = \sigma$ is at one potential minimum. *Left:* $k < V_0/2\sigma^2$ is slightly before the bifurcation, where the total potential still has two minima, *right:* $k > V_0/2\sigma^2$ slightly after the bifurcation, where the second minimum has disappeared. Another visualisation of the total potential (2.56) is provided in Fig. 2.12

is around $k \approx V_0/2\sigma^2$), a new stable solution in the form of a *limit cycle* bifurcates. This limit cycle corresponds to X-oscillations with period $\sim\tau$ between both sides of V_s [then, $X(t)$ and $X(t-\tau)$ are in opposite minima of V_s, switching back and forth]. The initial conditions alone decide which solution is visible. This bifurcation occurs at precisely the value where, when $X(t)$ and $X(t-\tau)$ are in *opposite* minima of V_s, i.e., $\{X(t) = \pm\sigma, X(t-\tau) = -X(t)\}$, the *total* potential $V_s(x) + V_d(x, x_\tau)$ has only one unique minimum. (For smaller values of k, the total potential has two minima.) A plot of the history-dependent total potential before (left panel) and after the bifurcation (right panel) is shown in Fig. 2.16. This simplified picture gives a good idea of the dynamics. However, the whole bifurcation scenario is actually not that simple due to the infinite-dimensional nature of the system (x and x_τ are not two independent phase space variables, as we have pretended in this description of the bifurcation).

When we quantify coherence resonance (specifically in Chap. 9), we will focus on parameter regimes, where no delay-induced oscillations would occur in the absence of noise, i.e., at k values *before* that bifurcation (left panel in Fig. 2.16). Thus, we will consider oscillations that are *delay- and noise-induced.* One can take the perspective that the noise makes this limit cycle "visible" before it actually bifurcates.

References

1. Schwabl F (2002) Statistical mechanics. Springer, Berlin
2. Gardiner CW (2002) Handbook of stochastic methods, 2 edn. Springer, Berlin
3. Hänggi P, Marchesoni F, Nori F (2005) Brownian motors. Ann Phys 14:51–70
4. Kapral R (2008) Multiparticle collision dynamics: simulation of complex systems on mesoscales. Adv Chem Phys 140:89
5. Ramaswamy S (2010) The mechanics and statistics of active matter. Annu Rev Condens Matter Phys 1:323–345

6. Brown R (1827) A brief account of microscopical observations. Philos Mag Ann Philos 4:161
7. Dhont JK (1996) An introduction to dynamics of colloids, vol 2. Elsevier, Amsterdam
8. Löwen H (2008) Colloidal dispersions in external fields: recent developments. J Phys Condens Matter 20:404201
9. Yethiraj A, van Blaaderen A (2003) A colloidal model system with an interaction tunable from hard sphere to soft and dipolar. Nature 421:513–517
10. Khadka U, Holubec V, Yang H, Cichos F (2018) Active particles bound by information flows. Nat Commun 9:3864
11. Berg J (2008) Out-of-equilibrium dynamics of gene expression and the Jarzynski equality. Phys Rev Lett 100:188101
12. Ovaskainen O, Meerson B (2010) Stochastic models of population extinction. Trends Ecol Evol 25:643–652
13. Majda AJ, Timofeyev I, Eijnden EV (1999) Models for stochastic climate prediction. PNAS 96:14687–14691
14. Bouchaud J-P, Cont R (1998) A Langevin approach to stock market fluctuations and crashes. Europ Phys J B-Conden Matter Complex Syst 6:543–550
15. Longtin A, Milton JG, Bos JE, Mackey MC (1990) Noise and critical behavior of the pupil light reflex at oscillation onset. Phys Rev A 41:6992
16. Verechtchaguina T, Schimansky-Geier L, Sokolov I (2004) Spectra and waiting-time densities in firing resonant and nonresonant neurons. Phys Rev E 70:031916
17. Van Zon R, Ciliberto S, Cohen E (2004) Power and heat fluctuation theorems for electric circuits. Phys Rev Lett 92:130601
18. Kish LB, Granqvist C-G (2012) Electrical Maxwell demon and Szilard engine utilizing Johnson noise, measurement, logic and control. PloS one 7:e46800
19. Gillespie DT (2000) The chemical Langevin equation. J Chem Phys 113:297–306
20. Zwanzig R (2001) Nonequilibrium statistical mechanics. Oxford University Press, Oxford
21. Risken H (1984) The Fokker-Planck equation. Springer, Berlin
22. Karatzsas I, Shreve SE (1991) Brownian motion and stochastic calculus. Graduate texts in Mathematics 113
23. Sekimoto K (2010) Stochastic energetics, vol 799. Springer, Berlin
24. Hida T, Kuo H-H, Potthoff J, Streit L (2013) White noise: an infinite dimensional calculus, vol 253. Springer Science & Business Media, Berlin
25. Bewersdorff J (2011) Statistik - wie und warum sie funktioniert: Ein mathematisches Lesebuch. Springer, Berlin
26. Kudryashov NA (2003) Nonlinear differential equations with exact solutions. ArXiv:0311058
27. Einstein A (1905) Über die von der molekularkinetischen Theorie der Wärme geforderte Bewegung von in ruhenden Flüssigkeiten suspendierten Teilchen. Ann Phys 322:549–560
28. Goychuk I (2009) Viscoelastic subdiffusion: from anomalous to normal. Phys Rev E 80:046125
29. Jeon J-H, Leijnse N, Oddershede LB, Metzler R (2013) Anomalous diffusion and power-law relaxation of the time averaged mean squared displacement in worm-like micellar solutions. New J Phys 15:045011
30. Metzler R, Jeon J-H, Cherstvy AG, Barkai E (2014) Anomalous diffusion models and their properties: non-stationarity, non-ergodicity, and ageing at the centenary of single particle tracking. Phys Chem Chem Phys 16:24128–24164
31. Kloeden P, Platen E (1992) Numerical solution of stochastic differential equations. Springer, Berlin
32. Chow CC, Buice MA (2015) Path integral methods for stochastic differential equations. JMN 5:8
33. Mannella R, McClintock PV (2012) Itô versus Stratonovich: 30 years later. Fluct Noise Lett 11:1240010
34. Vale RD (2003) The molecular motor toolbox for intracellular transport. Cell 112:467–480
35. Sokolov I (2010) Itô, Stratonovich, Hänggi and all the rest: The thermodynamics of interpretation. Chem Phys 375:359–363

36. Moon W, Wettlaufer J (2014) On the interpretation of Stratonovich calculus. New J Phys 16:055017
37. Sancho JM (2011) Brownian colloidal particles: Itô, Stratonovich, or a different stochastic interpretation. Phys Rev E 84:062102
38. Farago O, Grønbech-Jensen N (2014) Langevin dynamics in inhomogeneous media: re-examining the Itô-Stratonovich dilemma. Phys Rev E 89:013301
39. Yuan R, Ao P (2012) Beyond Itô versus Stratonovich. J Stat Mech: Theory Exp 2012:P07010
40. Seifert U (2012) Stochastic thermodynamics, fluctuation theorems and molecular machines. Rep Prog Phys 75:126001
41. Puglisi A, Villamaina D (2009) Irreversible effects of memory. EPL 88:30004
42. Hänggi P (1989) Path integral solutions for non-Markovian processes. Z Phys B 75:275–281
43. Munakata T, Rosinberg ML (2014) Entropy production and fluctuation theorems for Langevin processes under continuous non-Markovian feedback control. Phys Rev Lett 112:180601
44. Rosinberg ML, Munakata T, Tarjus G (2015) Stochastic thermodynamics of Langevin systems under time-delayed feedback control: second-law-like inequalities. Phys Rev E 91:042114
45. Abe S (2009) Generalized molecular chaos hypothesis and the H theorem: problem of constraints and amendment of nonextensive statistical mechanics. Phys Rev E 79:041116
46. Ehrenfest P, Ehrenfest T (1990) The conceptual foundations of the statistical approach in mechanics. Courier, Corporation
47. Zwanzig R (1973) Nonlinear generalized Langevin equations. J Stat Phys 9:215–220
48. Mori H (1965) Transport, collective motion, and Brownian motion. Progress Theoret Phys 33:423–455
49. Kubo R (1966) The fluctuation-dissipation theorem. Rep Prog Phys 29:255
50. Maes C, Safaverdi S, Visco P, Van Wijland F (2013) Fluctuation-response relations for nonequilibrium diffusions with memory. Phys Rev E 87:022125
51. Lee HK, Lahiri S, Park H (2017) Nonequilibrium steady states in Langevin thermal systems. Phys Rev E 96:022134
52. Hemmer PC, Maximon L, Wergeland H (1958) Recurrence time of a dynamical system. Phys Rev 111:689
53. Strogatz SH, Friedman M, Mallinckrodt AJ, McKay S et al (1994) Nonlinear dynamics and chaos: With applications to physics, biology, chemistry, and engineering. Comput Phys 8:532
54. Banasiak J (2008) Chaotic linear systems in mathematical biology. S Afr J Sci 104:173–179
55. Klages R, Dorfman J (1995) Simple maps with fractal diffusion coefficients. Phys Rev Lett 74:387
56. Moran B, Hoover WG, Bestiale S (1987) Diffusion in a periodic Lorentz gas. J Stat Phys 48:709–726
57. Machta J, Zwanzig R (1983) Diffusion in a periodic Lorentz gas. Phys Rev Lett 50:1959
58. Cecconi F, del Castillo-Negrete D, Falcioni M, Vulpiani A (2003) The origin of diffusion: the case of non-chaotic systems. Physica D 180:129–139
59. Franosch T et al (2011) Resonances arising from hydrodynamic memory in Brownian motion. Nature 478:85
60. Rouse PE Jr (1953) A theory of the linear viscoelastic properties of dilute solutions of coiling polymers. J Chem Phys 21:1272–1280
61. Berner J, Müller B, Gomez-Solano JR, Krüger M, Bechinger C (2018) Oscillating modes of driven colloids in overdamped systems. Nat Comm 9:1–8
62. Kheifets S, Simha A, Melin K, Li T, Raizen MG (2014) Observation of Brownian motion in liquids at short times: instantaneous velocity and memory loss. Science 343:1493–1496
63. Marchetti MC et al (2013) Hydrodynamics of soft active matter. Rev Mod Phys 85:1143
64. Reinken H, Klapp SH, Bär M, Heidenreich S (2018) Derivation of a hydrodynamic theory for mesoscale dynamics in microswimmer suspensions. Phys Rev E 97:022613
65. Maes C (2014) On the second fluctuation-dissipation theorem for nonequilibrium baths. J Stat Phys 154:705–722
66. Seifert U, Speck T (2010) Fluctuation-dissipation theorem in nonequilibrium steady states. EPL 89:10007

67. Bohec P et al (2012) Probing active forces via a fluctuation–dissipation relation. Arxiv:1203.3571
68. Cui B, Zaccone A (2018) Generalized Langevin equation and fluctuation-dissipation theorem for particle-bath systems in external oscillating fields. Phys Rev E 97:060102
69. Price H (1997) Time symmetry in microphysics. Philos Sci 64:S235–S244
70. Frank TD, Beek PJ, Friedrich R (2003) Fokker-Planck perspective on stochastic delay systems: exact solutions and data analysis of biological systems. Phys Rev E 68:021912
71. Wu T-Y (1975) Boltzmann's H theorem and the Loschmidt and the Zermelo paradoxes. Int J Theor Phys 14:289–294
72. Posch HA, Dellago C, Hoover WG, Kum O (1997) Microscopic time-reversibility and macroscopic irreversibility–still a paradox? In: Pioneering Ideas for the Physical and Chemical Sciences, 233–248. Springer, Berlin
73. Felderhof B (1978) On the derivation of the fluctuation-dissipation theorem. J Phys A: Math Gen 11:921
74. Noether E (1918) Invariante variationsprobleme. Nachrichten von der Gesellschaft der Wissenschaften zu Göttingenm, pp 235–257
75. Evans DJ, Searles DJ (2002) The fluctuation theorem. Adv Phys 51:1529–1585
76. Caticha A (2008) Lectures on probability, entropy, and statistical physics. ArXiv:08080012
77. Stoica G (2005) A stochastic delay financial model. Proc Am Math Soc 133:1837–1841
78. Voss HU, Kurths J (2002) Modelling and forecasting financial data: techniques of nonlinear dynamics, vol 2. Springer Science & Business Media, Berlin
79. Tambue A, Brown EK, Mohammed S (2015) A stochastic delay model for pricing debt and equity: numerical techniques and applications. Commun Nonlinear Sci Numer Simul 20:281–297
80. Callen JL, Khan M, Lu H (2013) Accounting quality, stock price delay, and future stock returns. Contemp Account Res 30:269–295
81. Petrillo A, Salvi A, Santini S, Valente AS (2018) Adaptive multi-agents synchronization for collaborative driving of autonomous vehicles with multiple communication delays. Trans Res Part C: Emer Tech 86:372–392
82. Waibel A, Hanazawa T, Hinton G, Shikano K, Lang KJ (1995) Backpropagation: Theory, Architectures and Applications. Lawrence Erlbaum Assoc. Mahwah
83. Cabral JR et al (2014) Exploring mechanisms of spontaneous functional connectivity in MEG: how delayed network interactions lead to structured amplitude envelopes of band-pass filtered oscillations. Neuroimage 90:423–435
84. Karmeshu Gupta, V. & Kadambari, K. (2011) Neuronal model with distributed delay: analysis and simulation study for gamma distribution memory kernel. Biol Cybern 104:369–383
85. Bechhoefer J (2005) Feedback for physicists: a tutorial essay on control. Rev Mod Phys 77:783
86. Schöll E, Schuster HG (eds) (2008) Handbook of chaos control. Wiley, Berlin
87. (2019). Weblink: https://www.nobelprize.org/ceremonies/the-nobel-prize-award-ceremony-2018/
88. Ashkin A (1970) Acceleration and trapping of particles by radiation pressure. Phys Rev Lett 24:156
89. (2019). Weblink: https://www.nobelprize.org/prizes/physics/2018/ashkin/facts/
90. Blickle V, Bechinger C (2012) Realization of a micrometre-sized stochastic heat engine. Nat Phys 8:143
91. Killian JL, Ye F, Wang MD (2018) Optical tweezers: a force to be reckoned with. Cell 175:1445–1448
92. Schirber M (2018) Focus: Nobel prize–lasers as tools. Physics 11:100
93. Kotar J, Leoni M, Bassetti B, Lagomarsino MC, Cicuta P (2010) Hydrodynamic synchronization of colloidal oscillators. PNAS 107:7669–7673
94. Kim KH, Qian H (2007) Fluctuation theorems for a molecular refrigerator. Phys Rev E 75:022102

95. Wördemann M (2012) Structured light fields: applications in optical trapping, manipulation and organisation. Springer Science & Business Media, Berlin
96. Balijepalli A, Gorman JJ, Gupta SK, LeBrun TW (2012) Significantly improved trapping lifetime of nanoparticles in an optical trap using feedback control. Nano Lett 12:2347–2351
97. Roldán É, Martínez IA, Parrondo JMR, Petrov D (2014) Universal features in the energetics of symmetry breaking. Nat Phys 10:457
98. Jahnel M, Behrndt M, Jannasch A, Schäffer E, Grill SW (2011) Measuring the complete force field of an optical trap. Opt Lett 36:1260–1262
99. Kaufman AM, Lester BJ, Regal CA (2012) Cooling a single atom in an optical tweezer to its quantum ground state. Phys Rev X 2:041014
100. Zhong M-C, Wei X-B, Zhou J-H, Wang Z-Q, Li Y-M (2013) Trapping red blood cells in living animals using optical tweezers. Nat Commun 4:1768
101. Ashkin A, Dziedzic JM (1987) Optical trapping and manipulation of viruses and bacteria. Science 235:1517–1520
102. Ashkin A (2007) The study of cells by optical trapping and manipulation of living cells using infrared laser beams. Gravitational and Space Research 4
103. Peterman EJ, Gittes F, Schmidt CF (2003) Laser-induced heating in optical traps. Biophys J 84:1308–1316
104. Kuo SC, Sheetz MP (1993) Force of single kinesin molecules measured with optical tweezers. Science 260:232–234
105. Bloom GS, Endow SA (1995) Motor proteins 1: kinesins. Protein Profile 2:1105–1171
106. Liphardt J, Dumont S, Smith SB, Tinoco I, Bustamante C (2002) Equilibrium information from nonequilibrium measurements in an experimental test of Jarzynski's equality. Science 296:1832–1835
107. Juniper MPN, Straube AV, Aarts DGAL, Dullens RPA (2016) Colloidal particles driven across periodic optical-potential-energy landscapes. Phys Rev E 93:012608
108. Martínez IA et al (2016) Brownian carnot engine. Nat Phys 12:67
109. Lavergne FA, Wendehenne H, Bäuerle T, Bechinger C (2019) Group formation and cohesion of active particles with visual perception-dependent motility. Science 364:70–74
110. Debiossac M, Grass D, Alonso JJ, Lutz E, Kiesel N (2019) Thermodynamics of continuous non-Markovian feedback control (2019). ArXiv:1904.04889
111. Wallin AE, Ojala H, Hæggström E, Tuma R (2008) Stiffer optical tweezers through real-time feedback control. Appl Phys Lett 92:224104
112. Braun M, Cichos F (2013) Optically controlled thermophoretic trapping of single nano-objects. ACS Nano 7:11200–11208
113. Braun M, Würger A, Cichos F (2014) Trapping of single nano-objects in dynamic temperature fields. Phys Chem Chem Phys 16:15207
114. Braun M, Bregulla AP, Günther K, Mertig M, Cichos F (2015) Single molecules trapped by dynamic inhomogeneous temperature fields. Nano Lett 15:5499–5505
115. Haeufle DFB et al (2016) External control strategies for self-propelled particles: optimizing navigational efficiency in the presence of limited resources. Phys. Rev. E 94:012617
116. Kim KH, Qian H (2004) Entropy production of Brownian macromolecules with inertia. Phys Rev Lett 93:120602
117. Levy N, Pressman A, Mussa-Ivaldi FA, Karniel A (2010) Adaptation to delayed force perturbations in reaching movements. PLoS one 5
118. Liu M, Wang K, Hong Q (2013) Stability of a stochastic logistic model with distributed delay. Math Comput Model 57:1112–1121
119. René A, Longtin A (2017) Mean, covariance, and effective dimension of stochastic distributed delay dynamics. Chaos 27:114322
120. Kyrychko Y, Blyuss K, Schöll E (2013) Amplitude and phase dynamics in oscillators with distributed-delay coupling. Philos Trans Royal Soc A 371:20120466
121. Smith HL (2011) An introduction to delay differential equations with applications to the life sciences, vol 57. Springer, New York

122. Bernard S, Bélair J, Mackey MC (2001) Sufficient conditions for stability of linear differential equations with distributed delay. Discrete Contin Dyn Syst Series B 1:233–256
123. Giuggioli L, McKetterick TJ, Kenkre VM, Chase M (2016) Fokker-Planck description for a linear delayed Langevin equation with additive Gaussian noise. J Phys A 49:384002
124. Hänggi P, Talkner P (1978) On the equivalence of time-convolutionless master equations and generalized Langevin equations. Phys Lett A 68:9–11
125. Hänggi P, Thomas H (1982) Stochastic processes: time evolution, symmetries and linear response. Phys Rep 88:207–319
126. Hernández-Machado A, Sancho J, San Miguel M, Pesquera L (1983) Joint probability distribution of nonMarkovian sde. EPJ B 52:335–343
127. Budini AA, Cáceres MO (2004) Functional characterization of generalized Langevin equations. J. Phys. A 37:5959
128. McKetterick TJ, Giuggioli L (2014) Exact dynamics of stochastic linear delayed systems: application to spatiotemporal coordination of comoving agents. Phys Rev E 90:042135
129. Diekmann O, Van Gils SA, Lunel SM, Walther H-O (2012) Delay equations: functional-, complex-, and nonlinear analysis, vol 110. Springer Science & Business Media, Berlin
130. Driver RD (2012) Ordinary and delay differential equations, vol 20. Springer Science & Business Media, Berlin
131. Hale JK, Lunel SMV (2013) Introduction to functional differential equations, vol 99. Springer Science & Business Media, Berlin
132. 1. Hale JK, Lunel SMV (2001) Effects of small delays on stability and control. In: Operator theory and analysis, pp 275–301. Springer, Berlin
133. Erneux T (2009) Applied delay differential equations, vol 3. Springer Science & Business Media, Berlin
134. Kim AV, Ivanov AV (2015) Systems with delays: analysis, control, and computations. Wiley, Berlin
135. Pyragas K (1995) Control of chaos via extended delay feedback. Phys Lett A 206:323–330
136. Pyragas K (1992) Continuous control of chaos by self-controlling feedback. Phys Lett A 170:421–428
137. Parmananda P et al (1999) Stabilization of unstable steady states and periodic orbits in an electrochemical system using delayed-feedback control. Phys Rev E 59:5266
138. Schneider I, Bosewitz M (2016) Eliminating restrictions of time-delayed feedback control using equivariance. Disc Cont Dyn Syst A 36:451–467
139. Schneider I (2013) Delayed feedback control of three diffusively coupled Stuart-Landau oscillators: a case study in equivariant hopf bifurcation. Philos Trans Royal Soc A 371:20120472
140. Longtin A (2010) Complex time-delay systems: theory and applications, edited by Fatihcan M. Atay. Springer Berlin
141. Buckwar E (2000) Introduction to the numerical analysis of stochastic delay differential equations. J Comput Appl Math 125:297–307
142. Luo J (2007) Fixed points and stability of neutral stochastic delay differential equations. J Math Anal Appl 334:431–440
143. Wu F, Mao X, Szpruch L (2010) Almost sure exponential stability of numerical solutions for stochastic delay differential equations. Numer Math 115:681–697
144. Klosek MM, Kuske R (2005) Multiscale analysis of stochastic delay differential equations. Multiscale Model Simul 3:706–729
145. Frank TD, Beek PJ (2001) Stationary solutions of linear stochastic delay differential equations: applications to biological systems. Phys Rev E 64:021917
146. Baker CT, Ford JM, Ford NJ (2004) Bifurcations in approximate solutions of stochastic delay differential equations. Int J Bifur Chaos 14:2999–3021
147. Küchler U, Mensch B (1992) Langevins stochastic differential equation extended by a time-delayed term. Stoch Stoch Rep 40:23–42
148. Guillouzic S, L'Heureux I, Longtin A (1999) Small delay approximation of stochastic delay differential equations. Phys Rev E 59:3970

149. Kramers HA (1940) Brownian motion in a field of force and the diffusion model of chemical reactions. Physica 7:284–304
150. Hänggi P, Talkner P, Borkovec M (1990) Reaction-rate theory: fifty years after Kramers. Rev Mod Phys 62:251
151. Landauer R (1961) Irreversibility and heat generation in the computing process. IBM J Res Dev 5:183–191
152. Tsimring LS, Pikovsky A (2001) Noise-Induced Dynamics in Bistable Systems with Delay. Phys Rev Lett 87:250602
153. Masoller C (2003) Distribution of residence times of time-delayed bistable systems driven by noise. Phys Rev Lett 90:020601
154. Du L, Mei D (2015) Global delay induced transition in a bistable system with multiplicative and additive noises. Indian J Phys 89:267–272
155. Piwonski T, Houlihan J, Busch T, Huyet G (2005) Delay-Induced Excitability. Phys Rev Lett 95:040601
156. Goulding D et al (2007) Kramers' law for a bistable system with time-delayed noise. Phys Rev E 76:031128
157. Guillouzic S, Heureux LI, Longtin A (2000) Rate processes in a delayed, stochastically driven, and overdamped system. Phys Rev E 61:4906
158. Reimann P et al (2001) Giant acceleration of free diffusion by use of tilted periodic potentials. Phys Rev Lett 87:010602
159. Emary C, Gernert R, Klapp SHL (2012) Minimal model for short-time diffusion in periodic potentials. Phys Rev E 86:061135
160. Arteaga MA et al (2007) Experimental evidence of coherence resonance in a time-delayed bistable system. Phys Rev Lett 99:023903
161. Gernert R, Emary C, Klapp SHL (2014) Waiting time distribution for continuous stochastic systems. Phys Rev E 90:062115
162. Loos SAM, Klapp SHL (2017) Force-linearization closure for non-Markovian Langevin systems with time delay. Phys Rev E 96:012106
163. Kappler J, Daldrop JO, Brünig FN, Boehle MD, Netz RR (2018) Memory-induced acceleration and slowdown of barrier crossing. J Chem Phys 148:014903
164. Geiss D, Kroy K, Holubec V (2019) Brownian molecules formed by delayed harmonic interactions. New J Phys 21:093014
165. Masoller C (2002) Noise-induced resonance in delayed feedback systems. Phys Rev Lett 88:034102
166. Balanov AG, Janson NB, Schöll E (2004) Control of noise-induced oscillations by delayed feedback. Physica D 199:1–12
167. Loos SAM, Klapp SHL (2019) Heat flow due to time-delayed feedback. Sci Rep 9:2491
168. Xiao T (2016) Heat dissipation and information flow for delayed bistable Langevin systems near coherence resonance. Phys Rev E 94:052109

Chapter 3
Fokker-Planck Equations

In the preceding chapter, we have introduced the Langevin equation, which describes the random processes studied in this thesis on a stochastic level. For Markovian systems, it is well known that Fokker-Planck equations (FPE) provide a complementary way of description, on the *probabilistic level*. These are deterministic equations, whose solutions are probability density functions. In the following, we will briefly introduce this concept, first focusing on the Markovian case. Towards the end of this chapter, we will review some previous results from the literature on Fokker-Planck (FP) descriptions for systems with time delay. We will keep this discussion short, because the entire Part II of this thesis is dedicated to the probabilistic treatment of delay systems.

3.1 Markovian Case

There is a unique FPE for each Markovian LE of type (2.16), where F might depend on time and the system state X [1], describing the temporal evolution of the probability densities. We have introduced the concept of probability density functions in Chap. 2 [see Eq. (2.7)]. Recall that $\int_y^z \rho_1(x,t)\,\mathrm{d}x$ gives the probability to find the system in a state $x \in [y,z]$ at time t. The FPE for the one-time PDF reads [1, 2]

$$\frac{\partial}{\partial t}\rho_1(x,t) = -\frac{\partial}{\partial x}\left[\frac{1}{\gamma}F(x,t)\rho_1(x,t)\right] + D_0\frac{\partial^2}{\partial x^2}\rho_1(x,t). \tag{3.1}$$

Given a normalized initial condition $\rho(x, t = 0)$ with $\int_\Omega \rho(x, t = 0)\mathrm{d}x = 1$, this equation gives the PDF at all $t > 0$ (assuming that the system is stable). The first term on the r.h.s is the *drift term* describing how the PDF shifts in time due to external forces. The second term on the r.h.s describes the diffusive spreading due to the white noise, and is called *diffusion term*. To illustrate the impact of the latter, we consider

© The Author(s), under exclusive license to Springer Nature Switzerland AG 2021

S. A. M. Loos, *Stochastic Systems with Time Delay*, Springer Theses,
https://doi.org/10.1007/978-3-030-80771-9_3

a Wiener process performed by the motion of a free particle with zero drift ($F = 0$), with an initial condition given by a delta distribution at x_0, describing that the particle is initially fixed at $X(0) = x_0$. This yields [1] (p. 99)

$$\rho_1(x, t) = \frac{1}{\sqrt{4\pi D_0 t}} e^{-\frac{(x-x_0)^2}{4D_0 t}}. \tag{3.2}$$

Thus, the PDF is a zero-mean[1] Gaussian distribution, reflecting the Gaussianity of the white noise and the linearity of the underlying LE. Furthermore, the variance increases linearly by $\langle [X(t) - x_0]^2 \rangle = 2D_0 t$, consistent with the mean square displacement calculated from the LE, see Eq. (2.18).

3.1.1 Natural Boundary Conditions

Throughout this thesis, we will consider natural boundary conditions, i.e.,

$$\lim_{x\to\pm\infty} \rho_1(x, t) = 0, \qquad \lim_{x\to\pm\infty} \partial_x \rho_1(x, t) = 0, \tag{3.3}$$

for $\Omega = \mathbb{R}$. Thus, the PDF and its spatial derivative both decay for large $|x|$. This boundary condition is appropriate in many scenarios where the overall extent of the system is large compared to the length scales of interest.

Here and in the following, we occasionally use ∂_t, ∂_x, and ∂_{xx}, as short notations for $\frac{\partial}{\partial t}$, $\frac{\partial}{\partial x}$, and $\frac{\partial^2}{\partial x^2}$, respectively. We will further denote improper integrals $\lim_{r\to\infty} \int_{-r}^{r}$ simply as $\int_{-\infty}^{\infty}$.

3.1.2 Joint Probability Densities

Similar to (3.1), one can also find FPEs for *multiple*-time PDFs. As we will later see, they play a particularly important role in the description of non-Markovian processes. Analogously to the one-time PDF [from Eq. (2.7)], one can also define, based on the stochastic process $X(s)$, $s \in [t_0, t_n]$, n-time (joint) PDFs (for any $n \in \mathbb{N}$), as

$$\begin{aligned} &\rho_n(x^{(n)}, t_n; x^{(n-1)}, t_{n-1}; \ldots; x^{(1)}, t_1) \\ &\quad = \langle \delta[X(t_n) - x^{(n)}] \delta[X(t_{n-1}) - x^{(n-1)}] \ldots \delta[X(t_1) - x^{(1)}] \rangle. \end{aligned} \tag{3.4}$$

This gives the *joint* PDF for multiple random *events* $\{x^{(i)}, t_i\}$ at different times t_i, with $i \in \{1, n\}$. For example,

[1] This is due to the spatial symmetry of the system.

$$\int_{x}^{y+\Delta y}\int_{y'}^{y'+\Delta y'}\int_{y}^{y+\Delta y} \rho_3\left(x^{(3)}, t_3; x^{(2)}, t_2; x^{(1)}, t_1\right) \mathrm{d}x^{(3)}\mathrm{d}x^{(2)}\mathrm{d}x^{(1)}$$

gives the joint probability to find the particle at times t_1 and t_3 within the interval $[y, y + \Delta y]$ *and* at time t_2 in $[y', y' + \Delta y']$. Since the overall probability to find the particle somewhere in Ω at all three times must be one, the normalization is $\int\int\int_\Omega \rho_n \mathrm{d}x^{(1)} \ldots \mathrm{d}x^{(n)} = 1$.

Marginalizing the joint n-time PDF w.r.t. one variable, yields the respective $(n-1)-$time PDF. For instance,

$$\int_\Omega \rho_2\left(x^{(2)}, t_2; x^{(1)}, t_1\right) \mathrm{d}x^{(2)} = \rho_1\left(x^{(1)}, t_1\right). \tag{3.5}$$

A further connection between joint and marginal PDF is given by the *conditional* PDF, ρ_c, defined as

$$\rho_2\left(x^{(2)}, t_2; x^{(1)}, t_1\right) = \rho_c\left(x^{(2)}; t_2 \middle| x^{(1)}, t_1\right) \rho_1\left(x^{(1)}, t_1\right), \tag{3.6}$$

here again using the case $n = 2$ as an example. $\int_y^{y+\Delta y} \rho_c\left(x^{(2)}; t_2 \middle| x^{(1)}, t_1\right) \mathrm{d}x^{(2)}$ tells us how likely it is the find the particle at time t_2 around y, given that it was at position $x^{(1)}$ at time t_1.

For Markovian processes it is however not necessary to consider n-time PDFs with $n > 2$. This is because of the following simplification. For the three-time PDF with $t_3 > t_2 > t_1$, Markovianity implies

$$\rho_c\left(x^{(3)}, t_3 \middle| x^{(2)}, t_2; x^{(1)}, t_1\right) = \rho_c\left(x^{(3)}, t_3 \middle| x^{(2)}, t_2\right). \tag{3.7}$$

The last expression in fact represents a common *definition of Markovianity*. In the present context, Eq. (3.7) further implies

$$\rho_3\left(x^{(3)}, t_3; x^{(2)}, t_2; x^{(1)}, t_1\right) = \rho_c\left(x^{(3)}, t_3 \middle| x^{(2)}, t_2\right) \rho_c\left(x^{(2)}, t_2 \middle| |x^{(1)}, t_1\right) \rho_1\left(x^{(1)}, t_1\right). \tag{3.8}$$

In the same way, the four-time PDF can be expressed using the conditional PDFs, and so on.

Indeed, together with the initial condition, the conditional PDF $\rho_c(x, t|x', t')$, which is also called "transition probability", provides the full information about the probabilistic evolution of a Markovian system,[2] in contrast to the one-time PDF alone. In turn, the conditional PDF $\rho_c(x^{(1)}, t^{(1)}|x^{(2)}, t^{(2)})$ is just the one-time PDF

---

[2] Which is as well immediately apparent in the so called Chapman-Kolmogorov equation [1] $\rho_c(x^{(3)}; t_3|x^{(1)}, t_1) = \int_\Omega \rho_c(x^{(3)}; t_3|y^{(2)}, t_2)\rho_c(y^{(2)}; t_2|x^{(1)}, t_1)\mathrm{d}y^{(2)}$.

$\rho_1(x^{(1)}, t^{(1)})$ from the FPE (3.1) for the special initial condition $\rho_1(x, t^{(2)}) = \delta(x - x^{(2)})$.

3.1.3 The Probability Current and Steady States

Now we will introduce another important probabilistic quantity, the probability current. This will also bring us to the notion of equilibrium on the level of the Markovian FPE.

As the total probability to find the particle is a conserved quantity, the FPE (3.1) can be rewritten in the form of a continuity equation

$$\partial_t \rho_1(x,t) = -\partial_x J(x,t), \tag{3.9}$$

with the *probability current*

$$J(x,t) = \left[\gamma^{-1} F(x,t) - D_0\, \partial_x\right] \rho_1(x,t). \tag{3.10}$$

We recall that steady states are defined by $\partial_t \rho_{1,\mathrm{ss}}(x,t) = 0$, which means the l.h.s. of the FPE is naturally zero. As immediately follows from Eq. (3.9), the probability current is then automatically constant in x and t. Steady states can be distinguished between two cases: $J = 0$, and $J \neq 0$, which are, for Markovian systems, associated with EQ and NESS, respectively [3]

$$J = 0 \Leftrightarrow \text{EQ} \tag{3.11}$$

$$J = \text{const.} \neq 0 \Leftrightarrow \text{NESS}. \tag{3.12}$$

To better understand this connection, it is interesting to look at the *particle current* $\langle \dot{X} \rangle$, which is closely related to J via $\langle \dot{X} \rangle = \int_\Omega J(x,t)\mathrm{d}x$ [3]. Thus, for Markovian systems in steady states, a nonzero probability current implies net *particle transport*, which is a directly measurable nonequilibrium property.

As a first example, we consider a situation where all external forces stem from a potential $F = -V_\mathrm{s}'$. Then, the system approaches a steady state, where the one-time PDF is given by the *Boltzmann distribution* [1]

$$\rho_{1,\mathrm{ss}}(x) = Z^{-1}\, e^{-V_\mathrm{s}(x)/(\gamma D_0)} \tag{3.13}$$

with normalization constant Z, as readily follows from the FPE (3.9). Furthermore, plugging the Boltzmann distribution (3.13) into Eq. (3.10) yields a vanishing probability current, implying zero particle transport and thermal equilibrium.

Second, a prototypical example for a one-dimensional, Markovian system in a NESS is a colloidal particle on a ring (i.e., a finite spatial domain with periodic boundaries) which is driven by a constant force F in one direction of the ring. Here, for symmetry-reasons $\rho_{1,\mathrm{ss}}$ is constant in x, which implies $\partial_x \rho_1 = 0$ and further $J \propto F$ [from Eq. (3.10)]. Thus, due to the preferred direction of motion aligned with the external drive F, this system exhibits a finite probability and particle current (with the same sign as F). Also see Ref. [4].

Multidimensional case We have used above a notation indicating *one*-dimensionality of the process, where the different random events are denoted $\{x^{(i)}, t_i\}$ with an index i referring to the specific event, i.e., an instant in time $t = t_i$ and a one-dimensional value of $X = x^{(i)}$. However, the formulae can as well be applied to multidimensional systems, where the system state itself is multi-dimensional, for example by replacing X by a vector $\underline{X} = \{X_0, X_1, \ldots, X_n\}$. This could be used to describe Brownian motion in three-spatial dimensions (then $n = 2$). Then, every stochastic event is a multi-dimensional object as well, e.g., $\left\{\underline{x}^{(n)}, t_n\right\} = \left\{\{x_0^{(i)}, x_1^{(i)}, x_2^{(i)}\}, t_i\right\}$.

The Markovian $(n+1)$-variate FPE reads

$$\partial_t \rho_{n+1}(\underline{x}, t) = -\sum_{j=0}^{n} \partial_{x_j} \underbrace{\left[\frac{F_j}{\gamma_j} - \frac{k_\mathrm{B}\mathcal{T}_j}{\gamma_j}\partial_{x_j}\right]\rho_{n+1}}_{=J_j(\underline{x},t)}, \tag{3.14}$$

with the $(n+1)$-point *one*-time PDF $\rho_{n+1}(\{x_0, x_1, \ldots, x_n\}, t)$, and the various $(n+1)$-dimensional probability currents $J_j(x_0, x_1, \ldots, x_n, t)$. The natural boundary conditions amount to $\lim_{x_i \to \pm\infty} \rho_1 = 0$ and $\lim_{x_i \to \pm\infty} \partial_{x_i} \rho_1 = 0, \forall i \in \{0, 1, \ldots, n\}$. We will consider multi-variate FPEs various times in Parts II and III.

3.2 Introduction to Fokker-Planck Descriptions of Systems with Time Delay

Above, we have seen that for Markovian systems, the full probabilistic information is encoded in the transition probabilities. Furthermore, the probability current (3.10) is a sufficient criterion to distinguish thermal equilibrium from nonequilibrium steady states. Both, the transition probabilities and the probability current are given by the FPE for the one-time PDF, which is a closed equation (3.1). Now we turn to the systems with time delay, where the situation completely changes, and is found to be literally "infinitely more" complicated. In this chapter, we review some literature results and hint to open problems. Later, in Part II, we present new results for the probabilistic treatment of systems with delay.

The FPE for the one-time PDF for the delay LE (2.43) was introduced in [5, 6], and reads

$$\frac{\partial}{\partial t}\rho_1(x,t|\phi) = -\frac{\partial}{\partial x}\left[\frac{1}{\gamma}\int_\Omega \mathrm{d}x_\tau F(x,x_\tau)\rho_2(x,t;x_\tau,t-\tau|\phi)\right] + D_0\frac{\partial^2}{\partial x^2}\rho_1(x,t|\phi), \tag{3.15}$$

with $x_\tau \in \Omega$ being a second spatial variable needed due to the two involved particle positions in the delay force. The initial condition ϕ is now a function on the interval $[-\tau, 0]$, and can either be given by a constant sharp trajectory $X(-\tau \leq t \leq 0)$, or a distribution of such paths. The initial condition dependency of the PDFs will be omitted in the following.

Comparing (3.15) to an ordinary FPE for Markovian systems (3.1), one notes the appearance of an integral over all possible past positions weighted with ρ_2. Thus, there is a "delay-averaged" drift term (instead of the usual drift term). This is a direct consequence of the delay in the underlying equation of motion. For this reason, Eq. (3.15) is not self-sufficient, as opposed to the Markovian FPE. Even worse, the corresponding FPE for ρ_2, which we will discuss in Chap. 5, is found to explicitly involve the three-time PDF, ρ_3, and so forth. In this way, an infinite hierarchy of coupled equations emerges, whose nth member depends on the $(n+1)$-time probability density $\rho_{n+1}(x,t;x_\tau,t-\tau;\ldots;x_{n\tau},t-n\tau)$.

Please note that we use here subscripts $i\tau$ for x to indicate different instants in time $t - i\tau$ of the stochastic event, which is somewhat inconsistent with Eq. (3.14), where subscripts denote dimensions (and superscripts denote times). However, as we explicitly give the time arguments as well, the notion should not be ambiguous.

Linear delay systems—time-convolutionless transformation The hierarchical structure of the FPE is in general a major issue in the analytical treatment. An exception is again the case of *linear* delay Langevin equations, where the (steady-state) PDFs are all *Gaussian*. This makes it possible to obtain rigorous results for the various n-time PDFs in the steady state (for arbitrary initial conditions) [7, 8], or even for the transient dynamics if the initial condition is Gaussian or delta-distributed [9]. We we will give explicit closed-form expressions for the steady-state PDFs in Sect. 5.2. Notably, in this special case, the infinite Fokker-Planck hierarchy can, in fact, be alternatively represented by a *finite* set of equations. This was explicitly shown in [10], where a time-convolutionless transform of the Langevin equation was employed. The latter is a technique to convert a time-nonlocal equation to a time-local one by means of Green's functions (which is not possible for nonlinear delayed LEs).

Side note: Hierarchical problems in probabilistic descriptions Similar hierarchy problems also occur in other contexts in statistical physics. A well-known example is the Bogoliubov-Born-Green-Kirkwood-Yvon (BBGKY) hierarchy, in which the time evolution of the one-particle density of an interacting multi-particle system depends upon the two-particle density and so on. For this problem, various closing strategies have been proposed, such as the simple mean-field (factorization) approximation and the more sophisticated dynamical density functional theory (involving an adiabatic approximation) [11, 12]. Another closure example is the mode coupling theory for glassy systems [13, 14]. We stress, however, that in all these examples the hierarchical structure emerges due to (conservative) particle–particle interactions. In contrast, the hierarchy of Fokker-Planck equations appearing in time-delayed systems – although having a similar structure – arises due to delay-induced "*temporal* interactions". Thus, the aforementioned strategies are not readily applicable in the present context. One might try to adopt some ideas, but to us, no obvious, useful connection occurred. This is a possible direction of future research.

3.2.1 Earlier Approximation Schemes

Part II of this thesis is dedicated to the apparent problems in the probabilistic treatment induced by the fact that (3.15) is not closed. Besides explicitly considering the higher members of the hierarchy, a main objective is to discuss and compare different strategies to find closed, approximate expressions for the one-time PDF of nonlinear delay LE. Here, we will briefly review the main ideas of the most prominent approaches proposed in earlier literature. In Part II, we will explicitly apply them to our example systems.

Small delay expansion The unarguably most popular scheme is the small delay approximation on the level of the LE, which was originally introduced in [5, 15]. This is essentially a Taylor expansion up to linear order of the total deterministic forces (and of the noise intensity, if multiplicative, i.e., system-state dependent noise is present), in powers of τ. For the delay LE (2.43) this yields the approximate Markovian LE [5, 15]

$$\dot{X}(t) = \gamma^{-1}\, F[X(t), X(t)] \left[1 - \tau \frac{\partial F(x, x_\tau)}{\partial x_\tau} \bigg|_{\substack{x=X(t) \\ x_\tau=X(t)}} \right] + \sqrt{2D_0}\, \xi(t), \tag{3.16}$$

where τ only appears as a parameter. Due to the Markovianity, one can further readily write down a corresponding FPE.

Considering its simplicity in terms of the resulting expressions and the underlying idea, it is no wonder that this approximation is so popular [9, 16–19]. However, it should be emphasized that this approach is appropriate for very small delay times only, which is, at the same time, a regime where the dynamics is often qualitatively very similar to a Markovian one. This means that many effects induced by the memory are not captured by this approximation.

Perturbation theory Another approximation scheme introduced in earlier literature involves a first-order perturbation-theoretical ansatz for the density on level of the FPE. We refer the interested Reader to Ref. [17], where this *perturbation theory* (PT) was introduced, and only outline the main idea here.

Within the PT approach, one makes the assumption $|F_\mathrm{d}| \ll |F_\mathrm{s}|$ for the total deterministic force appearing in the LE (recall $F[X(t), X(t-\tau)] = F_\mathrm{s}[X(t)] + F_\mathrm{d}[X(t), X(t-\tau)]$, Eq. (2.44)). This justifies that the delay force F_d is regarded as a small perturbation to the non-delayed, i.e., Markovian dynamics. As typical for a perturbation theory, an ansatz for the PDF with additive terms of increasing order in the perturbation is plugged into the Fokker-Planck equation.

The resulting first-order equation [17]

$$\frac{\partial}{\partial t}\rho_1(x,t) = -\frac{\partial}{\partial x}\left[\frac{1}{\gamma}\int_\Omega \mathrm{d}x_\tau F(x,x_\tau)\rho_c^{F_\mathrm{d}\equiv 0}(x_\tau,t-\tau|x,t)\right]\rho_1(x,t) + D_0\frac{\partial^2}{\partial x^2}\rho_1(x,t), \tag{3.17}$$

involves the conditional probability density with respect to the unperturbed, i.e., non-delayed system, here denoted $\rho_c^{F_\mathrm{d}\equiv 0}$. The latter follows a closed *Markovian* FPE, and consequentially, the combination of the resulting two FPEs is self-sufficient. In this way, one obtains an approximate one-time PDF.

Disadvantages of existing approximations One conceptual disadvantage of these approximations is that they both effectively render Markovian descriptions. Thus, they may not capture the essence of the process with memory. Furthermore, they both rely on assumptions which are only appropriate for small delays, or small delay forces. To tackle this problem, we will explore new schemes in Part II, going beyond the "quasi-Markovian" regime.

2-state reduction It is noteworthy that specifically for the bistable system with delay (see Sect. 2.5.1), there is yet another type of approximation known from earlier literature, namely, the 2-state reduction proposed by Tsimring and Pikovsky [20]. Please note that this approach is not on equal footing with the other approaches discussed here, as it does not yield an approximate one-time PDF. In contrast, the 2-state reduction is based on the corresponding Master equation, and aims for the position-autocorrelation function. We will elucidate this approach in more detail and apply it to our example model in Chap. 9. By construction, the 2-state reduction is appropriate in parameter regimes, where the dynamics is dominated by jumps between wells, as the intrawell fluctuations are fully neglected. It would be very interesting to extend this idea to other nonlinear systems, e.g., the washboard potential (which we have introduced in Sect. 2.5.2).

3.2.2 *Probability Current and Apparent Equilibrium of Time-Delayed Systems*

We have mentioned above that, for Markovian systems, the probability current J is an important quantity, which enables an easy distinction between EQ and NESS. Here we consider the probability current for systems with delay.

The current is defined analogously to the Markovian case [by writing (3.15) in the form (3.9)]. Due to the time delay, the probability current is a functional of the two-time PDF

$$J(x,t) = \left[\gamma^{-1}\int_{\Omega} dx_{\tau} F(x,x_{\tau})\rho_{c}(x_{\tau},t-\tau|x,t) + D_0\partial_x\right]\rho_1(x,t), \qquad (3.18)$$

which is typically very hard to calculate. It contains the delay-averaged drift term.

In regard to Markovian systems, we have argued above that J is constant in steady states, where $J = 0$ signals EQ. How is the situation in the presence of time delay? In particular, let us consider the three example systems (the linear, bistable and periodic one), introduced in the last sections of Chap. 2. An apparent shared feature of the systems is their *coordinate inversion symmetry*, i.e., the equations are invariant under the transformation $x \rightarrow -x$. In other words, the spatial symmetry of the systems is not broken, even in the presence of delay force. As a consequence, the probability current naturally vanishes in steady states. This can be easily seen from the FPE, which then reads $\partial_x J = 0$, yielding a constant current which would switch its sign under the coordinate inversion. Of course, a constant current with $J = -J$ must be zero. Besides the main example applications that we will consider throughout this thesis, in fact, various systems fall into this very class. Naively employing the aforementioned criterion of EQ established for Markovian systems, such systems with time delay appear to be in thermal equilibrium. We will prove this idea wrong in Part III.

3.2.3 *Side Note: Delay in Ensemble-Averaged Quantities*

In this thesis, we consider systems that are subject to a feedback force that depends on the time-delayed position $X(t-\tau)$. Thus, the delay appears in a *stochastic* quantity X in the LE. As we have mentioned above, the corresponding Fokker-Planck equation for the one-time PDF is not closed. In earlier research [21], we have also considered the situation that a time delay is instead introduced in an ensemble-averaged, and hence *deterministic* quantity. Specifically, we have considered feedback that depends on the delayed *mean* particle position $\langle X(t-\tau)\rangle$. In this case, one readily finds a closed FPE, which then involves the delayed probability density $\rho(t-\tau)$. Thus, this case is much easier to handle analytically. The underlying reason is that then, there is no direct interplay between delay and noise, and both can be treated separately. As we have shown in [21], such a delay force may indeed also yield interesting dynamics, e.g., induce a ratchet effect. However, this type of feedback is harder to realize in an experimental setup, and a LE with a $\langle X(t-\tau)\rangle$-dependent force term has much fewer applications. We therefore focus on systems with a time-delayed position-dependent force in this thesis.

References

1. Risken H (1984) The Fokker-Planck equation. Springer
2. Gardiner CW (2002) Handbook of stochastic methods, 2nd edn. Springer
3. Reimann P (2002) Brownian motors noisy transport far from equilibrium. Phys Rep 361
4. Seifert U (2012) Stochastic thermodynamics, fluctuation theorems and molecular machines. Rep Prog Phys 75:126001
5. Guillouzic S, L'Heureux I, Longtin A (1999) Small delay approximation of stochastic delay differential equations. Phys Rev E 59:3970
6. Frank TD, Beek PJ, Friedrich R (2003) Fokker-Planck perspective on stochastic delay systems: exact solutions and data analysis of biological systems. Phys Rev E 68:021912
7. Küchler U, Mensch B (1992) Langevins stochastic differential equation extended by a time-delayed term. Stoch Stoch Rep 40:23–42
8. Frank TD, Beek PJ (2001) Stationary solutions of linear stochastic delay differential equations: applications to biological systems. Phys Rev E 64:021917
9. Geiss D, Kroy K, Holubec V (2019) Brownian molecules formed by delayed harmonic interactions. New J Phys 21:093014
10. Giuggioli L, McKetterick TJ, Kenkre VM, Chase M (2016) Fokker-Planck description for a linear delayed Langevin equation with additive Gaussian noise. J Phys A 49:384002
11. Marconi UMB, Tarazona P (1999) Dynamic density functional theory of fluids. J Chem Phys 110:8032–8044
12. Menzel AM, Saha A, Hoell C, Löwen H (2016) Dynamical density functional theory for microswimmers. J Chem Phys 144:024115
13. Das SP (2004) Mode-coupling theory and the glass transition in supercooled liquids. Rev Mod Phys 76:785
14. Janssen LMC, Reichman DR (2015) Microscopic dynamics of supercooled liquids from first principles. Phys Rev Lett 115:205701
15. Guillouzic S, Heureux LI, Longtin A (2000) Rate processes in a delayed, stochastically driven, and overdamped system. Phys Rev E 61:4906
16. Longtin A (2010). In: Atay FM (ed) Complex time-delay systems: theory and applications. Springer, Berlin Heidelberg
17. Frank TD (2005) Delay Fokker-Planck equations, perturbation theory, and data analysis for nonlinear stochastic systems with time delays. Phys Rev E 71:031106
18. Mijalkov M, McDaniel A, Wehr J, Volpe G (2016) Engineering sensorial delay to control phototaxis and emergent collective behaviors. Phys Rev X 6:011008
19. Khadem SMJ, Klapp SH (2019) Delayed feedback control of active particles: a controlled journey towards the destination. Phys Chem Chem Phys 21:13776
20. Tsimring LS, Pikovsky A (2001) Noise-induced dynamics in bistable systems with delay. Phys Rev Lett 87:250602
21. Loos SAM, Gernert R, Klapp SHL (2014) Delay-induced transport in a rocking ratchet under feedback control. Phys Rev E 89:052136

Chapter 4
Stochastic Thermodynamics

"Information is physical"

– Rolf Landauer, 1991 [1]

Finally, we introduce another theoretical framework employed in this thesis: stochastic thermodynamics. The latter is a rather young theory, which aims at a generalization of thermodynamic notions towards small-scale systems dominated by thermal fluctuations and towards nonequilibrium systems [2]. On this scale, individual fluctuating trajectories can transiently violate the second law of thermodynamics, but in the limit of large systems (or long times), one shall always recover the well-established laws of macroscopic thermodynamics.

Because this theory is applicable to *nonequilibrium*, *stochastic* systems, it is a promising candidate to also describe stochastic processes with time delay. However, as we will see in the following, many fundamental concepts in stochastic thermodynamics are based on the *Markov assumption*.[1] In this chapter, we will first lay out the basic concepts in brief, focusing on Markovian, overdamped systems consisting of a single colloidal particle coupled to a heat bath. Towards the end of the chapter, we will turn to the question how stochastic thermodynamics applies to time-delayed systems, reviewing some previous results and forumlating expectations and apparent problems. These considerations provide the basis for our investigations in Part III of this thesis.

As in the preceding chapters, we will use a notation indicating that the system is one-dimensional, where X denotes the stochastic system state. However, the definitions can readily be applied to multidimensional systems as well, by simply replacing X by a vector $\underline{X} = \{X_0, X_1, ..., X_n\}$ ($n + 1$ being the spatial dimension). Further, we will again focus on steady states.

[1] The main problem is rooted in the notion of *total entropy production*, which lies at the heart of this theory.

© The Author(s), under exclusive license to Springer Nature Switzerland AG 2021

S. A. M. Loos, *Stochastic Systems with Time Delay*, Springer Theses,
https://doi.org/10.1007/978-3-030-80771-9_4

An educational introduction to this field, specifically to the energetics of stochastic systems, is given in Sekimoto's book [3]. Some ideas of our introduction are inspired by the latter. Further, the review article by Seifert [4] summarizes the elegant and consistent theoretical concepts, which have been developed in the last decades, focusing on the notion of fluctuating entropy.

Before we begin with introducing the main concepts, we give a brief historical overview, locating this theory within the overall framework of theoretical physics. This is a side note and not necessary to understand the remainder of this thesis.

4.1 Side Note: Some Historical Notes and Where Is Stochastic Thermodynamics "Located" in Theoretical Physics?

Macroscopic thermodynamics goes back to the seventeenth century.[2] Its importance for physics and engineering is undeniable and well known. This traditional, macroscopic thermodynamics is essential for the development of macroscopic machines, and made possible the industrial revolution. It describes many complex phenomena in the world on large scales, e.g., on length scales comparable to the size of a human. The laws of macroscopic thermodynamics are, however, purely empirical. Further, many concrete statements (like Carnot efficiency) only hold for systems that are close to *equilibrium*, or quasi-statically driven. In Sect. 2.3.6 we have mentioned the corresponding notion of macroscopic reversibility. A key concept is the formulation of a purely phenomenological state-function, called *entropy*, which can never decrease when a system is isolated (the second law).

Later, the framework of *statistical mechanics* was founded, which connects thermodynamic laws to *first principles* by unraveling the routes to the microscopic, Newtonian laws. These ideas base on a picture, where the world is made of particles (atoms or molecules) that are ruled by the laws of classical mechanics. A central result is the *statistical interpretation of entropy* by, e.g., Ludwig Boltzmann, whose meaning for modern physics and our recent understanding of the world is monumental. The (Shannon) entropy, which was before just an empirical concept, is now linked to microscopic states and their probabilities. Loosely speaking, the entropy is proportional to the logarithm of the number of possible microstates (i.e., microscopic configurations of the individual particles) that could all correspond to the observed macroscopic state (defined via volume, temperature, pressure,...) [6]. Due to this interpretation, the second law of thermodynamics loses the phenomenological character and gains mathematical rigor.[3] This theoretical advancement also introduces a

[2] An example are the works by Robert Boyle, who lived 1627–1691 [5].

[3] Here one should also mention the long debate it has cased, including *Lochschmidt's paradox*, i.e., how can an irreversible thermodynamics be derived from reversible, time-symmetric fundamental laws, and the ongoing question why the initial state of the universe had an exceptionally low entropy [7].

link between physics and information theory [8], e.g., the immediate relation between Shannon entropy and information discussed in the works by Jaynes [9].

Later on, a new way of describing observed phenomena was developed using *stochastic* models, which we have introduced in the last chapter. We recall that this description is a coarse-grained, mesoscopic one, sitting in between the microscopic, mechanical laws and the macroscopic ones. It introduced a new formulation of statistical mechanics [10]. Important theoretical concepts include the Langevin and Fokker-Planck equation [11] (also the Smoluchowski equation, and dynamical density functional theory), and the fluctuation-dissipation relations. Taking thermal fluctuations explicitly into account, it is suitable for systems that are dominated by "random" perturbations, typically due to the chaotic motion of molecules in the surrounding gas or fluid. The stochastic way of modeling (for, e.g., colloidal suspensions) is the best description we have to date to describe complex phenomena on the nano- to micrometer scale. This is the length scale where one can find macromolecules, viruses, individual and small clusters of cells, which is why stochastic modeling is also particularly important for the understanding of biological systems. Rigorous connections between stochastic modeling and the microscopic world, i.e., first principles, were later established. Important cornerstones are the projection method by Mori and Zwanzig (see Sect. 2.3.1), and toy models like the Rouse model [12, 13] (suitable to model viscoelastic fluids).

Despite these great successes, we are far away from a full physical understanding of the complex world that we are part of. Specifically, some important links between the aforementioned frameworks remain elusive, and there are still lots of open questions. For example, there are the two following issues: First, the generalizing of thermodynamic laws and the frameworks of statistical physics to systems *far from equilibrium*. After all, the real world is mostly out of thermal equilibrium, and many phenomena cannot even be considered as a relaxation towards it. Also the traditional concept of considering passive systems subject to (quasistatic) external driving is limited; and to capture the essence of complex processes, one often needs models that are intrinsically driven. This is particularly true for biological, i.e., living matter, and artificial, energy-consuming systems, like computers or robots. It is also a useful way to look at the feedback-controlled systems with time delay considered in this thesis. A second issue is that among all theories and connections mentioned above, an important link is missing, which is the generalization of traditional, macroscopic thermodynamics to the stochastic level of description.

Both problems are addressed by stochastic thermodynamics, which aims at finding laws and suitable definitions of thermodynamic properties on the stochastic level. This new theory is currently under development. Significant theoretical advances include the path integral representation of the total entropy production, stochastic energetics, and the fluctuation theorems, which we all briefly introduce in the following. In this framework, thermodynamic properties are investigated on the level of individual fluctuating trajectories. Further, basic premises are not limited to equilibrium situations,[4] making this theory particularly promising to gain deeper physi-

[4] Accept for the important assumption that the "bath" is in equilibrium!

cal understanding of complex, real-world systems, including time-delayed systems. Stochastic thermodynamics thus provides an essential piece of the puzzle to the overall understanding, and has immediate importance for, e.g., biological systems [14, 15] and computer science [16]. Moreover, it also provides important insight into quantum systems [17–19], and deepens the connection between physics and *information theory*, which was already introduced by Boltzmann and Jaynes and which might play an increasing role in the future [14].

4.2 Stochastic Energetics

Let us start introducing the concrete frameworks. In this Section, we review the energetic level of description given by the framework of Sekimoto [3]. We start by considering the energy (ex-)change between an (overdamped) particle and the bath consisting of the surrounding fluid particles, along a trajectory $\mathbf{X} = \{X(s)\}$, $s \in [t, t + \mathrm{d}t]$. The bath exerts a force on the particle through random kicks, namely, the Langevin force $\sqrt{2k_\mathrm{B}T\gamma}\,\xi(s)$. On the contrary, the particle also exerts a force onto the bath, $-\gamma\dot{X}(s)$, that is the friction[5] When yielding a particle displacement $\mathrm{d}X$, each of these forces is inevitably associated with energy exchange between particle and bath, called heat. Accordingly, the fluctuating heat δq flowing to the reservoir during the infinitesimal time $\mathrm{d}t$, is thus given by[6]

$$\delta q(t) = \left[\gamma \mathrm{d}X(t)/\mathrm{d}t - \sqrt{2k_\mathrm{B}T\gamma}\,\xi(t)\right] \circ \mathrm{d}X(t). \tag{4.1}$$

The $\circ$-symbol indicates usage of Stratonovich calculus. Note that as is commonly done, we assume the idealized heat bath to remain at all times in equilibrium, at a constant temperature, unaffected by possible external forces or the feedback. Further note that the sign convention of the heat employed here is consistent with [4], but different from [3].

Second, the particle may have an internal energy u. Since we consider the overdamped limit here, we do not properly resolve its kinetic energy. We further assume the particle to have no internal degrees of freedom. Consequentially, u is simply given by the potential energy, i.e., the position of the particle within an external potential V. This means the increment of internal energy $\mathrm{d}u$ reads

$$\mathrm{d}u(t) = V'[X(t)] \circ \mathrm{d}X(t). \tag{4.2}$$

Finally, in the presence of external, *non*-conservative forces $F_{\mathrm{non-con}}$, these may perform work on the particle given by

[5] Note that these two forces are often related to each other by a fluctuation-dissipation relation, see Sect. 2.3, Eq. 2.36.

[6] Please note the usage of δ symbols to express inexact differentials [21], which is because the heat and work are path-functions (not state-functions).

$$w(t) = F_{\text{non-con}} \circ \mathrm{d}X(t). \tag{4.3}$$

In total, we have identified three types of possible energy (ex-)change of a particle along a random, erratic trajectory of infinitesimal short time, $\mathrm{d}t$. Accordingly, the fluctuating energy changes during a time interval $[t_s, t_f]$ of finite length, are given by

$$u = \int_{t_s}^{t_f} \mathrm{d}u, \tag{4.4}$$

$$e = \int_{t_s}^{t_f} \delta e, \text{ with } e \in \{q, w\}, \tag{4.5}$$

while the *rates* are given by $\dot{u} = \mathrm{d}u/\mathrm{d}t$, $\dot{e} = \delta e/\mathrm{d}t$.

These definitions seem to be very reasonable and consistent with our physical intuition. When combining them with the LE (2.16), i.e., the equation of motion (or force-balance), one readily recovers the *first law of thermodynamics*

$$w(t) = \mathrm{d}u(t) + q(t). \tag{4.6}$$

This is the generalization of the first law to the stochastic level. Energy conservation must be valid along *each* fluctuating trajectory (of arbitrary length), as expected.

The description of the fluctuations of heat and work is one central objective of stochastic thermodynamics. An important contribution was the work [20], where the heat fluctuations in equilibrium and in a NESS of a linear, Markovian system were analytically calculated and experimentally confirmed, see Fig. 4.1 for an example.

Special case: A free particle If no external potential or non-conservative forces are present, i.e., the LE reads $\gamma\dot{X} = \sqrt{2k_\mathrm{B}T\gamma}\xi(t)$, all energy contributions trivially

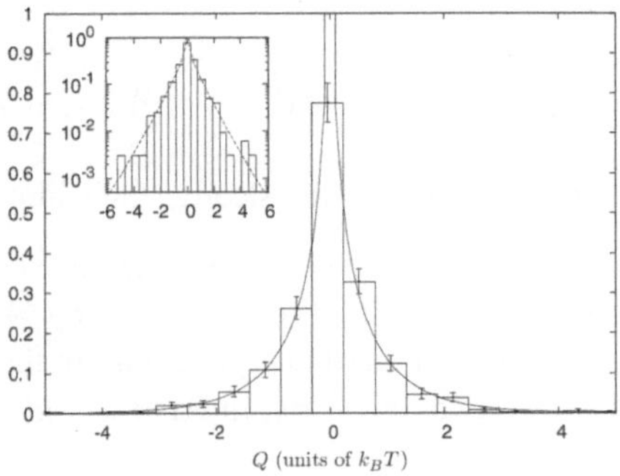

Fig. 4.1 A figure from Ref. [20], showing the probability distributions of heat Q generated along trajectories of length 0.5 s of a system described by a linear LE in thermal equilibrium. The lines plot the analytical expression given by Eq. (17) of [20], while the bars show experimental results from a colloidal particle in a harmonic optical trap generated by a laser. Inset: same data, with logarithmic y-axis

vanish, as follows from (4.1), (4.2), (4.3), (4.6). In such situations, the force exerted by the bath as a random kick does not "yield" energy exchange, but is *immediately* balanced out by friction. Thus, there are no heat, inner energy, or work fluctuations. This reflects the meaning of the Markov assumption on an energetic level (in the overdamped limit).

4.2.1 Steady States

Many systems approach steady states in the long-time limit ($t \to \infty$), where the transient initial condition effects have decayed away (see Sect. 2.2.2). In steady states, the ensemble average of the internal energy must be a conserved quantity, as is physically clear for stability reasons,[7] and as can also be seen from (4.2), which implies

$$\begin{aligned}\Delta U = \langle \Delta u \rangle = \left\langle \int_{t_s}^{t_f} \mathrm{d}u(t) \right\rangle = \left\langle \int_{t_s}^{t_f} V'\left[X(t)\right] \circ \mathrm{d}X(t) \right\rangle \\ = \left\langle V\left[X(t_f)\right]\right\rangle - \left\langle V\left[X(t_s)\right]\right\rangle \stackrel{\text{ss}}{=} 0. \end{aligned} \tag{4.7}$$

Here and in the following, we denote ensemble averages by capital letters: $\Delta U = \langle \Delta u \rangle$, $\delta W = \langle \delta w \rangle$, $\delta Q = \langle \delta q \rangle$. Thus, when considering steady-state ensemble averages, one is only interested in the heat and work, which must, according to the first law (4.6), fulfill

$$\delta W \stackrel{\text{ss}}{=} \delta Q. \tag{4.8}$$

Further, due to the time-invariance of steady states, the mean heat and work *linearly* increase with the length of the considered trajectory. For this reason, we will focus on the mean rates

$$\dot{Q} = Q/(t_f - t_s), \quad \text{and} \quad \dot{W} = W/(t_f - t_s). \tag{4.9}$$

In the following, we further distinguish the steady states into two different kinds.

Thermal equilibrium In the absence of non-conservative external forces, the work is trivially *zero* (without fluctuations) at all times, as given by (4.3). In such a situation, the system is expected to approach thermal equilibrium in the long-time limit, as no external driving prevents it. Then, the work, internal energy, and heat all have zero mean

$$\text{EQ}: \quad \delta W = \delta Q = 0. \tag{4.10}$$

[7] No system can remain stable when its internal energy constantly and for ever increases.

Internal energy and heat may nevertheless exhibit non-trivial fluctuations, see Fig. 4.1 for an example.

NESS When external non-conservative forces act on the particle, they prevent equilibration, because they continuously pump energy into the system. Ultimately, this energy is dissipated in the form of heat, as stated by the first law (4.6). Thus, a finite net heat flow is a signature of non-equilibrium. When the driving is time-independent, the system may nevertheless approach a stationary state, which is then a *nonequilibrium steady state* (NESS) characterized by a steady energy flow through the system[8]

$$\text{NESS}: \quad \delta W = \delta Q \neq 0. \tag{4.11}$$

This particular system state is very important for the framework of stochastic thermodynamics [4]. First of all, it is the normal operating regime of many nonequilibrium systems, and is hence relevant for experiments and applications. Further, it is technically somewhat easier to handle than other nonequilibrium states, as the PDFs are time-independent. Various exact relations are known for the fluctuation of entropy and energy (ex-)changes of systems in a NESS [4], including the existence of *fluctuation theorems* (which we will come to below), and exact expressions of the distributions in linear systems [20].

4.3 Fluctuating Entropy

So far, we have discussed the framework of stochastic energetics, where we encountered the generalization of the first law of thermodynamics to the fluctuating world. Now we consider the notion of fluctuating entropy, yielding a generalization of the second law. We will begin with the definition of the *total entropy production (EP) functional*, which is a central concept with deep physical implications. We will see that it is inextricably linked to the *thermodynamic arrow of time*. This definition will further automatically bring us to the fluctuating Shannon entropy and information.

The framework of stochastic thermodynamics reveals that the total EP of a process can literally be calculated via its *irreversibility*, i.e., the time-reversal symmetry breaking. Let us first give an illustration inspired by [22], and afterwards make things precise.

Imagine an observer sees a movie of a recorded process, for example the periodic motion a pendulum swinging back and forth, or the mixing of coffee and milk in a glass of latte macchiato [22]. The total EP measures how "sure" an observer can be of the fact that the movie is played forward, and not reversed. In the first case, a short movie of the oscillating pendulum does not enable the observer to clearly decide on

[8] A particular case of a NESS is approached in the absence of an external potential. Then, the inner energy (4.2) is trivially zero (has no fluctuations), and the fluctuations of heat and work, which still may have finite mean values, are *identical*, i.e., $\delta w = \delta q$ along each trajectory (4.6).

this question.[9] In sharp contrast, everybody could immediately tell[10] that the recorded movie of the latte macchiato, which would *de*mix if reversely played, is certainly played forwards. The difference between the two processes is that the mixing of coffee and milk yields a high increase of entropy, signaling a fast relaxation towards equilibrium (where coffee and milk are well mixed). On the contrary, the pendulum does not render a high increase in entropy, and hence, has a less pronounced arrow of time. In other words, the demixing of the coffee would violate "more clearly" the second law of thermodynamics (which we give below). The *decisiveness* of the imaginative observer is really what the total EP measures. For these macroscopic examples, the decisiveness appears to be a rather binary thing: The process is, or is not, reversible (corresponding to equilibrium or nonequilibrium). However, in situations where the observer considers small-scale processes with strong fluctuations (e.g., small numbers of colloids or molecules), it will usually not be so clear. For example, the process can transiently appear reversible, although it is on average irreversible (or *vice versa*). Thus, the total EP is a fluctuating quantity.

To be more precise, the total EP compares the following processes: The *forward* process is the actual, observable system state evolution. As thermodynamics tells us, this evolution is characterized by the system's attempt to relax towards equilibrium, possibly prevented by some external driving, which might be time-dependent. The *backward* process is an imagined process, where the time is running backwards and the system is driven by the same external driving protocol, but, if time-dependent, in reversed temporal order. Further, for a realization of the process, i.e., a single fluctuating trajectory $\mathbf{X} = \{X(t')\}, t' \in [t_s, t_f]$, the time-reversal operation amounts to replacing t by $-t$, and flipping the signs of velocities (which have "odd" parity under time-reversal [23]), but keeping the signs of positions X and accelerations $\ddot{X}$ the same (which are "even" quantities). In Markovian, memory-less systems, the backward process is typically as well realizable in experiments, and can further be accessed in simulations.

Comparing the chances to see a particular random trajectory in the forward process and the corresponding time-reversed trajectory in the backward process, is the very definition of total EP [4, 24]

$$\Delta s_{\text{tot}}[\mathbf{X}] = k_{\text{B}} \ln \frac{\mathcal{P}[\mathbf{X}]}{\hat{\mathcal{P}}[\hat{\mathbf{X}}]}, \tag{4.12}$$

involving the path probabilities $\mathcal{P}$ and $\hat{\mathcal{P}}$ for forward and backward process, respectively. The functional (4.12) assigns to a specific realization of the random process $\mathbf{X}$ a certain amount of total EP, which, in turn, depends on the path probabilities of the entire ensemble. As we will elaborate below, Δs_{tot} by construction fulfills an integral

[9] In fact, in a very long movie, the observer could also decide whether the movie of the pendulum was played forward, as the pendulum would then eventually stop due to friction, a process that produces entropy.

[10] Even without being aware of thermodynamics, as our brains are trained in a world which fulfills the second law.

fluctuation theorem [Eq. (4.22)], which is a statement about the symmetry properties of the fluctuations. Furthermore, the ensemble average of the total EP (4.12) automatically fulfills the *second law of thermodynamics*

$$\Delta S_{\text{tot}} \geq 0, \tag{4.13}$$

(as follows from the fluctuation theorem, see below). The capital letter is again used to denote the ensemble average. Equation (4.13) states that the mean total EP due to any spontaneous process can never be negative. This law breaks the symmetry between past and future and defines the thermodynamic arrow of time. The fact that (4.12) implies the second law (4.13) exemplifies that the definition of total EP employed in stochastic thermodynamics is a generalization of the concept of entropy to the fluctuating level, from which one can recover (on average) results from traditional thermodynamics.

At this point, we consider Markovian systems only, e.g., our delay LE (2.16) at $\tau = 0$ where $F[X(t), X(t-\tau)] \equiv F[X(t)]$. Using the conditional path probability

$$\mathcal{P}[\mathbf{X}] = \mathcal{P}[\mathbf{X}|x_s]\rho_1(x_s, t_s), \tag{4.14}$$

the total EP from definition (4.12) can be rewritten as the *entropy balance* equation

$$\Delta s_{\text{tot}}[\mathbf{X}] = \Delta s_{\text{sh}} + k_{\text{B}} \ln \frac{\mathcal{P}[\mathbf{X}|x_s]}{\hat{\mathcal{P}}[\hat{\mathbf{X}}|x_f]}, \tag{4.15}$$

involving the change of the *fluctuating Shannon entropy*

$$s_{\text{sh}}[X, t] = -k_{\text{B}} \ln[\rho_1(X, t)] \tag{4.16}$$

of the one-time probability density function ρ_1. Its mean value is

$$S_{\text{sh}} = \langle s_{\text{sh}} \rangle = -k_{\text{B}} \int_{\Omega} \rho_1(x, t) \ln[\rho_1(x, t)] \mathrm{d}x. \tag{4.17}$$

The change of Shannon entropy only depends on the start and end points[11] of the trajectory **X**.

Let us now consider the second term on the r.h.s. of Eq. (4.15). As mentioned in Chap. 2, $\mathcal{P}$ conditioned on the starting point $X(t_s) = x_s$ is proportional to the exponential of the Onsager-Machlup action [4, 23]. The corresponding $\mathcal{P}[\mathbf{X}|x_s]$ is provided in Eq. (2.28). In such a Markovian situation, the path probability of the *backward process* is analogously given by

[11] It can therefore be considered as a "temporal boundary term". Typically, it does not increase with the length of the trajectory and, thus, has a negligible contribution along very long trajectories.

$$\hat{\mathcal{P}}[\hat{\mathbf{X}}|x_f] = \mathcal{J}\, e^{-\int_{t_s}^{t_f} \left(-\dot{X}(t')-\gamma^{-1}F[X(t')]\right)^2/(4k_{\mathrm{B}}\mathcal{T}\gamma)\mathrm{d}t'}. \tag{4.18}$$

Importantly, the sign of velocities $\dot{X}$ is flipped in the backward process, while the position X keeps its sign, corresponding to odd and even time-reversal parity, respectively. Using the last expression and $\mathcal{P}$ from Eq. (2.28), one can immediately identify the remaining term in (4.15) as the *medium entropy production*, i.e.,

$$\Delta s_{\mathrm{m}} = k_{\mathrm{B}} \ln \frac{\mathcal{P}[\mathbf{X}|x_s]}{\hat{\mathcal{P}}[\hat{\mathbf{X}}|x_f]} = \frac{\delta q}{\mathcal{T}}, \tag{4.19}$$

which is proportional to the fluctuating heat, as defined in Sekimoto's framework given in Eq. (4.1). Contrary to the Shannon entropy change, this entropic contribution depends on the entire trajectory, i.e., the specific path that the particle takes, not only on the starting and end point.[12]

In total we have seen that, for a Markovian system (with position-dependent, time-dependent or constant) driving, the total EP is given by the change of Shannon entropy plus the medium entropy production. The first term quantitifies entropy production due to configurational changes of the system, while the second term quantifies entropy production in the medium associated with heat dissipation.

It is well known that the ensemble average of the total EP of a Markovian stochastic system can equivalently be derived by means of the Fokker-Planck equation [4] [instead of path integrals (4.15)], yielding the alternative expression for the rate

$$\dot{S}_{\mathrm{tot}} = \int \frac{\gamma J(x,t)^2}{k_{\mathrm{B}}\mathcal{T}\,\rho_1(x,t)}\,\mathrm{d}x \geq 0. \tag{4.20}$$

We will explicitly show this in Chap. 10.

4.3.1 Thermal Equilibrium & Nonequilibrium Steady States

In the previous Section, we have considered the energy flows characteristic for nonequilibrium steady states verses thermal equilibrium. After defining the total entropy production (4.15), we can now revisit these two types of steady states from an entropic perspective.

First we notice that in steady states, which are defined by $\partial_t \rho = 0$, the mean Shannon entropy (4.16) is, by construction, a conserved quantity. Thus, $\Delta S_{\mathrm{sh}} = 0$ and $\Delta S_{\mathrm{tot}} = \delta Q/\mathcal{T}$. Further, if equilibrium is approached, $\Delta S_{\mathrm{tot}} = 0$, finally giving the fundamental, thermodynamic definition of thermal equilibrium. This notion is consistent with the term equilibrium which we used above (e.g., when we discussed

---

[12] It can be considered a "temporal bulk term", which typically increases with the overall length of the trajectory.

the heat bath, or the probability current in Chap. 3). In all other cases, the LE yields $\Delta S_{\text{tot}} > 0$, as follows from (4.15), (4.19) and the second law. Thus,

$$\Delta S_{\text{tot}} = \delta Q/T \begin{cases} = 0, & \text{EQ}, \\ > 0, & \text{NESS} \end{cases} . \tag{4.21}$$

According to Eq. (4.21), a finite medium entropy production signals nonequilibrium. Further, the medium EP and the corresponding net heat flow are *strictly positive*[13] quantities. This means, in Markovian systems, external driving (such as a constant or position-dependent force, see [4] for examples), can only cause a net energy flow from the particle to the bath, and never a reversed one.[14] In other words such an ("uninformed", no feedback/control) external force ultimately yields a heating of the environment, and never cools it down. This is consistent with our physical intuition.[15]

One should note that in our theoretical model, where the bath is assumed to be infinitely big, see Sect. 2.3.1, such a heating or cooling can never yield a change of the bath's *temperature*, which is artificially kept fixed by the construction of the model.

Canonical ensemble We aim to mention that the Langevin equation generally describes a system at constant temperature, volume, and particle number, thus, the thermal equilibrium describes a canonical ensemble situation with a *free energy* $\mathcal{F} = U - TS_{\text{sh}}$ w.r.t. the colloid [4]. Here, the term "system" refers to the colloidal particle at position $X(t)$.

Let us recall that among the derivation of the LE from mechanical laws, the constant temperature was introduced by hand when the noise is "derived", see the given arguments between Eq. (2.35) and (2.36) in Sect. 2.3.1. This is noteworthy, as the starting point of this (Mori-Zwanzig) derivation was a system with constant energy, but fluctuating temperature. Thereby, the term "system" refers to the colloidal particle *plus* surrounding "bath particles".

Information gain The fluctuating Shannon entropy (4.16) has a direct connection to information theory. Specifically, s_{sh} is equivalent[16] to the *information* i gained by (i.e., the surprisal value of) a measurement of $x = X(t)$ [8]. Correspondingly, the mean Shannon entropy is equal to the expected information of one measurement, giving the *uncertainty* about the system [9]. Thus, it can be considered as the information that is needed to find out in which state the system is. The EQ state is the state with the maximal uncertainty under given external constraints (like constant temperature).

---

[13] Note that the sign of the heat actually depends on the used sign convention. However, in any case, the direction of the heat in a NESS is always directed such that energy is flowing on average to the heat bath.

[14] Be aware that this statement refers to ensemble averages only, *individual* random trajectories can transiently violate the second law and hence have a negative heat flow, as explicitly stated by the fluctuation theorems like (4.26).

[15] Heat is the energy exchange due to the friction between particle and solvent initiated by the external force.

[16] Here we refer to information measured in units of k_B. More generally $i \propto -\ln[\rho_1(X,t)]$.

4.4 Fluctuation Theorems

A particularly important result of stochastic thermodynamics are the various celebrated *fluctuation theorems* [25]. They represent general symmetry constraints for the distributions of thermodynamic quantities. Depending on the system at hand, there are fluctuation theorems for, e.g., the total EP, for the heat, and the work, which can often be derived from quite general assumptions. As a particularly important example, we here consider the *integral fluctuation theorem* (IFT)

$$\left\langle e^{-\Delta s_{\mathrm{tot}}/k_{\mathrm{B}}}\right\rangle = 1, \tag{4.22}$$

which is valid for the total EP along arbitrarily long trajectories, and holds in and outside of steady states or equilibrium. It can be derived from the definition (4.12), as follows

$$\begin{aligned}\left\langle e^{-\Delta s_{\mathrm{tot}}/k_{\mathrm{B}}}\right\rangle &= \int e^{-\Delta s_{\mathrm{tot}}[\mathbf{X}]/k_{\mathrm{B}}}\mathcal{P}[\mathbf{X}]\,\mathfrak{D}\mathbf{X}\\ &= \int e^{\ln\frac{\hat{\mathcal{P}}[\hat{\mathbf{X}}]}{\mathcal{P}[\mathbf{X}]}}\mathcal{P}[\mathbf{X}]\,\mathfrak{D}\mathbf{X}\\ &= \int \hat{\mathcal{P}}[\hat{\mathbf{X}}]\,\mathfrak{D}\mathbf{X} = 1.\end{aligned} \tag{4.23}$$

Thus, we only have to assume normalization of the paths, which is generally given.[17] This reasoning can readily be generalized to multidimensional systems (as we will do in Chap. 10). The IFT imposes a universal constraint on the symmetry of the total entropy fluctuations. Furthermore, the IFT (4.22) readily implies the second law of thermodynamics (4.13). This can be shown using the general relation

$$\ln(\sigma) \le \sigma - 1, \quad \forall \sigma > 0, \tag{4.24}$$

which can be applied to the present case, when expressing the ensemble average as sum over $N \to \infty$ realization of the random process, as follows

$$\begin{aligned}\langle -\Delta s_{\mathrm{tot}}/k_{\mathrm{B}}\rangle &= \sum_{i=1}^{N}\frac{-\Delta s_{\mathrm{tot,i}}/k_{\mathrm{B}}}{N} = \sum_{i=1}^{N}\frac{\ln(e^{-\Delta s_{\mathrm{tot,i}}/k_{\mathrm{B}}})}{N}\\ &\le \sum_{i}\frac{e^{-\Delta s_{\mathrm{tot,i}}/k_{\mathrm{B}}}-1}{N} = \left\langle e^{-\Delta s_{\mathrm{tot}}/k_{\mathrm{B}}}\right\rangle - 1 \overset{\mathrm{IFT}}{=} 0.\end{aligned} \tag{4.25}$$

More information about the entropy fluctuations are provided by the *detailed fluctuation theorem* [4, 26, 27]

[17] Please note that the backward path probabilities sum up to 1, also when integrated over all forward paths.

$$\frac{P(\Delta s_{\text{tot}})}{P(-\Delta s_{\text{tot}})} = e^{\Delta S_{\text{tot}}/k_{\text{B}}}, \tag{4.26}$$

which holds along arbitrarily long trajectories in a steady state (NESS or EQ) of a Markovian system. Loosely speaking, it states that observing "violations of the second law" is exponentially less likely than observing "fulfillment". In particular, the probability P of observing a particular *negative* value of total entropy production, $-\Delta s_{\text{tot}}$, along a fluctuating trajectory, is exponentially less likely then observing the corresponding positive value, $+\Delta s_{\text{tot}}$, where the exponent is given by the average value of produced total EP. Thus, in equilibrium (where $\Delta S_{\text{tot}} = 0$), negative and positive EP are equally likely, and the EP distribution is thus symmetric w.r.t. the origin. On the contrary, in a NESS, the mean total EP linearly increases with length of the trajectory. Therefore, the chances of observing negative EP exponentially decrease with the length of the trajectory.

There are various further fluctuation theorems. They are not only very important for the theoretical understanding, but also useful for applications [28–32], like the advancement of simulation techniques. For example, an immediate practical application of the IFT is that it can be used to test numerical integration schemes, by checking how well the generated random trajectories fulfill (4.22). Another prominent example of a fluctuation theorem is the *Jarzynski relation* [33]

$$\left\langle e^{-w/(k_{\text{B}}\mathcal{T})}\right\rangle = e^{-\Delta\mathcal{F}/(k_{\text{B}}\mathcal{T})}, \tag{4.27}$$

which enables reconstruction of the free-energy difference $\Delta\mathcal{F}$ between two EQ states upon experimental measurements of the fluctuating work w while driving the system from one of the EQ states to the other (along a non-equilibrium path). This relation has later been refined by Crooks [34], additionally taking the backwards process into account, in particular [4]

$$\frac{P(w)}{P(-w)} = e^{-(w-\Delta\mathcal{F})/\mathcal{T}}. \tag{4.28}$$

At this point, we aim to briefly revisit the discussion about causality, which we have started already in Chap. 2, Sect. 2.3.6.

4.4.1 Route to First Principles—Axiom of Causality

How deep is the physical insight provided by the integral fluctuation theorem?

While (4.23) is a "derivation" of the second law starting from the stochastic LE, one should emphasize that the LE itself already breaks time-reversal symmetry. Thus, this is not a derivation from first principles! Importantly, the route to first principles always requires at some point the introduction of an element of asymmetry by hand. This can be done by making suitable assumptions about the initial condition, or by imposing an *axiom of causality*. At this point, we resume the discussion from Sect. 2.3.6.

We recall that, in contrast to the underlying microscopic dynamics, the Langevin equation [that fulfills the FDR (2.36)] does not obey mechanical time-reversal symmetry. This is because, during the derivation, initial condition correlations must be neglected, or a causality principle must be assumed.

Indeed, the mathematical derivation of the fluctuation theorem from first principles, specifically the second law from [25], requires causality. To emphasize this, [25] explicitly shows that when assuming an anti-causal principle, one can derive—from the correct mechanical laws—a fluctuation theorem which implies an "anti-second law", i.e., decrease of entropy in the course of time (inconsistent with our observations of the real world). This clearly demonstrates the necessity of an axiom of causality to derive macroscopic irreversibility (thus, the second law), from time-reversible mechanical laws.

While the second law is hence rather imposed in the stochastic model than proven by (4.22), it should nevertheless be stressed that the symmetry properties uncovered by the fluctuation theorems are indeed nontrivial and important. In particular, fluctuation theorems represent one of the rare universal, rigorous results for *nonequilibrium* systems. By telling us, e.g., how unlikely transient "violations" of the second law are, they offer an intriguing view on thermal fluctuations.

As the fluctuation theorems are not readily applicable to time-delayed systems (which are, e.g., non-Markovian, have no notion of free energy,...), we will not go into further detail here, but refer the interested Reader to [4] and references therein.

4.5 Information

We have already mentioned a couple of times that concepts from information theory also play a crucial role in statistical physics, and especially in (stochastic) thermodynamics. This is particularly true in the context of control. In addition to energy and entropy, information represents a third perspective on thermodynamic processes [8]. In Sect. 4.3 we have already given the connection between Shannon entropy and information gain. Here, we will introduce some further information-theoretic quantities, which will be relevant for our investigations.

As opposed to the considerations up to this point, we now consider a system with multiple d.o.f. (subsystems), such that we can discuss the information flow between these subsystems. Lets say, for the sake of simplicity, we have a system with $n+1$ coupled d.o.f., $X_0, X_1, \ldots, X_n$. We again focus on the case of Markovian systems only. (The d.o.f. $X_{j\neq 0}$ will later represent the controller, as explained in detail in Chap. 10.)

But let us first introduce yet another, more established quantity in statistical physics, that is, the mutual information. In a second step, we will show their connection.

4.5.1 Mutual Information and Its Generalization

The mutual information between two d.o.f., say X_0 and X_1 is defined as

$$\mathcal{I}(x_0, x_1) = \int \rho_2(x_0, x_1) \ln \frac{\rho_2(x_0, x_1)}{\rho_1(x_0)\rho_1(x_1)} \mathrm{d}x_0 \mathrm{d}x_1, \tag{4.29}$$

with the 2-point joint probability distribution ρ_2. To have a shorter notation, we here omit the time-dependencies in the formulae.

In other words, the mutual information is the Kullback-Leibler divergence between the joint and the marginal PDFs. It quantifies the "amount of information" obtained about one d.o.f. through measurement of the other degree of freedom. It is, by construction, a *symmetric* quantity, i.e., the mutual information does not distinguish which of the two d.o.f. is the observed, and which is the unobserved one.

In steady states, the mutual information is automatically conserved, i.e., $\dot{\mathcal{I}} = 0$ (because $\partial_t \rho_2 = 0$). However, the individual information flows (defined below in Eq. 4.31), which are closely related by $\dot{\mathcal{I}} = \dot{I}_{\to 0} + \dot{I}_{\to 1}$, are not necessarily zero [35, 36]).

The definition of the mutual information itself, and its connection to information flows, is only applicable for situations where *two* subsystems exchange information ($n = 1$). In this thesis, we are also interested in situation where multiple degrees of freedom are involved. Thus, we will generalize this framework to arbitrary system sizes and topologies.

Generalization of the mutual information to more degrees of freedom In this thesis, we will consider a multivariate generalization of the mutual information between all d.o.f. defined as [37]

$$\mathcal{I}(\underline{x}) = \iint \rho_{n+1}(\underline{x}) \ln \frac{\rho_{n+1}(\underline{x})}{\rho_1(x_0)\rho_1(x_1)...\rho_1(x_n)} \mathrm{d}\underline{x}, \tag{4.30}$$

with the $(n+1)$-point joint pdf $\rho_{n+1}(\underline{x}, t)$ and $\underline{x} = \{x_0, x_1, ..., x_n\}$. For $n = 1$, this reduces to the usual mutual information (4.29). This multivariate generalization is (like the usual mutual information) a conserved quantity in steady states (because $\partial_t \rho_{n+1} = 0$). We note that there is not a unique way of generalizing the mutual information to systems with more than two subsystems, see [38–41].

4.5.2 Information Flow

We will consider the information flow to X_j from all other d.o.f.

$$\dot{I}_{\to j} = \iint \ln \frac{\rho_1(x_j)}{\rho_{n+1}(\underline{x})} \partial_{x_j} J_j \, \mathrm{d}\underline{x}. \tag{4.31}$$

While this quantity (or actually its bi-variate version [42]) is widely used to investigate discrete systems, it is less common to describe time- and space-continuous systems [36]. It should further be noted that there are other notions of information flows and information exchanges, see [42] for an educational overview. It should be noted that there are various other notions of information flows and information exchanges, which are more appropriate in other contexts, see [42] for an educational overview.

To derive the connection between information flow and the multivariate generalization of the mutual information, we start with considering the total derivative of $\mathcal{I}$ from Eq. (4.30), that is,

$$\dot{\mathcal{I}} = \int \underbrace{\partial_t \rho_{n+1}(\underline{x})}_{=(*)} \ln \frac{\rho_{n+1}(\underline{x})}{\rho_1(x_0)...\rho_1(x_n)} \, \mathrm{d}\underline{x} + \int \rho_{n+1}(\underline{x}) \left\{ -\frac{\partial_t \rho_{n+1}(\underline{x})}{\rho_{n+1}(\underline{x})} - \frac{\partial_t[\rho_1(x_0)\rho_1(x_1)...\rho_1(x_n)]}{\rho_1(x_0)\rho_1(x_1)...\rho_1(x_n)} \right\} \mathrm{d}\underline{x}. \quad (4.32)$$

To bring the last expression into another form, we utilize the multivariate FPE (3.14) $\partial_t \rho_{n+1} = -\sum_{j=0}^{n} \partial_{x_j} J_j$ to substitute $(*)$, and find

$$\begin{aligned}
\dot{\mathcal{I}} &= \sum_{j=0}^{n} \int \partial_{x_j} J_j(\underline{x}) \ln \frac{\rho_1(x_0)...\rho_1(x_n)}{\rho_{n+1}(\underline{x})} \mathrm{d}\underline{x} - \int \underbrace{\partial_t \rho_{n+1} \, \mathrm{d}\underline{x}}_{\to 0} \\
&\quad - \int \frac{\rho_{n+1}}{\rho_1(x_0)...\rho_1(x_n)} \underbrace{\partial_t[\rho_1(x_0)...\rho_1(x_n)] \, \mathrm{d}\underline{x}}_{\to 0} \\
&= \sum_{j=0}^{n} \int \partial_{x_j} J_j(\underline{x}) \ln \frac{\rho_1(x_0)...\rho_1(x_n)}{\rho_{n+1}(\underline{x})} \mathrm{d}\underline{x}. \qquad (4.33)
\end{aligned}$$

Let us now consider the individual summands. By application of basic properties of the logarithm and the natural boundary conditions, we find

$$\begin{aligned}
\int \partial_{x_j} J_j(\underline{x}) \ln \frac{\rho_1(x_0)\rho_1(x_1)..\rho_1(x_n)}{\rho_{n+1}(\underline{x})} \, \mathrm{d}\underline{x} &= \iint \ln \frac{\rho_1(x_j)\rho_1(x_{i\neq j})}{\rho_{n+1}(\underline{x})} \partial_{x_j} J_j \, \mathrm{d}\underline{x} \\
&\quad - \iint \ln \rho_1(x_i) \underbrace{\left[J_j\right]_{-\infty}^{\infty}}_{\to 0} \mathrm{d}\underline{x}_{\neq j} \\
&= \iint \ln \frac{\rho_1(x_j)}{\rho_{n+1}(\underline{x})} \partial_{x_j} J_j \, \mathrm{d}\underline{x} = \dot{I}_{\to j}. \quad (4.34)
\end{aligned}$$

Thus, the sum over all information flows gives the change of the (generalized multivariate) mutual information, i.e.,

$$\dot{\mathcal{I}} = \sum_{j=0}^{n} \dot{I}_{\to j}. \tag{4.35}$$

The last identity further implies that in steady states, the information flows between all components of the entire network nullify (thus, from an information-theoretical point of view, the network as a whole is "closed"). In Chap. 3, we have mentioned that EQ is associated with vanishing probability current, while a NESS has constant, nonzero J. Combining this with the definition (4.31), one can see that in EQ all individual information flows vanish, whereas *finite information flows are characteristic for a NESS*. This resembles the aforementioned net energy flows (heat and work), which likewise signal nonequilibrium.

In Chap. 10, we will derive rigorous relations between information flows and other thermodynamic quantities. Further, we will show that one can derive a generalized second law valid for the individual mesoscopic subsystems, based on this notion of information flow.

4.6 Previous Results, Expectations and Apparent Problems for Systems with Time Delay

After having introduced the basic concepts, focusing on Markovian systems, we now turn to systems subject to time-delayed feedback. We will review some previous results and point out open problems. In addition, we will formulate expectations, apparent technical challenges, and relevant research questions, in this way preparing the considerations presented in Part III.

While we will later in Part III consider both, systems with discrete and distributed time delay, we will here focus on the former case.

4.6.1 Energetics in the Presence of Delay

We start with considering the energetic perspective. For Markovian systems, we have introduced the framework of stochastic energetics in Sect. 4.2. From a formal perspective, the incorporation of time-delay delayed is straightforward: the first law and the fundamental definitions of the energy (ex-)changes (4.1)–(4.3) are still valid in the very form introduced here. After all, energy conservation must generally hold.

According to this framework, a position-dependent delay force $F[X(t-\tau)]$, which is *per se* a non-conservative force, is expected to perform work (4.3) on the system (see Fig. 4.2). However, as we have discussed in Chap. 3 in the context of probabilistic treatments, there are a significant differences between this history-dependent force and usual time-dependent or constant (non-conservative) Markovian forces. First, a formal difference is related to symmetry breaking and probability cur-

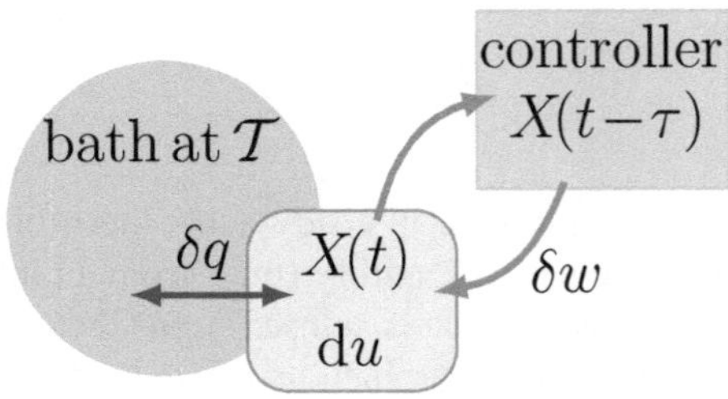

Fig. 4.2 Fluctuating energy (ex-)changes of the system $X(t)$: the controller can perform work δw on the system, heat δq might flow to the bath, and the internal energy of the system might change $\mathrm{d}u$. While in Markovian systems, the direction of the mean heat flow is dictated by the second law (4.21), a control with delay might induce a reversed net heat flow

rents. Specifically, the feedback force considered here has at each instant in time *on ensemble average* no preferred direction. In other words, it does not break the spatial symmetry of the system, and, therefore, it does not induce a net probability current. In turn, the absence of a probability current is a signature of thermal equilibrium, as is well known for Markovian processes (see Chap. 3). From this point of view, it is hence not immediately clear whether our feedback force will automatically yield nonequilibrium and introduce finite *average* energy flows.

Furthermore, in Sect. 4.3.1 we have seen that a simple external, position-dependent driving can never induce a negative mean heat flow in a NESS of a Markovian system [see (4.21)]. This is because of the second law, which forbids net energy extraction from a single heat bath in such a stationary situation. However, based on physical intuition, one may guess that it will make a substantial difference that our force is the result of a *control*, which involves something like a "decision"[18] between a measurement and according control action. It is a form of driving realizable by some external "intelligence", or controller; different from, lets say, some interaction forces between mutually coupled, mechanical degrees of freedom.[19] In principle, it is well known that an ("informed") control force can have a very different effect than usual ("uninformed") drivings. For example, control forces may enable *energy extraction of a single heat bath*, as illustrated by the "Maxwell demon" thought experiment [43–45]. This demon is a fictive creature which can, by tracking individual particle velocities and making decisions based on the measurement, "defeat the second law", i.e., extract useful work from a single heat bath. In our setting, energy extraction of a single heat bath would imply a net *negative heat current*. Another realization is a Szilárd engine [43, 46], which involves a position-dependent feedback. One can consider such devices "information to energy" converter [14]. Note that (in contrast

[18] We will later in Chap. 10 make this distinction rigorous, by connecting it to associated information flow.

[19] As an example, compare with the (friction) force applied by the surrounding heat bath oscillators derived in Sect. 2.3.1, which illustrates that not every history-dependent force is necessarily a "control".

to the aforementioned examples) the control considered here acts continuously and is, in this sense, closer to a so-called *autonomous* control device [36].

These considerations motivate us to investigate whether the feedback inevitably results in a net heat flow (despite the absence of a probability current), and second, whether it is possible to induce a *reversed heat flow* with a time-delayed feedback (which, although being very different from a Maxwell demon or Szilárd engine, likewise involves a measurement operation).

So far, no explicit general expressions for the thermodynamic quantities are worked out for time- and space-continuous systems with a continuously operating time-delayed feedback control. However, for the case of a *linear* delay LE, a couple of earlier studied can be found in the literature. We aim to highlight the theoretical contributions to this field by Rosinberg and Munakata [47–50]. This previous research has established that delayed feedback indeed induces a net heat flow [47] (despite the absence of a probability current). The delay indeed pushes the system out of equilibrium, and, in the absence of time-dependent drivings, the system may approach a NESS in the long-time limit [47]. Moreover, for underdamped systems, it has been shown that a delayed feedback loop can indeed induce a reversed (negative) heat flow. In contrast, the study [47] for overdamped systems suggests that the delay force can *not* induce a reversed heat flow. However, the arguments used there are based on an incomplete entropic treatment (see below). In addition, there are two further open problems concerning the NESS energy flows which we will address. First, the previous results for overdamped systems fail to correctly describe the limit of vanishing delay time. In [47], an offset of the heat rate at $\tau \to 0$ (where the heat flow should nullify) is detected, which is attributed to an Itô verses Stratonovich inconsistency. Second, the approaches taken in [47–50] are only applicable to linear systems.

This motivates us to revisit in this thesis the delay-induced steady-state heat flow, investigating linear and *nonlinear* delay systems (Chap. 9).

4.6.2 Entropic Description

Now we turn to the entropic treatment of our system in the presence of time-delayed feedback.

First, we aim to revisit the discussion from the previous Section, where we have mentioned the connection to "Maxwell demon"-devices, and the possibility of steady heat extraction from a single heat bath (see Sect. 4.6.1).

The long, historic discussion about the "Maxwell demon" has revealed that, from a formal perspective, the reason for the apparent "violation of the second law" implied by a negative steady-state heat flow, are *additional entropic terms* that need to be taken into account in the entropy balance (4.15) (in addition to the medium EP and the Shannon entropy change, which is on average zero in steady states). In particular, it has been established that measurement, storing, and deleting of information, which are associated with the control, have entropic cost, see Fig. 4.3. This is the essence

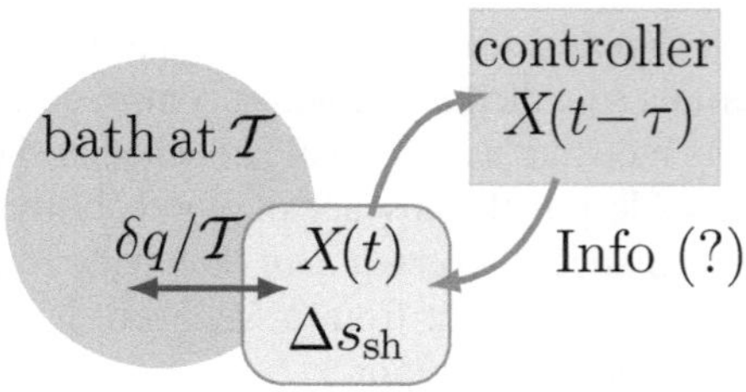

Fig. 4.3 Fluctuating entropy balance of the controlled system: the Shannon entropy s_{sh} may vary upon a change of the PDF or $X(t)$. The medium entropy might change due to a heat flow δq. Due to measurement and control action, we additionally expect an information-theoretic term in the entropy balance. Figure 4.2 shows the corresponding energetic picture. The unknown term "Info (?)" will be unraveled in Sect. 10.5.1

of the famous principle by Landauer [43, 51]. As a consequence, the total average entropy production ΔS_{tot} of controlled systems in a NESS may differ from the medium entropy $\Delta S_m = \delta Q/\mathcal{T}$ (contrary to the uncontrolled case described above), such that the second law does not impose nonnegativity on ΔS_m alone.

While this statement is generic, how to write down explicit formulae for the additional terms is not at all obvious. Specifically, the incorporation of continuously-operating, time-delayed feedback turns out to be highly nontrivial, as we will see in the following.

It is noteworthy that the treatment is much easier, when the controller only interacts with the system X at some (predefined) *solitary* instant in time. This is the scenario considered in various earlier studies, e.g., [52–54], an educational example being the feedback ratchet from Ref. [55]. The treatment of such systems is much more straightforward. In particular, one can chop the trajectories in a way that the control actions are at the boundaries only. On these time intervals, one has then a process that is Markovian, and can be treated by the standard approach described at the beginning of this chapter. For example, one can in principle evaluate the the total EP functional on the short trajectory pieces, and in a second step sum over all of the pieces.[20] To emphasize this difference, the type of control considered here was named *non-Markovian feedback control* in Ref. [56], and the considered continuous, stochastic system is denoted "autonomous system" [36, 49]. We aim to stress that for many physical or biological mechanisms such a continuous, non-Markovian feedback constitutes the more realistic model.

[20] In particular, the backward image of this process does not violate causality, contrasting the situation considered here, as will elaborate in the next Section.

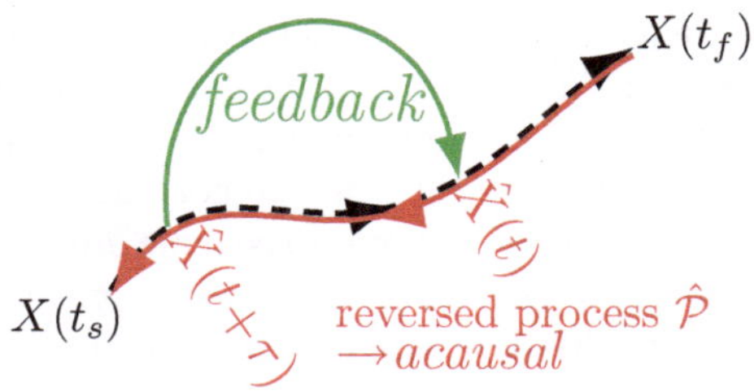

Fig. 4.4 Sketch of a trajectory (dotted line), and its time-reversed picture (red solid line). In the presence of a feedback force, which connects the instants in time t and $t - \tau$ (green arrow), the backward process becomes acausal, as it depends on its own future

4.6.3 The Acausality Issue

Above we have shown that the entropy balance readily follows from evaluating the logarithm of the ratio of path probabilities of forward and backward process (4.15). One crucial issue in the thermodynamic description of delayed systems, or more generally non-Markovian systems, is related to the time-reversed process. Recall that the backward process with probability $\hat{\mathcal{P}}$ shall be the time-reversed image of the process, where the driving is carried out in opposite direction in time (and the velocities are flipped).

For systems with time delay, where the driving involves $F_\mathrm{d}[X(t-\tau)]$, the forward process depends on the *previous* trajectory, i.e., the history. Thus, when flipping the time, one needs to think about how to flip the history-dependence: shall the time-reversed process depend on its own history (i.e., the future of the forward process), or shall it depend on its own future?

As argued by Rosinberg and Munakata [48, 49, 57], the time-reversal operation shall imply that the backwards process is not delayed, but *anticipating*, i.e., $F_\mathrm{d}[\hat{X}(t+\tau)]$. In other words, it is a stochastic process that depends on its own *future*, as sketched in Fig. 4.4. It is important to note that the latter is in principle unforseeable, due to the stochastic nature of the process. When following this idea, the backward process is *acausal*. As we have discussed in Sects. 2.3 and 4.4.1, such a process is in violation with the general formulation of statistical physics. From a technical point of view, the main problem is that the standard path integral formalism is not readily applicable, and instead one encounters a Jacobian $\mathcal{J}[\mathbf{X}]$ which is a highly nontrivial, path-dependent functional, see [48–50] (at least for any nonlinear system).

At this point it is further interesting to note that, in sharp contrast to ordinary differential equations, delayed differential equations have no unique backward solution,[21] even in the absence of noise (see Sect. 2.4.4). In the literature, we could not find a discussion about how this affects the total EP for delayed systems, but it might add an interesting argument to the general debate on this problem.

---

[21] Thus, the log-ratio of the probabilities to observe a certain forward and backward process is not zero in the absence of noise.

For Markovian systems, the Fokker-Planck equation represents an alternative route to calculate the EP [4]. We will demonstrate this in Chap. 10. However, because there is also no closed FPE for the systems with delay (see Chap. 3 and Part III), this alternative standard approach to the EP is as well not readily applicable.

More generally, entropic concepts appear to be quite difficult for non-Markovian systems. We think that this is indeed not only a minor problem of stochastic thermodynamics, given that the Markov assumption, which requires a clear timescale separation between "process" and "environment", is a very strong simplification only valid in a specific limit and inappropriate in many realistic systems. A framework that is restricted to such cases is of somewhat limited use to capture the essence of real-world systems.

Next, we will briefly consider a recent approach to the problem of calculating irreversibility measures alternative to the total EP for non-Markovian systems.

4.6.4 Short Comment on Effective Thermodynamics

Recent studies explore strategies to define reasonable irreversibility measures for non-Markovian systems based on the trajectories of X alone, similar to the total EP for Markovian systems (4.15). These approaches aim for an "effective thermodynamic" description, e.g., see Ref. [48–50, 58–61]. To this end, one may start with a Markovian description of the process. This can be done by considering the probabilities of the joint paths $\mathcal{P}[\mathbf{X}, \mathbf{Y}]$, where $\mathbf{Y}$ could represent the *entire relevant past* of the process $\mathbf{X}$ (and not only the starting point), see Ref. [49] for a more detailed description. Alternatively, $\mathbf{Y}$ could represent appropriate auxiliary variables, for example a variable that represents correlated fluctuations, as in [59]. We will give explicit examples for possible auxiliary variables for different non-Markovian systems in Part III. In the literature, different possibilities are discussed how to return to the non-Markovian case (without $\mathbf{Y}$).

As suggested in [49], one can first consider the log-ratio of probabilities of the joint paths, something like

$$R[\mathbf{X}, \mathbf{Y}] = k_B \ln \frac{\mathcal{P}[\mathbf{X}, \mathbf{Y}]}{\hat{\mathcal{P}}[\hat{\mathbf{X}}, \hat{\mathbf{Y}}]}. \tag{4.36}$$

Then, in a second step, R is integrated over all possible histories (i.e., tracing out $\mathbf{Y}$).

Alternatively, in [59] the authors decide to first integrate out $\mathbf{Y}$ already in the path probabilities (and making some assumptions about the correlations of the initial values of $\mathbf{X}$ and $\mathbf{Y}$), and then consider the log-ratio of

$$\widetilde{R}[\mathbf{X}] = k_B \ln \frac{\mathcal{P}[\mathbf{X}]}{\hat{\mathcal{P}}[\hat{\mathbf{X}}]}. \tag{4.37}$$

The different strategies yield different irreversibility measures.

We have shown above that the IFT of the total EP can be derived from quite general assumptions. Indeed, the IFT is quite robust and also holds for other functionals that are similar to the total EP (this is also true for some other fluctuation theorems, see [4]). Hence, it is not too surprising that also effective irreversibility measures like R, or an integral over R (where $\mathbf{Y}$ is traced out), or $\widetilde{R}$, fulfill an IFT. The fulfillment is explicitly shown in [48, 49, 59]. Hence, the existence of some fluctuation theorems alone is not a sufficient criterion to confirm the meaningfulness of an reversibility measure.

While these effective thermodynamics approaches are interesting and promising, they still comprise many open questions and ambiguities. For example: at which point should one marginalize, i.e., trace out $\mathbf{Y}$? Besides conceptual issues, there are also technical challenges. First, a generalization to other types of memory is not straightforward. So far, many studies are focused on a special memory kernel, often exponential or delta-correlated memory; or are only available for non-Markovianity introduced by colored noise. Second, the generalization to *nonlinear* systems is extremely difficult. For example, the approach in [59] relies on the Green's function method, which is only applicable for linear systems. Also, the entropy-like functional from [48, 49], could so far not be calculated for any nonlinear system. This is somewhat problematic, as in real-world systems, nonlinear models are often necessary to capture many important effects [62]. It is a great challenge for future research in the field of stochastic thermodynamics to sort out these issues.

Our approach to the problem (pursued in Chap. 10) In contrast to approaches that try to define meaningful irreversibility measures based on $\mathbf{X}$ alone (like the "effective thermodynamics" descriptions), we will explore a different strategy in Chap. 10, in line with the earlier works [23, 63]. In particular, we explicitly keep all relevant d.o.f. (i.e., consider the full EP of the joint process $\{\mathbf{X}, \mathbf{Y}\}$), and see what we can learn from that. This approach allows us to use the standard definition of total EP, and is in a sense more general w.r.t. the mathematical details of the control (memory kernel, nonlinear forces). In turn, it makes crucial assumptions on the physical picture behind the control, and is, in this sense, much more restrictive.

From the viewpoint of effective thermodynamics, one can also regard our investigation as a first step. Building up on our findings, one might, in a second step, construct again an effective thermodynamics, by some sort of marginalization. This is beyond the scope of this thesis, but provides an interesting perspective for future research.

Moreover, without further reasoning, the approach employed in Chap. 10 is appropriate for the following physical picture: Assume that the delayed force stems from physical *external subsystems*, which generate the feedback loop and interacts with the particle. We call them "external", as they are only implicitly accounted for in our original non-Markovian description [in the form of a delayed force in the LE (2.43)]. One can also say that they are *hidden*. These "external subsystems" might belong to a specifically designed agent, a feedback controller, or, more generally, could be a part of a more complex biological system (for example part of a chemical reaction network). In any case, we assume that this external system as well consists of physical components, which cannot perform forces without energetic and entropic cost, i.e., we assume that the external agent is a physical system. Furthermore, we assume that these external subsystems interact with the particle in an instantaneous (Markovian) manner, (e.g., the involved measurement operation is instantaneous). The delay comes through a coupling of multiple subsystems to each other in a specific way (in the form of a unidirectional ring).

We will explain all of this in detail in Chap. 10. There we will show that such a setup is only possible with *non-reciprocal* coupling among the "controller subsystems".

Role of hidden degrees of freedom Indeed, it is well known that hidden degrees of freedom, i.e., physical subsystems which interact with the studied systems but are not explicitly accounted for in the employed model, can have a significant impact on the EP, see for example [64, 65]. This provides an additional motivation for the study discussed in Chap. 10.

From a more general point of view, this problem is connected to the question of correct coarse-graining procedures, which, ideally, preserve as much of the thermodynamic properties as possible. We will only very briefly comment on this issue in Part III, but refer the interested Reader to [2, 4, 61, 64–68].

4.7 Side Note: Active Particles & Non-reciprocal Interactions

By investigating the random trajectories of a feedback-controlled, non-Markovian particle, one notices differences to the trajectories of a normal, passive colloidal system subject to e.g. external "uninformed" forces, $F[X, t]$. This is for example evident by the correlation functions, which reveal the nontrivial memory in the process (compare Figs. 2.4 and 6.3). However, the differences become much more evident, when we consider the thermodynamic properties (see Part III). An important distinction is the nonexistence of a thermal equilibrium state. As we have already mentioned before, one may thus consider the feedback-controlled colloids as a "novel" type of particle, in comparison to the passive particles usually studied in traditional statistical mechanics. Indeed, there is another "novel" type of particle, which is nowadays studied by many researchers by means of statistical physics and specifically stochastic thermodynamics; that is *active systems*.

Active systems A historic curiosity is that Robert Brown in his original paper from 1827 [69], in which Brownian motion was described for the first time, speculated that the observed motion of the pollen might be due to their liveliness, and called them "active molecules". He therefore dedicated his investigation mainly on how the motion differs between a living and dead plant, e.g., by burning them. Of course, the observed stochastic motion is also present in the pollen of dead plants, and is in this case independent of this question.

In contrast, the stochastic motion of bacteria, sperm cells, and many other, actually living objects is indeed very different from *passive* Brownian motion (of, e.g., a pollen, or dead bacteria), because these systems actively "swim", i.e., propel themselves. The *self-propulsion* is possible because these organisms convert other forms of energy, like chemical energy stored in ATP, into motion (just as we humans do). So far, we have introduced the Langevin equation as a model for passive particles who may be subject to external forces (e.g., feedback forces). Besides this, Langevin equations can indeed also be used to model such an active motion. Active motion is often characterized by *persistence* [70–72]. In regard to the diffusive behavior this means, it has a ballistic regime on intermediate timescales, while the motion is typically like normal diffusion on very short and on very long timescales. This behavior is described by all of the three common models [73]: active Brownian particles, active Ornstein-Uhlenbeck particles, and run-and-tumble motion.

Besides the aforementioned examples, such active colloids can also be individual parts of a larger organism, like the hair bundles of auditory cells [74, 75]. Further, they don't have to be (parts of) living creatures, but can also be artificial, an obvious example being robots. Other well-know realizations of artificial active systems are Janus particles with unihemispherical coating [76], and vibrating screws or rods [77]. Studying the statistical properties of active motion is subject of ongoing research in the field of *active matter* [78].

A defining criterion of active particles is an inherent (independent) energy input [79–82]. Thus, these systems are intrinsically out of equilibrium. While this statement is generic, the development of a thermodynamic description of active systems is subject of an ongoing debate, and the objective of numerous latest research articles in the field of stochastic thermodynamics, see, e.g., [59, 60, 76, 81, 83–90]. Here, key problems lie in the search for appropriate definitions of irreversibility measures and entropy production, and the interpretation of heat flux and work. A further open question is the time-reversal parity of the involved variables [60], which is relevant for the total EP functional (4.12). These problems are further connected to a more general question in active matter, that is, the search for suitable models to describe active systems.

4.7.1 Active Ornstein-Uhlenbeck Particles

The simplest and therefore very popular model for an active swimmer is the *Active Ornstein-Uhlenbeck particle* (AOUP) [58–60, 84, 85]. It can be expressed using two d.o.f. following the equations

$$\gamma \dot{X}(t) = Y(t) + F + \sqrt{2k_B \mathcal{T} \gamma}\, \xi(t), \tag{4.38a}$$

$$\tau_p \dot{Y}(t) = -Y(t) + \xi_p(t), \tag{4.38b}$$

with the two, zero-mean white noises: $\langle \xi(t)\xi(t')\rangle = \delta(t-t')$, $\langle \xi_p(t)\xi_p(t')\rangle = \delta(t-t')$, $\langle \xi(t)\xi_p(t')\rangle = 0$. Thereby, Y is a d.o.f. that models the effect of the flagella of a bacterium, or the asymmetric flow field around a Janus colloid, in a very simplistic, one-variable manner. However, X which corresponds to the particle position, is the d.o.f. of main interest. One can thus also employ a one-variable description focusing on X, obtained by projecting the second equation with the d.o.f. Y into the X equation. To this end, one formally solves the equation for Y [assuming $Y(0) = 0$], which yields $Y(t) = \int_0^t e^{-(t-t')/\tau_\mathrm{p}} \xi_Y(t')\mathrm{d}t' =: \nu(t)$. Then one substitutes the result into the first equation of (4.38) obtaining

$$\gamma \dot{X}(t) = \nu(t) + F + \sqrt{2k_\mathrm{B}T\gamma}\,\xi(t),$$
$$\text{with } \langle \nu(t)\nu(t+T)\rangle = \int_0^t e^{-(t-t')/\tau_\mathrm{p}} \xi_\mathrm{p}(t')\mathrm{d}t' =: C_\nu(T), \tag{4.39a}$$

and $\langle \nu(t)\rangle = 0$. We have renamed Y to ν, to emphasize the different characters of the variable, which is a dynamical d.o.f. in (4.38) and a colored Gaussian noise in (4.39). This is a simple, non-Markovian description of an active particle's motion with persistence.

It is noteworthy that this very model is in fact also used to describe a passive (i.e., "dead") colloid in an *active bath*, i.e., a heat bath consisting of self-propelled particles. However, here we are going to take the aforementioned, alternative viewpoint of an active colloidal swimmer in a passive bath.

Both Eqs. (4.39) and (4.38) equally describe the dynamics of a *microswimmer* at position X. The auxiliary d.o.f. Y in (4.38) introduces correlated fluctuations in the X dynamics with characteristic timescale $1/\tau_\mathrm{p}$, and in this way models the effect of the flagella of a bacterium, or the asymmetric flow field around a Janus colloid, responsible for the propulsion force on the swimmer [76, 91]. Thus, here the memory models the persistence of the motion. Importantly, in (4.39) there is no direct connection between memory in noise correlations and in the friction, hence, no FDR (see Sect. 2.3.1). This signals *nonequilibrium*, as one shall indeed expect for an active system. However, the broken FDR alone does not quantify the distance from equilibrium, nor does it provide further insights into the thermodynamic properties of this nonequilibirum system. This motivates us to take a closer look at this model as we do in Part III. Another motivation is the connection between active matter and feedback-controlled systems.

4.7.2 Connection Between Active Matter and Time-Delayed Systems

We have already mentioned one important, common feature of active and feedback systems, i.e., they are intrinsically out of equilibrium already on the level of individual constituents. Conventionally, active systems are mostly associated with "energy consumption" of individual constituents [79, 80, 88], and are hence rather studied from an energetic perspective (or from a purely dynamical perspective), while feedback-controlled systems are considered to be "information fueled" [55]. Therefore, the second type of system is usually studied from an entropic and information-theoretical perspective.

However, it has already been recognized in earlier literature that also active systems involve nontrivial information flows [59, 92]. On the other hand, one clearly expects that a feedback controller "consumes energy" in order to operate. Thus, these systems might not be too different after all. Recent experiments and theoretical works indeed expose the connections between both types of systems, see e.g. [93–96]. It is also noteworthy, that some active systems, like *E.coli* bacteria, use *intrinsic feedback loops* to create a motion towards food sources, for example chemotaxis [93, 97, 98].

In fact, there is another—less obvious—shared feature of active and feedback systems, that is, the presence of memory. While the discussion of memory, or delay, in feedback loops has a long tradition [48–50, 99–103] (where it stems e.g. from finite signal processing speed), this viewpoint is less common in the context of active motion. However, the presence of history-dependence becomes apparent in the form of *persistence* of active swimmers [70–72] as one can see above in Eq. (4.39). Also the viscoelastic properties of active gels [82], or the giant Kovacs-like memory effect in the Vicsek model [104] give a hint to the importance of memory in active systems.

We believe that a fundamental understanding of the two classes of systems from *both* perspectives, the energetic and the entropic/information-theoretic one, is very important for the thermodynamic understanding and their statistical properties. An apparent advantage of combining these fields are the various established results for the collective behavior of active systems on larger scales [71, 96]. In this spirit, we investigate an approach which naturally unifies both perspectives, as we will discus in detail in Chap. 10. A key ingredient is *non-reciprocal coupling*.

4.7.3 Non-reciprocal Interactions

The set of Eqs. (4.38) defines a two-dimensional network with *unidirectional* coupling, because X is influenced by Y, whereas Y does not "see" X, i.e.,

$$\begin{aligned} X &\leftarrow Y, \\ X &\nrightarrow Y. \end{aligned} \tag{4.40}$$

This unidirectional coupling is a special type of *non-reciprocal* (NR) coupling, which denotes interactions that cannot be represented by an interaction Hamiltonian and *violate Newtons third law actio = reactio*.

Such interactions also arise in "intelligent" systems [105], like pedestrian dynamics, or in collections of self-driving cars, but can also be found in chemical [106, 107] and quantum systems [108–112]. This type of interaction commonly arises in active matter [113]. Recent experiments [114] further realize non-reciprocal interactions between colloidal particles. At this point one may also note that in the community of nonlinear dynamics and network science, studying the interplay between non-reciprocal (symmetry-broken) coupling and dynamical states is a well-established research field. For example, the existence of chimera states [115], a special type of clustering, was linked to symmetry-broken coupling [116], and shown to persist in the presence of discrete delay [117] and Gamma-distributed memory [118].

As we have seen in Chap. 2, the mechanical laws are the foundations of the stochastic models. In fact, various derivations in statistical physics rely on them. Thus, in the absence of Newton's third law, one needs to revisit many well-known results. Only recently, researchers have started to investigate the consequences of NR coupling on statistical properties of many particle systems [106, 119]. However, the thermodynamic implications of the non-reciprocity have, to the best of our knowledge, not been considered in earlier literature.

In Chap. 10, we will examine in detail the heat and information flows induced by NR coupling, considering the AOUP model, but also more general types of network topologies. For example, the very same set of equations as (4.38) with reversed unidirectional coupling (i.e., $X \rightarrow Y$ and $X \not\leftarrow Y$), was recently suggested as a model for a *cellular sensor* [120, 121]. More importantly in the context of this thesis, systems with time-delayed feedback can also be represented as networks with NR coupling among X and auxiliary variables.

In the following two Parts of this thesis, we present novel results for the probabilistic description (Part II) and thermodynamic treatment (Part III). Therewith, we aim for two theoretical concepts, which are cornerstones of statistical physics, already established for passive particles, which are not subject to feedback ("uninformed" particles).

References

1. Landauer R et al (1991) Information is physical. Phys Today 44:23–29
2. Esposito M (2012) Stochastic thermodynamics under coarse graining. Phys Rev E 85:041125
3. Sekimoto K (2010) Stochastic energetics, vol 799. Springer
4. Seifert U (2012) Stochastic thermodynamics, fluctuation theorems and molecular machines. Rep. Prog. Phys. 75:126001
5. Hunter MCW, Davis EB (1999) The works of Robert Boyle, vol 7. Pickering & Chatto London
6. Schwabl F (2002) Statistical mechanics. Springer, Berlin
7. Carroll SM, Chen J (2005) Does inflation provide natural initial conditions for the universe? Int J Mod Phys D 14:2335–2339
8. Katz A (1967) Principles of statistical mechanics: the information theory approach. WH Freeman
9. Jaynes ET (1957) Information theory and statistical mechanics. Phys Rev 106:620
10. Hänggi P, Marchesoni F, Nori F (2005) Brownian motors. Ann der Phys 14:51–70
11. Risken H (1984) The Fokker-Planck equation. Springer
12. Rouse PE Jr (1953) A theory of the linear viscoelastic properties of dilute solutions of coiling polymers. J Chem Phys 21:1272–1280
13. Maes C, Safaverdi S, Visco P, Van Wijland F (2013) Fluctuation-response relations for nonequilibrium diffusions with memory. Phys Rev E 87:022125
14. Parrondo JMR, Horowitz JM, Sagawa T (2015) Thermodynamics of information. Nat Phys 11:131–139

15. Barato AC, Seifert U (2015) Thermodynamic uncertainty relation for biomolecular processes. Phys Rev Lett 114:158101
16. Wolpert DH (1992) Memory systems, computation, and the second law of thermodynamics. Int J Theor Phys 31:743–785
17. Esposito M, Lindenberg K, Van den Broeck C (2009) Thermoelectric efficiency at maximum power in a quantum dot. EPL 85:60010
18. Strasberg P, Schaller G, Brandes T, Esposito M (2013) Thermodynamics of a physical model implementing a Maxwell demon. Phys Rev Lett 110:040601
19. Morikuni Y, Tasaki H (2011) Quantum Jarzynski-Sagawa-Ueda relations. J Stat Phys 143:1–10
20. Imparato A, Peliti L, Pesce G, Rusciano G, Sasso A (2007) Work and heat probability distribution of an optically driven Brownian particle: theory and experiments. Phys Rev E 76:050101
21. Neumann C (1875) Vorlesungen über die mechanische Theorie der Wärme. Teubner
22. Aaronson S, Carroll SM, Ouellette L (2014) Quantifying the rise and fall of complexity in closed systems: The coffee automaton. ArXiv:1405.6903
23. Puglisi A, Villamaina D (2009) Irreversible effects of memory. EPL 88:30004
24. Lebowitz JL, Spohn H (1999) A Gallavotti-Cohen-type symmetry in the large deviation functional for stochastic dynamics. J Stat Phys 95:333–365
25. Evans DJ, Searles DJ (2002) The fluctuation theorem. Adv Phys 51:1529–1585
26. García-García R, Domínguez D, Lecomte V, Kolton AB (2010) Unifying approach for fluctuation theorems from joint probability distributions. Phys Rev E 82:030104
27. Esposito M, Van den Broeck C (2010) Three detailed fluctuation theorems. Phys Rev Lett 104:090601
28. Tasaki H (2000) Jarzynski relations for quantum systems and some applications. *arXiv preprint cond-mat/0009244*
29. Berg J (2008) Out-of-equilibrium dynamics of gene expression and the Jarzynski equality. Phys Rev Lett 100:188101
30. Wolf S, Stock G (2018) Targeted molecular dynamics calculations of free energy profiles using a nonequilibrium friction correction. J Chem Theory Comput 14:6175–6182
31. Williams SR, Evans DJ, Searles DJ (2011) Nonequilibrium Umbrella Sampling and the Functional Crooks Fluctuation Theorem. J Stat Phys 145:831–840
32. Sivak DA, Chodera JD, Crooks GE (2013) Using nonequilibrium fluctuation theorems to understand and correct errors in equilibrium and nonequilibrium simulations of discrete Langevin dynamics. Phys Rev X 3:011007
33. Jarzynski C (1997) Nonequilibrium equality for free energy differences. Phys Rev Lett 78:2690
34. Crooks GE (1999) Entropy production fluctuation theorem and the nonequilibrium work relation for free energy differences. Phys Rev E 60:2721
35. Allahverdyan AE, Janzing D, Mahler G (2009) Thermodynamic efficiency of information and heat flow. JSTAT 2009:P09011
36. Horowitz JM, Esposito M (2014) Thermodynamics with continuous information flow. Phys Rev X 4:031015
37. Loos SAM, Klapp SHL (2020) Irreversibility, heat and information flows induced by non-reciprocal interactions. New J Phys 22:123051
38. McGill W (1954) Multivariate information transmission. Trans IRE Prof Group Inf Theory 4:93–111
39. Te Sun H (1980) Multiple mutual informations and multiple interactions in frequency data. Inf Control 46:26–45
40. Ting HK (1962) On the amount of information. Theory Prob Appl 7:439–447
41. Srinivasa S (2005) A review on multivariate mutual information. Univ. of Notre Dame, Notre Dame. Indiana 2
42. Horowitz JM, Sandberg H (2014) Second-law-like inequalities with information and their interpretations. New J Phys 16:125007

43. Sagawa T (2012) Thermodynamics of information processing in small systems. Prog Theor Phys 127:1–56
44. Koski JV, Maisi VF, Sagawa T, Pekola JP (2014) Experimental observation of the role of mutual information in the nonequilibrium dynamics of a Maxwell Demon. Phys Rev Lett 113:030601
45. Maxwell JC (1871) Theory of heat. Dover Publications. inc
46. Roldán É, Martínez IA, Parrondo JMR, Petrov D (2014) Universal features in the energetics of symmetry breaking. Nat Phys 10:457
47. Munakata T, Iwama S, Kimizuka M (2009) Linear stochastic system with delay: Energy balance and entropy production. Phys Rev E 79:031104
48. Munakata T, Rosinberg ML (2014) Entropy production and fluctuation theorems for Langevin processes under continuous non-Markovian feedback control. Phys Rev Lett 112:180601
49. Rosinberg ML, Munakata T, Tarjus G (2015) Stochastic thermodynamics of Langevin systems under time-delayed feedback control: Second-law-like inequalities. Phys Rev E 91:042114
50. Rosinberg ML, Tarjus G, Munakata T (2017) Stochastic thermodynamics of Langevin systems under time-delayed feedback control. II. Nonequilibrium steady-state fluctuations. Phys Rev E 95:022123
51. Landauer R (1961) Irreversibility and heat generation in the computing process. IBM J Res Dev 5:183–191
52. Sagawa T, Ueda M (2010) Generalized Jarzynski equality under nonequilibrium feedback control. Phys Rev Lett 104:090602
53. Sagawa T, Ueda M (2012) Nonequilibrium thermodynamics of feedback control. Phys Rev E 85:021104
54. Esposito M, Schaller G (2012) Stochastic thermodynamics for "Maxwell demon" feedbacks. EPL 99:30003
55. Cao FJ, Feito M (2009) Thermodynamics of feedback controlled systems. Phys Rev E 79:041118
56. Debiossac M, Grass D, Alonso JJ, Lutz E, Kiesel N (2019) Thermodynamics of continuous non-Markovian feedback control. ArXiv:1904.04889
57. Rosinberg M, Tarjus G, Munakata T (2018) Influence of time delay on information exchanges between coupled linear stochastic systems. Phys Rev E 98:032130
58. Caprini L, Marconi UMB, Puglisi A, Vulpiani A (2019) The entropy production of Ornstein-Uhlenbeck active particles: a path integral method for correlations. J Stat Mech Theor Exp 2019:053203
59. Dabelow L, Bo S, Eichhorn R (2019) Irreversibility in active matter systems: Fluctuation theorem and mutual information. Phys Rev X 9:021009
60. Shankar S, Marchetti MC (2018) Hidden entropy production and work fluctuations in an ideal active gas. Phys Rev E 98:020604(R)
61. Polettini M, Esposito M (2017) Effective thermodynamics for a marginal observer. Phys Rev Lett 119:240601
62. Strogatz SH, Friedman M, Mallinckrodt AJ, McKay S et al (1994) Nonlinear dynamics and chaos: with applications to physics, biology, chemistry, and engineering. Comput Phys 8:532
63. Crisanti A, Puglisi A, Villamaina D (2012) Nonequilibrium and information: the role of cross correlations. Phys Rev E 85:061127
64. Mehl J, Lander B, Bechinger C, Blickle V, Seifert U (2012) Role of hidden slow degrees of freedom in the fluctuation theorem. Phys Rev Lett 108:220601
65. Kawaguchi K, Nakayama Y (2013) Fluctuation theorem for hidden entropy production. Phys Rev E 88:022147
66. Altaner B, Vollmer J (2012) Fluctuation-preserving coarse graining for biochemical systems. Phys Rev Lett 108:228101
67. Puglisi A, Pigolotti S, Rondoni L, Vulpiani A (2010) Entropy production and coarse graining in markov processes. J Stat Mech: Theory Exp P05015
68. Strasberg P, Esposito M (2017) Stochastic thermodynamics in the strong coupling regime: an unambiguous approach based on coarse graining. Phys Rev E 95:062101

69. Brown R (1827) A brief account of microscopical observations. Philos Mag Ann Philos 4:161
70. Narinder N, Bechinger C, Gomez-Solano JR (2018) Memory-induced transition from a persistent random walk to circular motion for achiral microswimmers. Phys Rev Lett 121:078003
71. Nagai KH, Sumino Y, Montagne R, Aranson IS, Chaté H (2015) Collective motion of self-propelled particles with memory. Phys Rev Lett 114:168001
72. Scholz C, Jahanshahi S, Ldov A, Löwen H (2018) Inertial delay of self-propelled particles. Nat Commun 9:5156
73. Fodor É, Marchetti MC (2018) The statistical physics of active matter: from self-catalytic colloids to living cells. Phys A 504:106–120
74. Ricci A, Crawford A, Fettiplace R (2000) Active hair bundle motion linked to fast transducer adaptation in auditory hair cells. J Neurosci 20:7131–7142
75. Barral J, Dierkes K, Lindner B, Jülicher F, Martin P (2010) Coupling a sensory hair-cell bundle to cyber clones enhances nonlinear amplification. PNAS 107:8079–8084
76. Speck T (2019) Thermodynamic approach to the self-diffusiophoresis of colloidal Janus particles. Phys Rev E 99:060602(R)
77. Blair DL, Neicu T, Kudrolli A (2003) Vortices in vibrated granular rods. Phys Rev E 67:031303
78. Bechinger C et al (2016) Active particles in complex and crowded environments. Rev Mod Phys 88:045006
79. Ramaswamy S (2010) The mechanics and statistics of active matter. Ann Rev Condens Matter Phys 1:323–345
80. Ramaswamy S (2017) Active matter. J Stat Mech Theor Exp 2017:054002
81. Fodor É et al (2016) How far from equilibrium is active matter? Phys Rev Lett 117:038103
82. Marchetti MC et al (2013) Hydrodynamics of soft active matter. Rev Mod Phys 85:1143
83. Takatori SC, Brady JF (2015) Towards a thermodynamics of active matter. Phys Rev E 91:032117
84. Mandal D, Klymko K, DeWeese MR (2017) Entropy production and fluctuation theorems for active matter. Phys Rev Lett 119:258001
85. Bonilla L (2019) Active Ornstein-Uhlenbeck particles. Phys Rev E 100:022601
86. Marconi UMB, Puglisi A, Maggi C (2017) Heat, temperature and clausius inequality in a model for active Brownian particles. Sci Rep 7:46496
87. Speck T (2016) Stochastic thermodynamics for active matter. EPL 114:30006
88. Nardini C et al (2017) Entropy production in field theories without time-reversal symmetry: quantifying the non-equilibrium character of active matter. Phys Rev X 7:021007
89. Pietzonka P, Seifert U (2017) Entropy production of active particles and for particles in active baths. J Phys A: Math Theor 51:01LT01
90. Ehrich J, Kahlen M (2019) Approximating microswimmer dynamics by active Brownian motion: energetics and efficiency. Phys Rev E 99:012118
91. Cates ME (2012) Diffusive transport without detailed balance in motile bacteria: does microbiology need statistical physics? Rep Prog Phys 75:042601
92. Micali G, Endres RG (2016) Bacterial chemotaxis: information processing, thermodynamics, and behavior. Curr Opin Microbiol 30:8–15
93. Ito S, Sagawa T (2015) Maxwells demon in biochemical signal transduction with feedback loop. Nat Commun 6:7498
94. Khadka U, Holubec V, Yang H, Cichos F (2018) Active particles bound by information flows. Nat Commun 9:3864
95. Aguilar-Hidalgo D et al (2018) Critical point in self-organized tissue growth. Phys Rev Lett 120:198102
96. Mijalkov M, McDaniel A, Wehr J, Volpe G (2016) Engineering sensorial delay to control phototaxis and emergent collective behaviors. Phys Rev X 6:011008
97. Sourjik V, Wingreen NS (2012) Responding to chemical gradients: bacterial chemotaxis. Curr Opin Cell Biol 24:262–268
98. Bourret RB, Stock AM (2002) Molecular information processing: lessons from bacterial chemotaxis. J Biol Chem 277:9625–9628

99. Loos SAM, Klapp SHL (2017) Force-linearization closure for non-Markovian Langevin systems with time delay. Phys Rev E 96:012106
100. Loos SAM, Klapp SHL (2019) Heat flow due to time-delayed feedback. Sci Rep 9:2491
101. Loos SAM, Klapp SHL (2019) Fokker-Planck equations for time-delayed systems via Markovian embedding. J Stat Phys 177:95–118
102. Longtin A (2010). In: Atay FM (ed) Complex time-delay systems: theory and applications. Springer, Berlin
103. Van Vu T, Hasegawa Y (2019) Uncertainty relations for time-delayed Langevin systems. Phys Rev E 100:012134
104. Kürsten R, Sushkov V, Ihle T (2017) Giant kovacs-like memory effect for active particles. Phys Rev Lett 119:188001
105. Ivlev A, Bartnick J, Heinen M, Löwen H (2014) Statistical mechanics for non-reciprocal forces. ArXiv:1403.2417
106. Agudo-Canalejo J, Golestanian R (2019) Active phase separation in mixtures of chemically interacting particles. Phys Rev Lett 123:018101
107. Saha S, Agudo-Canalejo J, Golestanian R (2020) Scalar active mixtures: The nonreciprocal cahn-hilliard model. Phys Rev X 10:041009
108. Fang K et al (2017) Generalized non-reciprocity in an optomechanical circuit via synthetic magnetism and reservoir engineering. Nat Phys 13:465
109. Metelmann A, Clerk AA (2015) Nonreciprocal photon transmission and amplification via reservoir engineering. Phys Rev X 5:021025
110. Bernier NR et al (2017) Nonreciprocal reconfigurable microwave optomechanical circuit. Nat commun 8:1–8
111. Wanjura CC, Brunelli M, Nunnenkamp A (2019) Topological framework for directional amplification in driven-dissipative cavity arrays. ArXiv:1909.11647
112. Chiacchio ER, Nunnenkamp A (2019) Dissipation-induced instabilities of a spinor bose-einstein condensate inside an optical cavity. Phys Rev Lett 122:193605
113. You Z, Baskaran A, Marchetti MC (2020) Nonreciprocity as a generic route to traveling states. Proc Natl Acad Sci 117:19767–19772
114. Lavergne FA, Wendehenne H, Bäuerle T, Bechinger C (2019) Group formation and cohesion of active particles with visual perception-dependent motility. Science 364:70–74
115. Loos SAM, Claussen JC, Schöll E, Zakharova A (2016) Chimera patterns under the impact of noise. Phys Rev E 93:012209
116. Premalatha K, Chandrasekar VK, Senthilvelan M, Lakshmanan M (2015) Impact of symmetry breaking in networks of globally coupled oscillators. Phys Rev E 91:052915
117. Zakharova A et al (2016) Controlling chimera patterns in networks: interplay of structure, noise, and delay. Springer
118. Kyrychko Y, Blyuss K, Schöll E (2013) Amplitude and phase dynamics in oscillators with distributed-delay coupling. Philos Trans R Soc A 371:20120466
119. Durve M, Saha A, Sayeed A (2018) Active particle condensation by non-reciprocal and time-delayed interactions. Eur Phys J E 41:49
120. Bo S, Del Giudice M, Celani A (2015) Thermodynamic limits to information harvesting by sensory systems. J Stat Mech Theor Exp 2015:P01014
121. Hartich D, Barato AC, Seifert U (2016) Sensory capacity: An information theoretical measure of the performance of a sensor. Phys Rev E 93:022116

Part II
Probabilistic Descriptions for Systems with Time Delay

The true Logic of this world is the Calculus of Probabilities
—James Clerk Maxwell, 1871 [336]

This Part II is dedicated to the probabilistic description of time-delayed stochastic systems. In Chap. 2, we have already introduced Fokker-Planck equations, and mentioned that for time-delayed systems, this type of description is given by an *infinite hierarchy*. Here we will elucidate this issue in more detail and explore different strategies to tackle the apparent problems in the probabilistic treatment arising due to this hierarchical structure. We will further explicitly inspect higher members of this hierarchy, which have so far received only little attention in earlier research.

On the one hand, we will review earlier derivation strategies that have been used to derive the first, not self-sufficient member, and see if they can be extended towards the higher members. Only one of these approaches results in a complete FP description in the form of the infinite hierarchy, namely the one basing on *Novikov's theorem*. We use this theorem to explicitly derive the second member, and find that it involves some unknown functional derivatives. Moreover, we will introduce a new approach via a Markovian embedding, and, in this way, derive a complete probabilistic description with an alternative representation of the hierarchy.

On the other hand, we will discuss and compare different ways to find closed, approximate equations. We will consider approaches from earlier literature, and further develop new ones, particularly, the force-linearization closure, and an approximation on the level of the second FP hierarchy member.

Chapter 5
Infinite Fokker-Planck Hierarchy

In this chapter, we will examine the infinite Fokker-Planck hierarchy. We will explicitly consider the higher members, which play a crucial role to understand the non-Markovian dynamics. Later in Chap. 6, we will present a new derivation of the FP description using a Markovian embedding technique [1–6]. To better understand the technical and conceptual difference of our approach, we will here also review some earlier work, in particular, Refs. [7–9].

5.1 Derivation of Fokker-Planck Hierarchy from Novikov's Theorem

We begin with reviewing the approach [7] based on Novikov's theorem (5.3). While [7] only considers the first member, we here further present a generalization towards the higher members of the hierarchy. We have already reported the resulting equation for the two-time PDF[1] $\rho_2(x,t;x_\tau,t-\tau)$ in Ref. [10].

Derivation of the second member We start with the definition $\rho_2(x,t;x_\tau,t-\tau) = \langle\delta[x-X(t)]\delta[x_\tau-X(t-\tau)]\rangle$ [as given in Part I in Eq. (3.4)], and consider its temporal derivative

$$\begin{aligned}\frac{\partial}{\partial t}\rho_2 =& \left\langle\left\{\frac{\partial}{\partial t}\delta[x-X(t)]\right\}\delta[x_\tau-X(t-\tau)]\right\rangle + \left\langle\delta[x-X(t)]\,\frac{\partial}{\partial t}\left\{\delta[x_\tau-X(t-\tau)]\right\}\right\rangle\\ =& -\frac{\partial}{\partial x}\left\langle\delta[x-X(t)]\delta[x_\tau-X(t-\tau)]\,\frac{\partial X(t)}{\partial t}\right\rangle\\ & -\frac{\partial}{\partial x_\tau}\left\langle\delta[x-X(t)]\delta[x_\tau-X(t-\tau)]\,\frac{\partial X(t-\tau)}{\partial t}\right\rangle.\end{aligned} \tag{5.1}$$

[1] We recall that a list of all abbreviations and symbols is given on pages VII and VIII, respectively.

© The Author(s), under exclusive license to Springer Nature Switzerland AG 2021

S. A. M. Loos, *Stochastic Systems with Time Delay*, Springer Theses,
https://doi.org/10.1007/978-3-030-80771-9_5

Substituting $\partial X(t)/\partial t$ and $\partial X(t-\tau)/\partial t$ with the LE (2.43), we obtain

$$\frac{\partial}{\partial t}\rho_2(x,t;x_\tau,t-\tau) = -\frac{\partial}{\partial x}\Big\langle \delta[x-X(t)]\,\delta[x_\tau - X(t-\tau)]\,\gamma^{-1}F\,[X(t),X(t-\tau)]\Big\rangle \quad \text{(A1)}$$
$$-\frac{\partial}{\partial x_\tau}\Big\langle \delta[x-X(t)]\,\delta[x_\tau - X(t-\tau)]\,\gamma^{-1}F\,[X(t-\tau),X(t-2\tau)]\Big\rangle \quad \text{(A2)}$$
$$-\sqrt{2D_0}\,\frac{\partial}{\partial x}\Big\langle \delta[x-X(t)]\,\delta[x_\tau - X(t-\tau)]\,\xi(t)\Big\rangle \quad \text{(B1)}$$
$$-\sqrt{2D_0}\,\frac{\partial}{\partial x_\tau}\Big\langle \delta[x-X(t)]\delta[x_\tau - X(t-\tau)]\,\xi(t-\tau)\Big\rangle. \quad \text{(B2)}$$

Let us now try to simplify the terms on the r.h.s. The terms (A1), (A2) solely involve deterministic forces (and no random Langevin forces). Hence, the ensemble average can readily be calculated, yielding

$$\text{(A1)} = -\frac{\partial}{\partial x}\iint_\Omega \delta(x-y)\,\delta(x_\tau - y_\tau)\,\gamma^{-1}F(y,y_\tau)\rho_2(y,t;y_\tau,t-\tau)\mathrm{d}y\mathrm{d}y_\tau$$
$$= -\frac{\partial}{\partial x}\left\{\gamma^{-1}F(x,x_\tau)\rho_2(x,t;x_\tau,t-\tau)\right\}$$

and

$$\text{(A2)} = -\frac{\partial}{\partial x_\tau}\left\{\int_\Omega \gamma^{-1}F(x_\tau,x_{2\tau})\rho_3(x,t;x_\tau,t-\tau;x_{2\tau},t-2\tau)\mathrm{d}x_{2\tau}\right\}.$$

In order to evaluate (B1), (B2), we need to deal with the correlations between the noise and the particle position, X. To this end, we introduce the functional of the random processes

$$\Lambda[\xi] := \delta[x - X(t)]\,\delta[x_\tau - X(t-\tau)], \quad (5.2)$$

which emphasizes by its form that X can be considered a functional of the random process ξ. At this point, we can employ Novikov's theorem.

Novikov's theorem Novikov's theorem connects the variational derivative of any functional $\Lambda[\xi]$ w.r.t. the Langevin force ξ with the correlations between that functional and the Langevin force. For the zero-mean Gaussian white noise with $\langle\xi(t)\xi(t')\rangle = \delta(t-t')$, $\forall t,t'$, it reads [11]

$$\big\langle \Lambda[\xi]\xi(t)\big\rangle = \left\langle \frac{\delta\Lambda[\xi]}{\delta\xi(t)}\right\rangle. \quad (5.3)$$

A short derivation of (5.3) is given in the Appendix A.2 by means of path integrals. Using this theorem, and simple rules for variational derivatives, we find

$$
\begin{aligned}
(\mathrm{B1}) &= -\sqrt{2D_0}\,\frac{\partial}{\partial x}\left\langle\frac{\delta\Lambda[\xi]}{\delta\xi(t)}\right\rangle \\
&= -\sqrt{2D_0}\,\frac{\partial}{\partial x}\left\{\left\langle\frac{\delta\Lambda[\xi]}{\delta X(t)}\frac{\delta X(t)}{\delta\xi(t)}\right\rangle+\left\langle\frac{\delta\Lambda[\xi]}{\delta X(t-\tau)}\underbrace{\frac{\delta X(t-\tau)}{\delta\xi(t)}}_{\to 0}\right\rangle\right\},
\end{aligned}
$$

$$
\begin{aligned}
(\mathrm{B2}) &= -\sqrt{2D_0}\,\frac{\partial}{\partial x_\tau}\left\langle\frac{\delta\Lambda[\xi]}{\delta\xi(t-\tau)}\right\rangle \\
&= -\sqrt{2D_0}\,\frac{\partial}{\partial x_\tau}\left\{\left\langle\frac{\delta\Lambda[\xi]}{\delta X(t)}\frac{\delta X(t)}{\delta\xi(t-\tau)}\right\rangle+\left\langle\frac{\delta\Lambda[\xi]}{\delta X(t-\tau)}\frac{\delta X(t-\tau)}{\delta\xi(t-\tau)}\right\rangle\right\}.
\end{aligned}
$$

The second term of (B1) vanishes since it is *acausal*: The particle position cannot depend on the future noise. We have discussed this causality argument in Sects. 2.3 and 4.4. We will apply it several times in the following. Further, using the integral form of the LE (2.43), i.e., $X(t')=\int_0^{t'}\left\{\gamma^{-1}F\left[X(t'),X(t'-\tau)\right]+\sqrt{2D_0}\xi(t')\right\}\mathrm{d}t'+X(0)$, we generally find

$$
\begin{aligned}
\frac{\delta X(t-\tau)}{\delta\xi(t-\tau)} = \frac{\delta X(t)}{\delta\xi(t)} &= \frac{\delta}{\delta\xi(t)}\int_0^t\left[\frac{F[X(s),X(s-\tau)]}{\gamma}+\sqrt{2D_0}\xi(s)\right]\mathrm{d}s \\
&= \sqrt{2D_0}\int_0^t\delta(s-t)\,\mathrm{d}s=\sqrt{\frac{D_0}{2}},
\end{aligned}
\tag{5.4}
$$

independently of the specific form of F. This means, the effect of varying the noise realization on the resulting path X, is given by the diffusion constant alone, and is not influenced by the energy landscape. This finding severely simplifies (B1) and (B2).

On the contrary to (5.4), the functional derivative of $X(t)$ w. r. t. the *earlier* noise $\xi(t-\tau)$, which also appears in (B2), cannot be treated in this manner. In particular, evaluating $\delta X(t)/\delta\xi(t-\tau)$ requires the formal solution of the delayed LE (2.43), which is not known for general F (an exception being linear systems, which we discuss in Sect. 5.2.1). For general (nonlinear) cases, we need to keep this term explicitly in the equation.

For the remaining terms in (B1), (B2), we can immediately write down

$$
\frac{\delta\Lambda[\xi]}{\delta X(t)} = -\frac{\partial}{\partial x}\delta[x-X(t)]\delta[x_\tau-X(t-\tau)] \tag{5.5}
$$

$$
\frac{\delta\Lambda[\xi]}{\delta X(t-\tau)} = -\frac{\partial}{\partial x_\tau}\delta[x-X(t)]\delta[x_\tau-X(t-\tau)], \tag{5.6}
$$

which finally yields the FPE[2]

[2] We have assumed that the variational derivative w. r. t. the delayed noise ξ is deterministic, which should generally be the case for additive white noise considered here.

$$
\begin{aligned}
\frac{\partial}{\partial t}\rho_2(x,t;x_\tau,t-\tau) =& -\frac{\partial}{\partial x}\left[\frac{1}{\gamma}F(x,x_\tau)\rho_2(x,t;x_\tau,t-\tau)\right] \\
&- \frac{\partial}{\partial x_\tau}\left[\int_\Omega \frac{1}{\gamma}F(x_\tau,x_{2\tau})\rho_3(x,t;x_\tau,t-\tau;x_{2\tau},t-2\tau)\mathrm{d}x_{2\tau}\right] \\
&+ D_0\left[\frac{\partial^2}{\partial x^2}+\frac{\partial^2}{\partial x_\tau^2}\right]\rho_2(x,t;x_\tau,t-\tau) \\
&+ \sqrt{2D_0}\,\frac{\partial^2}{\partial x_\tau \partial x}\left[\frac{\delta X(t)}{\delta\xi(t-\tau)}\bigg|_{\substack{X(t)=x\\ X(t-\tau)=x_\tau}}\rho_2(x,t;x_\tau,t-\tau)\right].
\end{aligned}
\tag{5.7}
$$

Equation (5.7) is the second member of the Fokker-Planck hierarchy for a system with a delayed (nonlinear) force $F(x, x_\tau)$ and additive Gaussian white noise. It involves the three-time PDF, indicating the emergence of an infinite hierarchy of FPEs. In contrast to the first member, it further contains a variational derivative which needs to be separately calculated for the specific force. In Sect. 5.2.1, we give explicit results for these missing terms specifically for linear systems [where $F(x, x_\tau)$ is linear in x and x_τ], and further elucidate the meaning of this term.

We note that Eq. (5.7) resembles a Markovian FPE of a two-dimensional (or a two-particle) system in the variables x and x_τ, with the exception of the diffusion term which involves derivatives w. r. t. *both* variables. This term induces in the solutions correlations between particle positions at different times, which we describe in detail in Sect. 5.2.

First member from Novikov's theorem The first FP member can be derived in an analogous way, as shown in Ref. [7]. To this end, one starts with the temporal derivative of ρ_1, instead of ρ_2 [like in Eq. (5.1)]. Moreover, as one should expect, integrating (5.7) over the delayed variable, i.e., $\int_\Omega \mathrm{d}x_\tau$ on both sides of (5.7), as well yields the first member of the FP hierarchy, confirming the consistency of the derived equations. In order to show this, one may use $\int_\Omega \rho_2(x, x_\tau)\mathrm{d}x_\tau = \rho_1(x)$, and apply the natural boundary conditions.

5.1.1 Alternative Approach with Two Time Arguments

An alternative derivation of the FPE for the one-time PDF ρ_1 was introduced in [8, 9]. We will here review and extended it towards the next member of the hierarchy. This derivation bases on a description with *two* time arguments. In particular, an integer $j = 0, 1, 2, \ldots$ is counting the number of intervals of length τ that have passed since the beginning $t = 0$, and a continuous variable $z \in [0, \tau]$ measures the time within the current interval.

Then, auxiliary phase-space variables $X_{j\tau}(z) := X(z + j\tau)$ are introduced, which, by construction, all follow a LE of the identical form

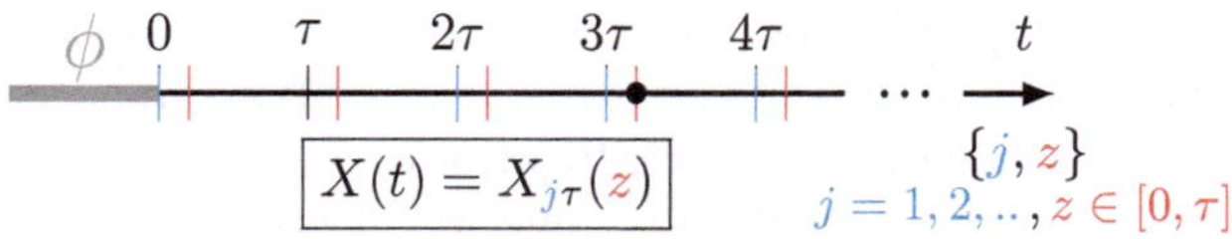

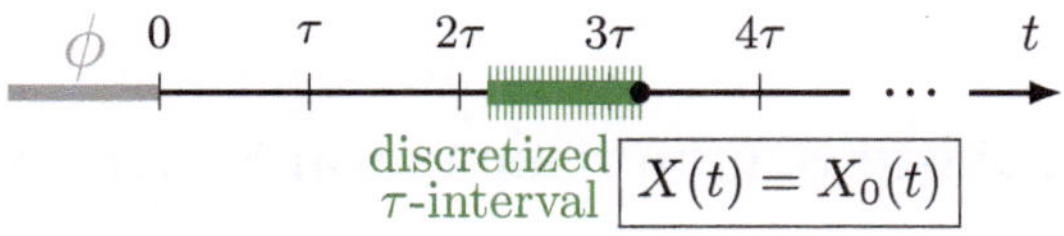

Fig. 5.1 Sketch of the different ways to introduce auxiliary variables, *upper panel*: within the approach based on two time arguments [8, 9] (mentioned here), *lower panel*: within the Markovian embedding approach discussed later in Sect. (6.1). These figures stem from [12]

$$\mathrm{d}X_{j\tau}(z) = F[X_{j\tau}(z), X_{(j-1)\tau}(z)]\mathrm{d}z + \sqrt{2D_0}\,\mathrm{d}W_{j\tau}(z). \tag{5.8}$$

The steady-state dynamics can be accessed if $j \to \infty$. A sketch of the construction of the auxiliary variables is given in Fig. 5.1, upper panel. In the probability space belonging to the phase space $\{X_{0\tau}(z), \ldots, X_{N\tau}(z)\}$, the process is Markovian, and the corresponding FPE is given by

$$\begin{aligned}\partial_z \rho_{N+1} =& \sum_{j=1}^{N} \Big[-\partial_{x_{j\tau}} \left\{ F(x_{j\tau}, x_{(j-1)\tau})\rho_{N+1} \right\} + D_0 \partial^2_{x_{j\tau}} \rho_{N+1} \Big] \\ & - \partial_{x_{0\tau}} \Big[F(x_{0\tau}, x_{1\tau})\rho_{N+1} \Big] + D_0 \partial^2_{x_{0\tau}} \rho_{N+1},\end{aligned} \tag{5.9}$$

with $\rho_{N+1}(x_{0\tau}, x_{1\tau}, \ldots, x_{N\tau}, z) = \langle \delta[x_{0\tau} - X(z)]\delta[x_{1\tau} - X(z+\tau)] \ldots \delta[x_{N\tau} - X(z + N\tau)] \rangle$. By integrating the equation over the entire domain of all variables but the first one, this many-variable FPE can be used to derive the well-known FPE for ρ_1. By integrating over all variables but the first and last one, further yields a FPE for ρ_2, reading

$$\partial_t \rho_2 = -\partial_x \left[F(x, x_\tau)\rho_2 \right] - \partial_{x_\tau} \int_\Omega \left[F(x, x_\tau)\rho_3 \right] \mathrm{d}x_{2\tau} + D_0 \left[\partial^2_x + \partial^2_{x_\tau} \right] \rho_2, \tag{5.10}$$

with $\rho_2(x, t; x_\tau, t-\tau)$ and $\rho_3(x, t; x_\tau, t-\tau; x_{2\tau}, t-2\tau)$. As opposed to the corresponding equation from Novikov's theorem, Eq. (5.7), this equation lacks of the term with the unknown functional derivative.

We think that this is because the stochastic set of Eqs. (5.8) is, in fact, only equivalent to the delayed process, if one further imposes *additional constraints* $X_j(\tau)=X_{j+1}(0),\ \forall j$, ensuring that the end point of the stochastic trajectories on each $[j\tau, (j+1)\tau]$ matches the starting point on $[(j+1)\tau, (j+2)\tau]$. Translating these constraints to the level of PDFs is a non-trivial task on its own. However, the comparison between Eq. (5.7) from Novikov's theorem and (5.10) suggests that the terms with the unknown functional derivatives in (5.7) indeed represent such terms ensuring the additional constraints on the trajectories.

While appearing quite elegant, this two-time arguments approach seems not to be appropriate to derive higher members, due to the need for additional terms which do not come automatically.

5.2 Exact Probabilistic Solutions for Linear Systems with Time Delay

Now we again turn to the purely linear case, which we have introduced in Sect. 2.4.5, where more analytical results can be obtained. Importantly for the present context, the missing terms in the second member of the FP hierarchy (5.7) can be analytically calculated for this case, as well as the steady state solutions, i.e., the one-time PDF and the two-time PDF.

5.2.1 Derivation of the Second Member of the Fokker-Planck Hierarchy

We aim to calculate the variational derivatives of the particle position with respect to the delayed Langevin force, in order to specify the FPE for ρ_2 from Eq. (5.7) to the linear case

$$F(x, x_\tau) = -\alpha x - \beta x_\tau. \tag{5.11}$$

The corresponding LE is given in (2.51). For this purpose, we will need a formal solution of the delayed Langevin equation. To obtain the latter, we will employ the method of steps.

Method of steps The linear LE with delay can be formally solved in an iterative manner, considering each interval $[(n-1)\tau, n\tau]$ after another, with increasing $n \in \mathbb{N} \cup \{0\}$. This technique is well known for delay differential equations and is called *method of steps* [13]. We call the formal solution on the nth interval $\mathcal{Y}_n$, such that

$$X(t) = \mathcal{Y}_n(t), \quad \text{for } t \in [(n-1)\tau, n\tau].$$

We recall that, in contrast to deterministic equations, these solutions explicitly contain the random force ξ, hence, they can only be regarded as "formal" solutions. Please further note that the intervals are all closed, such that they overlap at the discrete times $t = n\tau$. However, this is not problematic because $\mathcal{Y}_n(n\tau) = \mathcal{Y}_{n+1}(n\tau)$ [see Eq. (5.13)].

Starting with the initial condition ϕ on the zeroth interval $t \in [-\tau, 0]$

$$n = 0: \quad \mathcal{Y}_0(t) = \phi(t), \quad \forall t \in [-\tau, 0], \tag{5.12}$$

we find for the first interval: $\mathrm{d}\mathcal{Y}_1(t)/\mathrm{d}t = -\alpha\mathcal{Y}_1(t) - \beta\phi(t-\tau) + \sqrt{2D_0}\,\xi(t)$. Since ϕ is a given function, this is a stochastic *non-delayed* differential equation with the formal solution

$$n = 1: \quad \mathcal{Y}_1(t) = \phi(0)e^{-\alpha t} + \int_0^t e^{-\alpha(t-t')}[\sqrt{2D_0}\,\xi(t') - \beta\phi(t'-\tau)]\mathrm{d}t'.$$

Hence $X(t) = \mathcal{Y}_1(t), \ \forall t \in [0, \tau]$. In the same manner, we can express $\mathcal{Y}_2$ with the help of $\mathcal{Y}_1$, and so on. Consequently, given that the solution $\mathcal{Y}_n$ is known, the non-delayed stochastic differential equation for $\mathcal{Y}_{n+1}$ reads:

$$\frac{\mathrm{d}}{\mathrm{d}t}\mathcal{Y}_{n+1}(t) = -\alpha\mathcal{Y}_{n+1}(t) - \beta\mathcal{Y}_n(t-\tau) + \sqrt{2D_0}\,\xi(t),$$

with the (iterative) formal solution [13]

$$\mathcal{Y}_{n+1}(t) = \mathcal{Y}_n(n\tau)e^{-\alpha(t-n\tau)} + \int_{n\tau}^t e^{-\alpha(t-t')}[\sqrt{2D_0}\,\xi(t') - \beta\mathcal{Y}_n(t'-\tau)]\mathrm{d}t'. \tag{5.13}$$

Variational derivatives With the help of Eq. (5.13) we can readily calculate the variational derivative of the particle position w. r. t. the instantaneous, or w. r. t. the delayed Langevin force, $\delta X(t)/\delta\xi(t)$ and $\delta X(t)/\delta\xi(t-\tau)$, which appear in Eq. (5.7). We have already calculated the first one in (5.4) for the general case of (nonlinear) forces. We nevertheless reconsider it here, as a consistency check.

To this end, we write the arbitrary time t as $t = n\tau + \Delta t$, with $\Delta t < \tau$ without loss of generality. Second, we plug in the formal solution $X(t) = \mathcal{Y}_{n+1}(n\tau + \Delta t)$ for $t \in [n\tau, n\tau + \tau]$, such that

$$\begin{aligned} \frac{\delta X(t)}{\delta\xi(t)} &= \frac{\delta\mathcal{Y}_{n+1}(n\tau+\Delta t)}{\delta\xi(n\tau+\Delta t)} = \underbrace{\frac{\delta\mathcal{Y}_n(n\tau)}{\delta\xi(n\tau+\Delta t)}}_{\to 0} e^{-\alpha\Delta t} \\ &+ \int_{n\tau}^{n\tau+\Delta t} e^{-\alpha(n\tau+\Delta t-t')}\left[\sqrt{2D_0}\underbrace{\frac{\delta\xi(t')}{\delta\xi(n\tau+\Delta t)}}_{(\times)} - \beta\underbrace{\frac{\delta\mathcal{Y}_n(t'-\tau)}{\delta\xi(n\tau+\Delta t)}}_{\to 0}\right]\mathrm{d}t'. \end{aligned} \tag{5.14}$$

The first and last terms both vanish since they are *acausal*.[3] The remaining term $(\times)$ gives a delta-distribution, and thus we have[4]

[3] See Sects. 2.3 and 4.4 for a brief discussion concerning the meaning of causality in this context.

[4] We use $\int^{t_0} f(t)\delta(t-t_0)\mathrm{d}t = f(t_0)/2$. To see where this comes from, one may regard the delta-distribution as the limit of a normalized Gaussian distribution with variance ϵ, for $\epsilon \to 0$.

$$\frac{\delta X(t)}{\delta\xi(t)} = \sqrt{2D_0}\int_{n\tau}^{n\tau+\Delta t} e^{-\alpha(n\tau+\Delta t-t')}\delta(n\tau+\Delta t-t')\mathrm{d}t' = \sqrt{\frac{D_0}{2}}, \tag{5.15}$$

in agreement with Eq. (5.4), where we have shown that this results also holds for the general (nonlinear) case.

For the variational derivative w. r. t. the delayed Langevin force, we analogously utilize (5.13) which yields

$$\frac{\delta X(t)}{\delta\xi(t-\tau)} = \frac{\delta\mathcal{Y}_{n+1}\big(n\tau+\Delta t\big)}{\delta\xi(n\tau-\tau+\Delta t)} = \underbrace{\frac{\delta\mathcal{Y}_n\big(n\tau\big)}{\delta\xi\big(n\tau-\tau+\Delta t\big)}}_{(*)} e^{-\alpha\Delta t}$$
$$+\int_{n\tau}^{n\tau+\Delta t} e^{-\alpha(n\tau+\Delta t-t')}\left[\sqrt{2D_0}\,\delta([n\tau-\tau+\Delta t]-t') - \beta\underbrace{\frac{\delta\mathcal{Y}_n(t'-\tau)}{\delta\xi\big(n\tau-\tau+\Delta t\big)}}_{(**)}\right]\mathrm{d}t'. \tag{5.16}$$

Again applying the causality argument, $(**)$ is found to have a nonzero value at solely one single point of the integral, i.e., when $t' = n\tau+\Delta t$ (upper bound). Since, furthermore, the integrand is finite at this unique point, in particular, $\delta X(t)/\delta\xi(t) = \sqrt{D_0/2}$ [Eq. (5.15)], the entire integral vanishes. In the remaining term $(*)$, we plug in, once again, the formal solution Eq. (5.13) (now for $t \in [n\tau-\tau, n\tau]$), and find

$$(*) = \underbrace{\frac{\delta\mathcal{Y}_{n-1}\big(n\tau-\tau\big)}{\delta\xi\big(n\tau-\tau+\Delta t\big)}}_{\to 0} e^{\alpha\tau}$$
$$+\int_{n\tau-\tau}^{n\tau} e^{-\alpha(n\tau-t')}\left[\underbrace{\frac{\sqrt{2D_0}\,\delta\xi(t')}{\delta\xi\big(n\tau-\tau+\Delta t\big)}}_{(\#)} - \beta\underbrace{\frac{\delta\mathcal{Y}_n(t'!-\tau)}{\delta\xi\big(n\tau-\tau+\Delta t\big)}}_{\to 0}\right]\mathrm{d}t',$$

where two terms again vanish due to their acausality. In contrast, the first term in the integral $(\#) = \delta(n\tau-\tau+\Delta t-t')$ gives a finite contribution. We thus obtain

$$\frac{\delta X(t)}{\delta\xi(t-\tau)} = \sqrt{2D_0}\,e^{-\alpha\tau}. \tag{5.17}$$

Hence, this functional derivative is [just like (5.15)] independent of whether the delay force is present, or not ($\beta = 0$ or $\beta \neq 0$). This is somewhat surprising, because it means that the presence of delay does not influence how the trajectories change, when the (past) noise process is varied. We think, however, that this independence should be specific to the linear case.

From (5.7), (5.17), we find the FPE for the linear case

$$\begin{aligned}\frac{\partial}{\partial t}\rho_2(x,t;x_\tau,t-\tau) =& -\frac{\partial}{\partial x}\left[(-\alpha x-\beta x_\tau)\,\rho_2(x,t;x_\tau,t-\tau)\right]\\ &-\frac{\partial}{\partial x_\tau}\left[\int_\Omega(-\alpha x_\tau-\beta x_{2\tau})\,\rho_3(x,t;x_\tau,t-\tau;x_{2\tau},t-2\tau)\,\mathrm{d}x_{2\tau}\right]\\ &+D_0\left[\frac{\partial^2}{\partial x^2}+\frac{\partial^2}{\partial x_\tau^2}\right]\rho_2(x,t;x_\tau,t-\tau)\\ &+\sqrt{2D_0}\,e^{-\alpha\tau}\frac{\partial^2}{\partial x\,\partial x_\tau}\rho_2(x,t;x_\tau,t-\tau),\end{aligned} \tag{5.18}$$

which we have also reported in [10]. The occurrence of the three-time PDF in (5.18) confirms that the linear system is described by an *infinite* hierarchy [as already discussed for the general case, see Eq. (5.7)]. It is interesting to note that, for the *linear* case, however, a corresponding closed set of FP equations can be found, as shown in [14] using a time-convolutionless transform.

The fact that the linear case can equivalently be described by a closed set of (two) equations, in a sense, suits to the fact that the (steady-state) solutions are Gaussian distributions, which are in turn fully specified by the first two moments. We will discus the Gaussian solutions in the next Sect. 5.2.2.

Markovian limit To understand the meaning and origin of the different terms in (5.18), it is instructive to consider Markovian limits. To this end, we take the limit $\beta \to 0$ of (5.18).[5] We first note that τ, which represents a meaningful parameter defining the time lag in the delay force if $\beta > 0$, is, in this Markovian limit, only an (arbitrary) time difference. It specifies where we evaluate the two-time PDF, but has no further meaning. Second, we observe that the coupling to the higher members of the hierarchy vanishes, as expected. This is because the integral in the drift term can then be readily evaluated, yielding $\alpha x_\tau \int_\Omega \rho_3(x,t;x_\tau,t-\tau;x_{2\tau},t-2\tau)\,\mathrm{d}x_{2\tau} = \alpha x_\tau \rho_2(x,t;x_\tau,t-\tau)$. Finally, we realize that the diffusion terms stay unchanged in this limit. In particular, the mixed, $\partial_x\partial_{x_\tau}$ diffusion term, which quantifies some sort of correlation between the two instants in time (t and $t-\tau$), is *not* specific to the non-Markovian case. Thus, this term describes a type of "memory" which is also present in the absence of the delayed force (recall that the position-autocorrelation takes some time to relax to zero even in Markovian systems, representing a temporal correlation, see e.g. Fig. 2.4). For large time differences τ, this term completely vanishes as $e^{-\alpha\tau} \to 0$, and the variables x and x_τ decouple. This is in sharp contrast to the non-Markovian case, where $x, x_\tau, x_{2\tau}, \ldots$, are all additionally coupled via the drift term.

[5] Please note that the limit $\tau \to 0$ as well yields a Markovian system, but this limit affects, at the same time, the arguments of the PDF.

5.2.2 Steady-State Solutions

Considering the steady states (i.e., the limit $t \to \infty$), the FPE for ρ_2 (5.18) and ρ_1 (3.15) have (multivariate) Gaussian PDF solutions [15] of the form

$$\rho_{\mathrm{n,ss}} = \frac{1}{\sqrt{(2\pi)^n \det\{\mathbf{D_n}\}}} e^{-(1/2)(\boldsymbol{x}-\langle \boldsymbol{x}\rangle)\mathbf{D_n}^{-1}(\boldsymbol{x}-\langle \boldsymbol{x}\rangle)}, \tag{5.19}$$

with $\boldsymbol{x} = (x, x_\tau, \ldots, x_{(n-1)\tau})^{\mathrm{T}}$, $\mathbf{D_1} = C(0)$, and

$$\mathbf{D_2} = \begin{pmatrix} C(0) & C(\tau) \\ C(\tau) & C(0) \end{pmatrix}, \quad \mathbf{D_3} = \begin{pmatrix} C(0) & C(\tau) & C(2\tau) \\ C(\tau) & C(0) & C(\tau) \\ C(2\tau) & C(\tau) & C(0) \end{pmatrix}. \tag{5.20}$$

The Gaussianity of the PDFs is a direct consequence of the linearity of the underlying LE with Gaussian noise, and can be shown on the basis of the LE alone [15].

The covariance matrices $\mathbf{D_n}$ involve the steady-state temporal position-autocorrelation function between the systems state at time t and time $t + z$ (2.10), i.e., $C(z) := \langle X(t)X(t + z)\rangle_{\mathrm{ss}}$. Thus, the n-time PDF is fully known, once $C(z)$ with $z = 0, \tau, \ldots$ up to $z = n\tau$ are known.

In Sect. 2.4.5, we have given exact expressions for $C(0)$ and $C(\tau)$ as functions of α, β, τ and D_0. We recall

$$C(0) = D_0 \frac{1 + (\beta/\omega)\sinh(\omega\tau)}{\alpha + \beta\cosh(\omega\tau)},$$
$$C(z) = C(0)\cosh(\omega z) - (D_0/\omega)\sinh(\omega|z|),$$

with $\omega = \sqrt{\alpha^2 - \beta^2} \in \mathbb{C}$, as given in Eqs. (2.52), (2.53). This readily implies the exact one- and two-time steady-state PDFs.

To obtain the three-time steady-state PDF, we derive an explicit expression for $C(2\tau)$. We do this by generalization of the calculation from Ref. [13].

Extension to larger time differences Now we extend the result (2.52) towards higher z. Due to the symmetry property $C(z) = C(-z)$, we only need to consider non-negative z. One can directly deduce from the LE a differential equation for $C(z)$ for any $z > 0$ [13]

$$\frac{\mathrm{d}C(z)}{\mathrm{d}z} = \left\langle X(t)\frac{\mathrm{d}X(u)}{\mathrm{d}u}\bigg|_{u=t+z}\right\rangle_{\mathrm{ss}} = -\alpha C(z) - \beta C(z-\tau) + \sqrt{2D_0}\langle X(t)\xi(t+z)\rangle. \tag{5.21}$$

For $z > 0$, the last term vanishes, because it is *acausal*, thus

$$\mathrm{d}C(z)/\mathrm{d}z = -\alpha C(z) - \beta C(z-\tau). \tag{5.22}$$

This equation can be solved iteratively using the *method of steps*.

To be more specific, we plug in for the term $C(z-\tau)$ the solution on the interval $z \in [0, \tau]$, which is given in Eqs. (2.52), (2.53). This yields a closed equation for the correlation function on the interval $z \in [\tau, 2\tau]$, with the solution

$$C(z) = \frac{2D_0}{\beta[\alpha + \beta\cosh(\omega\tau)]}\Big(e^{\alpha(\tau-|z|)}[\beta\cosh(\omega\tau) + \alpha] - \alpha\cosh[\omega(\tau - z)] \\ - \frac{\beta}{2}\cosh[\omega(2\tau - z)] + \frac{\beta^2 - 2\alpha^2}{2\omega}\sinh[\omega(\tau - |z|)] \\ - \frac{\alpha\beta}{2\omega}\sinh[\omega(2\tau - |z|)]\Big). \tag{5.23}$$

We have reported this solution in [10]. Again, when ω becomes imaginary if $|\alpha| < \beta$, the hyperbolic functions convert to trigonometric ones, as in the case of $C(0 \le z \le \tau)$. Equation (5.23) generalizes the calculation of Ref. [13] towards larger time differences z. In the same manner, one can extend the result further towards higher

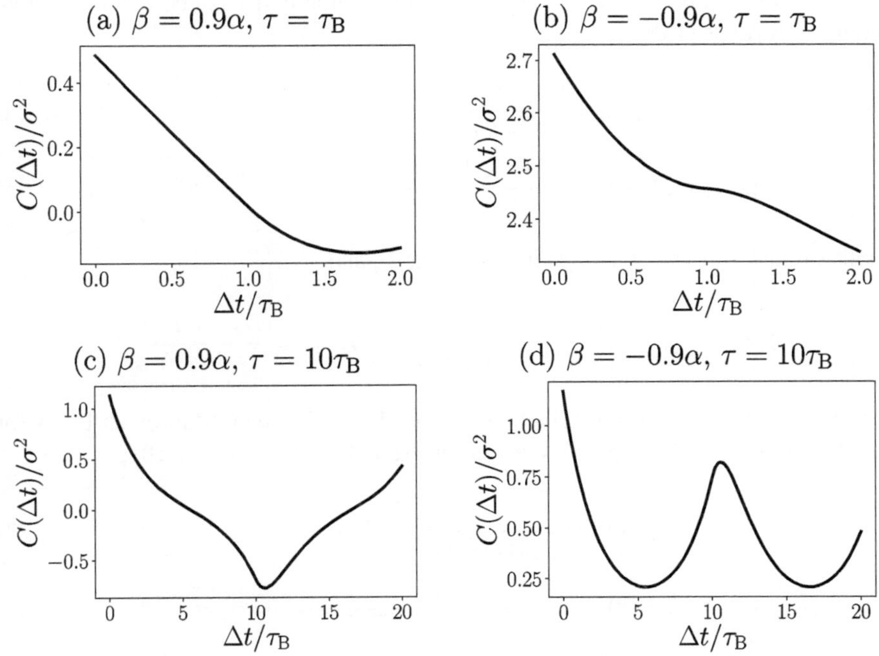

Fig. 5.2 Position-autocorrelation function on $t \in [0, 2\tau]$ from Eqs. (5.23), (2.53), of the linear system $\gamma\dot{X} = -\alpha X(t) - \beta X(t-\tau) + \sqrt{2D_0}\,\xi(t)$ with $\alpha = \gamma/\tau_B$, and different values of β, τ. *Upper panels,* (a+b): $\tau = \tau_B$, *lower panels,* (c+d): $\tau = 10\tau_B$. *Left panels,* (a+c): $\beta/\alpha = 0.9$, i.e., positive feedback, *Right panels,* (b+d): $\beta/\alpha = -0.9$, i.e., negative feedback. These results also already displayed in Figs. 2.10 and 2.11

z. Figure 5.2 shows the analytically calculated autocorrelation function for different values of α, β and τ.

To avoid confusion, we note that in Ref. [13], the there derived expression (2.53) is erroneously stated to be valid for all $z > 0$. Along their derivation, the equation with number (18) in [13] turns out to be not valid for $|z| > \tau$, although they claim otherwise. A correlation described by Eq. (2.53) $\forall z \in \mathbb{R}$ would, in fact, be unphysical, as it yields unbounded correlations, even in situations where the system is stable and confined (e.g., for $\beta < \alpha$).

Green's function method An alternative strategy to obtain the correlation functions is via the Green's function method. We explain how this works and derive the needed formula in Appendix A.3. In principle, for various linear systems this strategy yields analytical expressions, which can, e.g., be numerically integrated. However, explicit *closed-form solutions* are only available for specific cases, where some required integrals are known (as well as the back-transformation of the Green's function to real space). See Refs. [16, 17] for some explicit results. For the LE (2.51) considered here, we could not find general closed-form expression for the position-autocorrelation function on this way.

Using (5.19) and (2.53), the one-time and two-time PDFs can be written in quite compact form as [13]

$$\rho_{1,\mathrm{ss}}(x) = \frac{1}{\sqrt{2\pi K(\tau)}} e^{-\frac{x^2}{2K(\tau)}} \tag{5.24}$$

$$\rho_{2,\mathrm{ss}}(x, t; x_\tau, t-\tau) = \frac{d_1(\tau)\sqrt{1-d_2(\tau)}}{\pi} e^{-d_1(\tau)[x^2 + x_\tau^2 - 2x\, x_\tau d_2(\tau)]} \tag{5.25}$$

with variance $K(\tau) = C(0)$ given in Eq. (2.52), and

$$d_1(\tau) = \frac{1}{2} \frac{\beta^2\, K(\tau)}{[\beta K(\tau)]^2 - [D_0 - \alpha K(\tau)]^2}, \quad d_2(\tau) = \frac{D_0 - \alpha K(\tau)}{\beta K(\tau)}. \tag{5.26}$$

In the limit $\beta \to 0$, these results converge to the well-known Markovian PDFs for the Ornstein-Uhlenbeck process, given for example in Ref. [18], p. 101,

$$\rho_{1,\mathrm{ss}}^{(0)}(x) = \frac{1}{\sqrt{2\pi (D_0/\alpha)}} e^{-\frac{x^2}{2D_0/\alpha}}, \tag{5.27}$$

$$\rho_{2,\mathrm{ss}}^{(0)}(x, t; x_\tau, t-\tau) = \frac{1}{2\pi (D_0/\alpha)\sqrt{1 - e^{-2\alpha\tau/\gamma}}} e^{-\frac{x^2 + x_\tau^2 - 2x\, x_\tau e^{-2\alpha\tau/\gamma}}{2(D_0/\alpha)(1 - e^{-2\alpha\tau/\gamma})}}. \tag{5.28}$$

Figure 5.3 shows the two-time PDF (5.25) of linear systems with delay force, as well as, the Markovian limit (5.27). For very short τ, all depicted PDFs appear as (uprising) diagonals, indicating high probabilities to find $X(t)$ in the vicinity of $X(t-\tau)$. For very large τ, only the Markovian process always approaches a radial-symmetric ρ_2 [this can directly be seen from Eq. (5.27)], due to a vanishing correlation $C(\tau) \to 0$ for increasing τ. More specifically, the symmetric PDF is

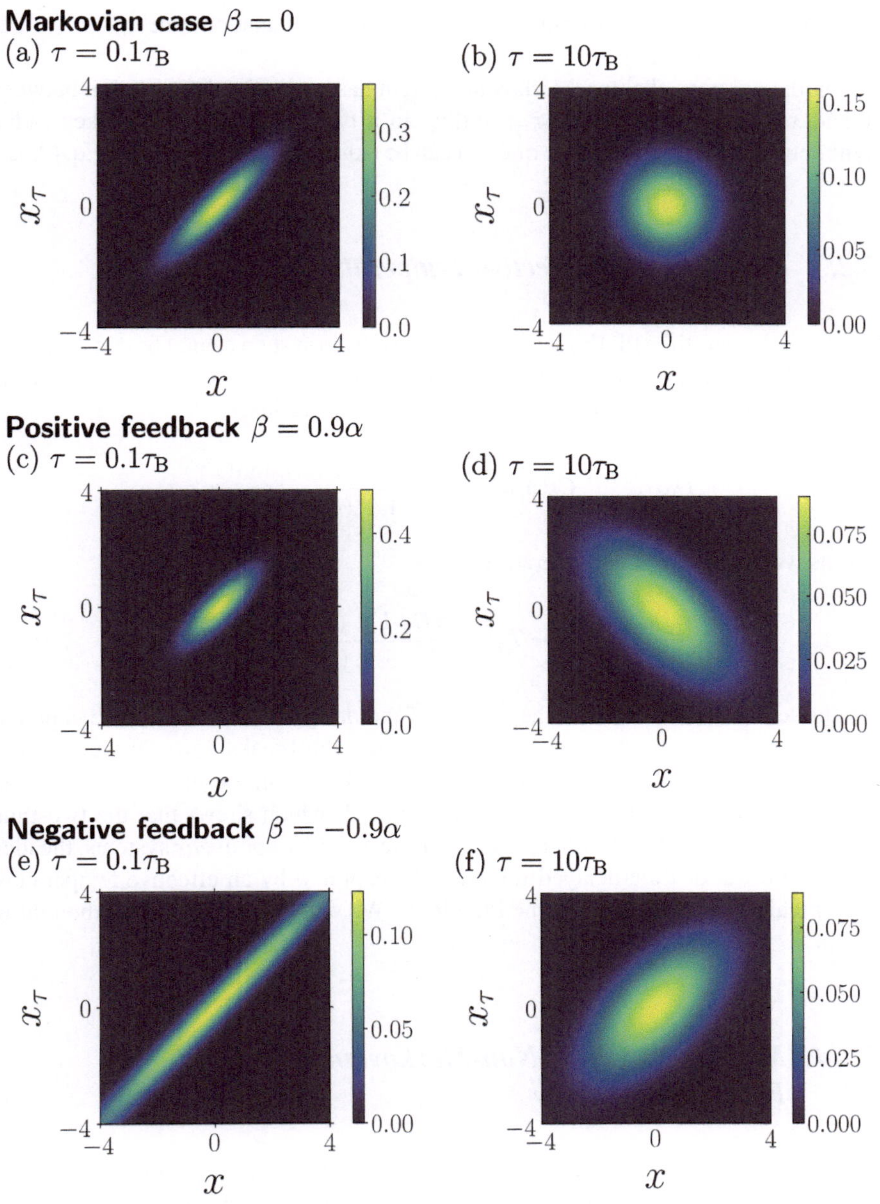

Fig. 5.3 Plots of the analytical results (5.25) for the steady-state two-time probability density function $\rho_2(x, t; x_\tau, t - \tau)$ of the linear LE (2.51) with $\alpha = \gamma/\tau_B$ and different values of β and τ. *Left / right panels:* Short / large delay times; *Upper / central / lower panels:* $\beta = 0$ / $\beta = 0.9\alpha$ / $\beta = -0.9\alpha$. The Markovian system always has a radial-symmetric PDF for large τ, the non-Markovian system has nontrivial correlations between x and x_τ for arbitrary τ. x and x_τ are in units of σ, the PDF (color code) is in units of $1/\sigma^2$

approached once τ is large compared to the intrawell relaxation time[6] $\tau_{\mathrm{ir}} \approx \gamma/\alpha$, as one can see from (5.28).

On the contrary, the non-Markovian system has a nontrivial correlation between x and x_τ for arbitrarily large τ, resulting in a two-time PDF with broken radial symmetry. The respective value of $C(\tau)$ can be extracted from Fig. 5.2, or Eq. (2.52).

5.2.3 *The Notion of Effective Temperature*

Based on the one-time PDF alone, the non-Markovian system cannot be distinguished from a Markovian one at a different temperature. This can be seen by comparing (5.24) with (5.27). The delay force only changes the variance

$$D_0/\alpha \;\rightarrow\; C(0) = (D_0/\alpha)\,\frac{1+(\beta/\omega)\sinh(\omega\tau)}{1+(\beta/\alpha)\cosh(\omega\tau)},$$

just as well as a changed temperature

$$\frac{\gamma D_0}{k_{\mathrm{B}}} = \mathcal{T} \;\rightarrow\; \mathcal{T}_{\mathrm{eff}} = \frac{\gamma D_0}{k_{\mathrm{B}}}\,\frac{1+(\beta/\omega)\sinh(\omega\tau)}{1+(\beta/\alpha)\cosh(\omega\tau)}$$

would do. Thus, in this sense the impact of the delay in the force can be captured by an *effective temperature*. However, this viewpoint breaks down, if we consider the higher order PDFs, which are crucially different than the one in a Markovian system. To illustrate this point, we recall Fig. 5.3, which shows that the two-time PDF significantly differ. Furthermore, when we consider *nonlinear* systems, the idea that the impact of time-delayed force can be captured by an effective temperature even breaks down on the one-time PDF level. We will come back to this thought in Chap. 7.

5.2.4 *Markovian Versus Non-Markovian Two-Time Probability Density*

We turn back to the aforementioned Markovian limit $\beta \rightarrow 0$ [discussed below Eq. (5.18)], and reconsider the meaning of the $e^{-\alpha\tau}$ decaying "memory" (induced by the mixed diffusion term) on the basis of the exact solutions. To this end, we study the Kolmogorov-like distance

$$\mathcal{D}(\tau) = \frac{1}{2}\iint \mathrm{d}x\mathrm{d}x_\tau\,\left|\rho_{2,\mathrm{ss}}(x,x_\tau;\tau) - \rho_{\mathrm{ss}}(x)\rho_{\mathrm{ss}}(x_\tau)\right|. \tag{5.29}$$

[6] The timescales have been introduced in Sect. 2.6. We further recall that a list of all symbols can be found on page VIII.

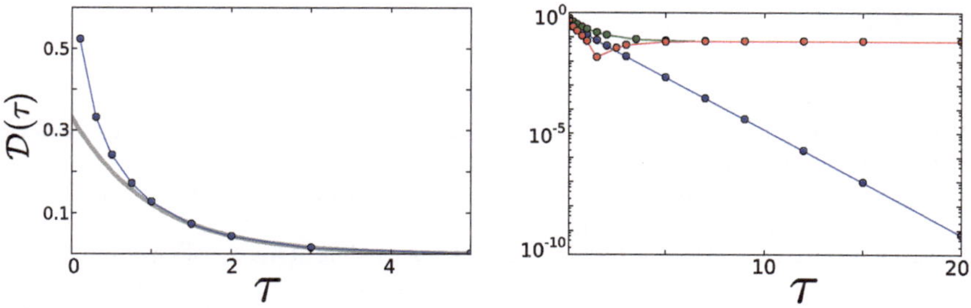

Fig. 5.4 Kolmogorov-like distance $\mathcal{D}$, as defined in (5.29), between two-time PDF $\rho_2(x, x_\tau)$ and the product of the one-time PDFs $\rho_1(x)\rho_1(x_\tau)$ at $\alpha = \gamma D_0/\sigma^2$. *Left panel:* Markovian case ($\beta = 0$, blue lines) plus an exponential fit: $(1/3)e^{-\alpha\tau/\gamma}$ as a guide to the eye (gray line). *Right panel:* Again the Markovian case ($\beta = 0$, blue lines), here in a logarithmic plot, supplemented by the results for non-Markovian cases ($\beta = -0.4\,\alpha$ green line, and $\beta = 0.4\,\alpha$ red line)

which was previously also used in [19] to quantify (non-)Markovianity. This article [19] further provides a systematic comparison between different non-Markovianity measures, and between quantum and classical non-Markovian dynamics. We find that it decays to zero in the Markovian case, as can be seen in Fig. 5.4. For $\tau \gg 0$, the decay is exponential, with the same exponent as the term that scales the mixed diffusion term in Eq. (5.18). Thus, the Markovian process has exponentially decaying correlations between the times t and $t - \tau$, such that the two-time PDF $\rho_2(x, t; x_\tau, t - \tau)$ decomposes to the product of the two identical one-time PDF for large time differences τ.

On the contrary, the distance is not decaying to zero if $\beta > 0$. The non-Markovian system is thus characterized by a correlation between t and $t - \tau$, for arbitrary τ, as one would expect. As one can further see from Fig. 5.4, the distance instead approaches a constant finite value. This is because in fact, both, the one-time and two-time PDF *saturate*, when τ is larger than the other characteristic timescales of the process (which is here the intrawell relaxation time, discussed in Chap. 7), see Eq. (5.24). This saturation is also observed in nonlinear systems, and we will come back to it in Sect. 7.4.

Interestingly, the approached saturation value of $\mathcal{D}$ identical for $+\beta$ and $-\beta$. This observation also holds for other values of α and D_0 (not shown here).

Consistency check of results We have plugged in the PDFs ρ_2 and ρ_3 (which were derived on the basis of the LE alone) into the FPE (5.18), and found that they indeed solve the equation. This confirms the consistency of the different approaches and of our results.

At this point, one might actually question the usefulness of the FPE, which is technically more involved than the LE, but still seems not necessary to find solutions. However, as we will see in the following, the FPE is a valuable tool to obtain exact relations (especially for nonlinear cases). It further represents a starting point for approximation schemes, as discussed in Chap. 7. Also the probabilistic solutions of the linear delay system will prove themselves useful in this regard.

References

1. Puglisi A, Villamaina D (2009) Irreversible effects of memory. EPL 88:30004
2. Crisanti A, Puglisi A, Villamaina D (2012) Nonequilibrium and information: the role of cross correlations. Phys Rev E 85:061127
3. Siegle P, Goychuk I, Talkner P, Hänggi P (2010) Markovian embedding of non-Markovian superdiffusion. Phys Rev E 81:011136
4. Siegle P, Goychuk I, Hänggi P (2011) Markovian embedding of fractional superdiffusion. EPL 93:20002
5. Bao J-D, Hänggi P, Zhuo Y-Z (2005) Non-Markovian Brownian dynamics and nonergodicity. Phys Rev E 72:061107
6. Villamaina D, Baldassarri A, Puglisi A, Vulpiani A (2009) The fluctuation-dissipation relation: how does one compare correlation functions and responses? J Stat Mech Theory Exp 2009:P07024
7. Frank TD (2005) Delay Fokker-Planck equations, Novikov's theorem, and Boltzmann distributions as small delay approximations. Phys Rev E 72:011112
8. Frank TD (2004) Analytical results for fundamental time-delayed feedback systems subjected to multiplicative noise. Phys Rev E 69:061104
9. Zheng Y, Sun X (2017) Governing equations for Probability densities of stochastic differential equations with discrete time delays. Discrete Contin Dyn Syst Ser B 22
10. Loos SAM, Klapp SHL (2017) Force-linearization closure for non-Markovian Langevin systems with time delay. Phys Rev E 96:012106
11. Novikov EA (1965) Functionals and the random-force method in turbulence theory. Sov Phys JETP 20:1290–1294
12. Loos SAM, Klapp SHL (2019) Fokker-Planck equations for time-delayed systems via Markovian embedding. J Stat Phys 177:95–118
13. Frank TD, Beek PJ, Friedrich R (2003) Fokker-Planck perspective on stochastic delay systems: exact solutions and data analysis of biological systems. Phys Rev E 68:021912
14. Giuggioli L, McKetterick TJ, Kenkre VM, Chase M (2016) Fokker-Planck description for a linear delayed Langevin equation with additive Gaussian noise. J Phys A 49:384002
15. Küchler U, Mensch B (1992) Langevins stochastic differential equation extended by a time-delayed term. Stoch Stoch Rep 40:23–42
16. Geiss D, Kroy K, Holubec V (2019) Brownian molecules formed by delayed harmonic interactions. New J Phys 21:093014
17. Munakata T, Iwama S, Kimizuka M (2009) Linear stochastic system with delay: energy balance and entropy production. Phys Rev E 79:031104
18. Risken H (1984) The Fokker-Planck equation. Springer
19. Vacchini B, Smirne A, Laine E-M, Piilo J, Breuer H-P (2011) Markovianity and non-Markovianity in quantum and classical systems. New J Phys 13:093004

Chapter 6
Markovian Embedding—A New Derivation of the Fokker-Planck Hierarchy

"A clever person solves a problem. A wise person avoids it."

– supposedly by Albert Einstein [1, 2]

In this Chapter, we derive the FP hierarchy by using a Markovian embedding technique, which is inspired by the *linear chain trick* [3–5]. Markovian embeddings are already a well-established tool to treat stochastic systems with memory, for example in the context of generalized Langevin equations [6–11]. The following considerations are based on Ref. [12], where we have introduced this derivation. Here, we add additional noise terms to the auxiliary variables (different from [12]), which might provide a conceptual advantage for future analytic treatments.

Main Idea We introduce $n \in \mathbb{N}$ auxiliary variables $X_{j=1,2,\ldots,n}$ whose dynamics is given by a set coupled, linear differential equations without delay. These degrees of freedom will generate the memory. They are coupled to another degree of freedom, X_0, which follows a (nonlinear) stochastic differential equation, which is as well delay-free, but represents the original delayed process. In the limit $n \to \infty$, this set of equations generates the same stochastic process as the delayed equation, where $X_0(t) \to X(t)$. Thus, the delay in the original equation is replaced by a coupling with (infinitely many) additional degrees of freedom.

This $(n+1)$-dimensional Markovian system is described by a closed, *self-sufficient* FPE. By marginalization, i.e., projection onto the one-dimensional subspace X_0, and a subsequent limiting procedure, we derive the well-known delayed FPE for the one-time PDF. Furthermore, by projection onto higher-dimensional subspaces, we find higher FPEs, yielding an infinite hierarchy which has a different form than the one obtained by Novikov's theorem [13]. Explicitly considering the second

© The Author(s), under exclusive license to Springer Nature Switzerland AG 2021

S. A. M. Loos, *Stochastic Systems with Time Delay*, Springer Theses,
https://doi.org/10.1007/978-3-030-80771-9_6

member, we show that the two hierarchies can actually be converted into each other. Interestingly, the limiting procedure contains one step where position-like variables are converted into velocity-like variables.

In a nutshell, the idea is to first employ a Markovian embedding on the level of the Langevin equation (6.1), and then, again integrate out the auxiliary variables from the embedding, on the level of the resulting Markovian FPE. Thus, the trick is to introduce and remove auxiliary variables at the right moments.

Meaning of Additional Noise Terms and Generalization of Ref. [12] Contrary to Ref. [12], where we have introduced this idea, we here generalize the approach by adding noise terms $\sim \xi_j$ to the auxiliary variables $X_{j>0}$, scaled with a temperature $\mathcal{T}'$. The limit $\mathcal{T}' \to 0$ corresponds to the case considered in [12]. In particular, we show that the impact of the additional noise terms on X_0 *vanishes* in the limit $n \to \infty$. Therefore, starting from the $n+1$ LEs *with*, or *without*, the additional noise terms, yields the *same* FPE for ρ_1 (despite the additional source of randomness).

The benefits of this generalization include technical, as well as physical aspects. First, there is a technical advantage of adding noise terms to all d.o.f.. In this way, we "regularize" the noise. More specifically, by adding the noise terms to $X_{j>0}$ we avoid coupling between deterministic and stochastic degrees of freedom. This also implies that the diffusion matrix in the multivariate Fokker-Planck equation (6.27) has full rank. This solves, for the specific case at hand, the long-standing problem of singular diffusion matrices, see [14]. For example, this matrix can now be inverted. It might be interesting to check, whether one can also approach other cases like the ones from [14] with an analogous approach (i.e., adding noise terms to render a diffusion matrix with full rank, and then consider some limit where the impact of the additional noise term vanishes.) Typically this is a very useful property in the search of exact results (e.g., see [15], p. 84ff).

Second, we will in Part III reconsider the network (6.1) from a thermodynamic perspective. In the framework of stochastic thermodynamics, it is problematic to have variables without noise, because some key quantities are not meaningful in the noise-free "limit".

Third, there is a physical argument to add the noise terms $\sim \xi_j$, which takes into account possible underlying microscopic dynamics. Specifically, the additional noise terms allow for an alternative interpretation of the variables in (6.1). In particular, one can take the viewpoint that all equations in (6.1) are (overdamped) equations of motion, expressing a *force balance*, like a Langevin equation. From that perspective, it is reasonable to balance each "friction force" $\sim -\gamma \dot{X}_j$ with a corresponding noise $\sim \xi_j$. In fact, friction and random force usually have the same origin, that is, a coupling with an (integrated out) heat bath (see Sect. 2.3 for an example). Thus, often one does not come without the other in a stochastic equation.

6.1 Markovian Embedding—A Different View on Memory

We consider the set of dynamical equations given by

$$\dot{X}_0(t) = \quad \gamma^{-1} F[X_0(t), X_n(t)] \quad + \sqrt{2k_B \mathcal{T}_0/\gamma}\, \xi_0(t) \tag{6.1a}$$

$$\dot{X}_j(t) = (n/\tau)[X_{j-1}(t) - X_j(t)] + \sqrt{2k_B \mathcal{T}'/\gamma'}\, \xi_j(t), \tag{6.1b}$$

with $j \in \{1, 2, ..., n\}$, $t \geq 0$, $X_j \in \mathbb{R}$ and independent Gaussian white noise terms $\langle \xi_i(t)\xi_j(t')\rangle = \delta_{ij}\delta(t - t')$. This is essentially a generalization of the so-called linear chain trick [3–5]. Importantly, the n auxiliary variables generally follow *linear* Eqs. (6.1b), while the delayed process itself may involve a nonlinear force, F. The Eqs. (6.1) define a unidirectionally coupled ring network, as depicted in Fig. 6.1. The initial conditions are set to $X_{j\in\{1,..,n\}}(0) \equiv 0$. As will become clear below, this amounts to the initial condition $X(t < 0) \equiv 0$ in the non-Markovian LE (see Sect. 6.1.4).

As we explicitly show in the following, *projection* of (6.1b) onto the variable X_0 yields

$$\gamma \dot{X}_0(t) = F\left[X_0(t), \int_{-\tau}^{t} K_n(t - t')X_0(t')\, \mathrm{d}t' + \nu_n(t)\right] + \sqrt{2k_B \mathcal{T}_0 \gamma}\, \xi_0(t), \tag{6.2}$$

with the n-dependent Gamma-distributed memory kernel

$$K_n(t - t') = \left(\frac{n}{\tau}\right)^n \frac{(t - t')^{n-1}}{(n - 1)!}\, e^{-n(t-t')/\tau}. \tag{6.3}$$

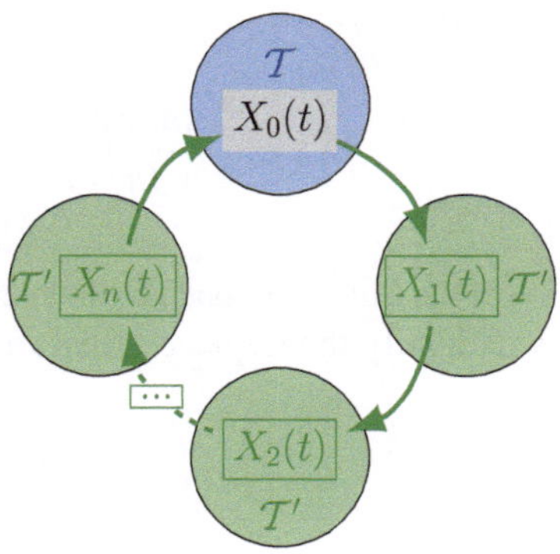

Fig. 6.1 Schematic visualization of the $(n + 1)$-dimensional Markovian system (6.1). In the limit $n \to \infty$, the dynamics of X_0 is identical to the time-delayed process (2.43). The variables $X_{j>0}$ represent the memory and form the feedback loop

and the n-dependent colored noise ν_n with $\langle \nu_n \rangle = 0$, and

$$C_\nu(T) = \frac{k_B T'}{\gamma'} \sum_{p=0}^{n-1} \sum_{l=0}^{p} \frac{2^{l-2p}(2p-l)!}{p!(p-l)!l!} \left(\frac{n}{\tau}\right)^{l-1} e^{-nT/\tau} T^l. \tag{6.4}$$

6.1.1 *Projection, Memory Kernel & Colored Noise*

Here we carry out the projection of the $X_{j>0}$ dynamics onto X_0. In this way, we derive the memory kernel and colored noise from Eq. (6.2). The projecting method is technically similar to the Mori-Zwanzig projection (see Sect. 2.3.1). However, in contrast to the derivation of a generalized LE, we here do not start from microscopic, mechanically coupled particles, but instead from the mesoscopic auxiliary d.o.f..

Projection We transform the equations for $X_{j\in\{1,2,...,n\}}$ into frequency space, and make use of their linearity (we recall that their linearity is irrespective of the question whether the equation for X_0 is linear). First, we apply the Laplace-transformation[1] $\mathcal{L}[X_j(t)](s) = \int_0^\infty X_j(t)e^{-st}\mathrm{d}s$ and obtain for each $j \in \{1, 2, ..., n\}$

$$\begin{aligned} s\hat{X}_j(s) &= (n/\tau)\left[\hat{X}_{j-1}(s) - \hat{X}_j(s)\right] + \sqrt{2k_B T'/\gamma'}\,\hat{\xi}_j(s) \\ \Rightarrow \hat{X}_j(s) &= \frac{(n/\tau)\hat{X}_{j-1}(s) + \sqrt{2k_B T'/\gamma'}\,\hat{\xi}_j(s)}{s + (n/\tau)}. \end{aligned} \tag{6.5}$$

Here we have already incorporated the initial conditions $X_{j\in\{1,..,n\}}(0) \equiv 0$, which simplify the expression (6.5). One should further note that the Laplace-transformation of the noise is in fact not well-defined. However, we can safely ignore this issue here, as we will transform back at the end. Iteratively substituting the solutions (6.5) into each other yields

$$\hat{X}_n = \frac{(n/\tau)^n}{[s + (n/\tau)]^n}\hat{X}_0 + \sum_{j=1}^{n} \frac{(n/\tau)^{j-1}\sqrt{2k_B T'/\gamma'}\,\hat{\xi}_{n-j+1}}{[s + (n/\tau)]^j}. \tag{6.6}$$

Now we transform back to the real space via inverse Laplace-transformation. In (6.6) we identify the Laplace-transformation of the Gamma distribution

$$\mathcal{L}\left[K_j(t)\right](s) = \frac{(n/\tau)^j}{[s + (n/\tau)]^j} \tag{6.7}$$

[1] As opposed to Fourier-transformations, the Laplace-transformation is useful to treat initial-value problems, while Fourier-transformations have advantages in steady states, where the initial condition effects have already decayed. Both transformations could be used here.

with the Gamma-distributed kernels $K_j(t) = \frac{n^j}{\tau^j(j-1)!} t^{j-1} e^{-nt/\tau}$. Furthermore, making use of the convolution theorem as well as the linearity of the Laplace-transformation, we readily find

$$X_n(t) = \int_0^t K_n(t-t')X_0(t')\,\mathrm{d}t' + \nu_n(t) \tag{6.8}$$

with the Gaussian colored noise

$$\nu_n(t) = \sum_{j=1}^{n} \int_0^t \frac{\tau}{n} K_j(t-t') \sqrt{\frac{2k_B T'}{\gamma'}}\, \xi_{n-j+1}(t')\,\mathrm{d}t'. \tag{6.9}$$

Hence, we have found the expressions for the memory kernel and colored noise in the equation of motion of X_0 (6.2), generated by the variables $X_{j>0}$.

Here we have used a special choice of initial conditions for the variables $X_{j>0}$, which simplified the derivation. For general initial conditions, there are additional terms in the projected equation for X_n, which read $X_n(0)e^{-nt/\tau} + \frac{n}{\tau}X_{n-1}(0)t\, e^{-nt/\tau} + \ldots$, as we have explicitly shown in Ref. [12] (where we have presented the projection in real space). However, these terms decay exponentially with time. Thus, for the long time limit, which we mostly focus on in this thesis, these terms are anyway irrelevant.

Memory Kernel The memory kernel is given by a Gamma distribution (6.3). This is a consequence of the linear coupling. Figure 6.2 depicts the kernel and the noise correlations for different numbers of auxiliary variables, i.e., different values of n. For $n = 1$ [Fig. 6.2a], the kernel is a simple exponential decay. For $n > 1$ [Fig. 6.2b, c], the kernel has a maximum around its mean value $\mu = \int_0^\infty K_n(\Delta t)\Delta t\,\mathrm{d}(\Delta t) = \tau$, which is independent of the value of n. By increasing n, the kernel gets sharper [and eventually approaches a delta-distribution (6.20), see below].

Noise Correlations The correlations $C_\nu(\Delta t) = \langle \nu_n(t)\nu_n(t+\Delta t)\rangle$ of (6.9) can be calculated exactly for an arbitrary $\Delta t > 0$, as follows. First, we perform the trivial

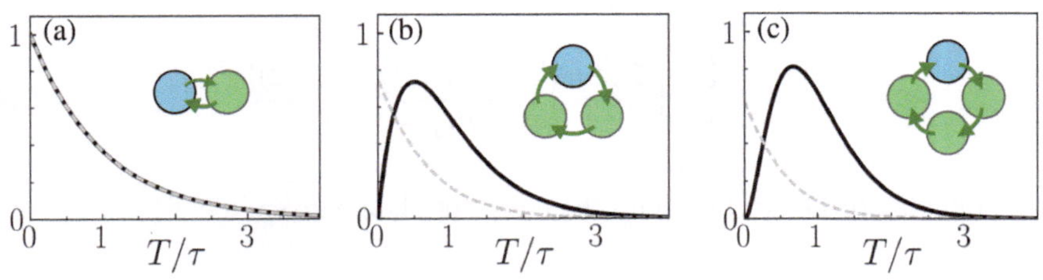

Fig. 6.2 The memory kernel (solid black lines) from Eq. (6.3) and the noise correlations (dashed gray lines) from Eq. (6.4), from $n = 1, 2, 3$ auxiliary variables shown in panel (**a**), (**b**), and (**c**), respectively. The insets show sketches of the respective networks (6.1), similar to Fig. 6.1. Here, $k_B T' = \sigma^2 \tau \gamma'$

integration over the δ-distributions from the white noise correlations $\langle \xi_{n-j+1}(t') \xi_{n-i+1}(t'') \rangle = 2\gamma' k_B \mathcal{T}' \delta_{i,j} \delta(t' - t'')$, which readily yields

$$\frac{\gamma' n^2}{k_B \mathcal{T}' \tau^2 2} C_\nu(\Delta t) = \sum_{j=1}^{n} \int_0^t K_j(t-t') K_j(t-t'+\Delta t)\, dt'$$
$$= \sum_{j=1}^{n} \frac{(n/\tau)^{2j}}{(j-1)!^2} \int_0^t (t'^2 + t'\Delta t)^{j-1} e^{-n(\Delta t + 2t')/\tau}\, dt'. \quad (6.10)$$

Performing the substitution $u = 2t'n/\tau$, we further obtain

$$\frac{\gamma' n^2}{k_B \mathcal{T}' \tau^2 2} C_\nu(\Delta t) = e^{-n\Delta t/\tau} \sum_{j=1}^{n} \frac{(n/\tau) 2^{1-2j}}{(j-1)!^2} \underbrace{\int_0^{2tn/\tau} \left(u^2 + u \frac{2n}{\tau} \Delta t \right)^{j-1} e^{-u}\, du}_{=(*)} . \quad (6.11)$$

The integral $(*)$ can be simplified by using the binomial theorem, which yields

$$(*) = \sum_{l=0}^{j-1} \binom{j-1}{l} \int_0^{2tn/\tau} u^{2(j-1-l)} \left(u \frac{2n\Delta t}{\tau} \right)^l e^{-u}\, du$$
$$= \sum_{l=0}^{j-1} \frac{(2n\Delta t)^l\, (j-1)!}{\tau^l\, l!\, (j-l-1)!} \int_0^{2tn/\tau} u^{2j-l-2} e^{-u}\, du. \quad (6.12)$$

Since we focus on steady states and are mainly interested in the noise correlations after the initial condition effects have decayed,[2] we now take the limit $t \to \infty$. Then, we perform the integration using the definition of the Gamma-function

$$\int_0^\infty x^p e^{-x} dx = \Gamma(p+1) = p!, \quad (6.13)$$

and obtain

$$(*) = \sum_{l=0}^{j-1} \left(\frac{2n\Delta t}{\tau} \right)^l \frac{(j-1)!(2j-l-2)!}{l!(j-l-1)!}. \quad (6.14)$$

Combining (6.11), (6.14) yields the noise correlation given in (6.4).

[2] One could treat transient dynamics similarly by using the incomplete Gamma function instead.

Dynamics of X_j With the very same arguments as used above to derive (6.2), one can show that each $X_{j>0}$ follows a similar equation. Specifically, one finds

$$X_j(t) = \int_0^t K_j(t-t')X_0(t')\,\mathrm{d}t' + \nu_j(t) \tag{6.15}$$

with $K_j(t-t') = \frac{n^j}{\tau^j}\frac{(t-t')^{j-1}}{(j-1)!}e^{-n(t-t')/\tau}$, and the Gaussian colored noise

$$\nu_j(t) = \sum_{l=1}^{j}\int_0^t \frac{\tau}{n}K_l(t-t')\sqrt{\frac{2k_\mathrm{B}T'}{\gamma'}}\,\xi_{j-l+1}(t')\,\mathrm{d}t'. \tag{6.16}$$

6.1.2 Limit n → ∞

Here we study the limit of the dynamical equation for the variable X_0, Eq. (6.2).

Limit of Memory Kernel In the limit $n \to \infty$, the memory kernel (6.2) collapses onto a delta peak at τ (6.20). To show this, we consider several properties of the kernel. First, it is clearly a non-negative function $K \geq 0$. Second, its mean value

$$\mu = \int_0^\infty K_n(\Delta t)\Delta t\,\mathrm{d}(\Delta t) = \tau, \tag{6.17}$$

independent of the value of n. The same holds for its weight, i.e., the integral over the kernel $\int_0^\infty K_n(\Delta t)\,\mathrm{d}(\Delta t) = 1$, which is unity. In contrast, its variance

$$\int_0^\infty K_n(\Delta t)^2\,\mathrm{d}(\Delta t) - \mu^2 = \frac{\tau^2}{n}, \tag{6.18}$$

vanishes for $n \to \infty$. Finally, for large n, its value at $\Delta t = \tau$ can be estimated by using the Stirling formula $n! \approx \sqrt{2\pi n}(n/e)^n$, yielding

$$K_n(\tau) = \frac{n^n\,e^{-n}}{(n-1)!\tau} \approx \frac{n!}{(n-1)!\sqrt{2\pi n}\tau} = \sqrt{\frac{n}{2\pi}}\frac{1}{\tau} \to \infty. \tag{6.19}$$

In total, these considerations give rise to the limit

$$\lim_{n\to\infty} K_n(\Delta t) = \delta(\Delta t - \tau). \tag{6.20}$$

Colored Noise Now we turn to the limit of the noise correlations. To this end, we consider the weight of the correlation C_ν [from Eq. (6.4)], which is given by

$$\int_0^\infty C_\nu(\Delta t)\mathrm{d}(\Delta t) = \frac{k_\mathrm{B}\mathcal{T}'}{\gamma'}\frac{\tau}{n}\sum_{p=0}^{n-1}\sum_{l=0}^{p}\frac{n^l\tau^1}{2^{2p-l}p!}\frac{(2p-l)!}{(p-l)!l!}\int_0^\infty e^{-n\Delta t/\tau}\Delta t^l\,\mathrm{d}(\Delta t)$$
$$= \frac{k_\mathrm{B}\mathcal{T}'}{\gamma'}\frac{\tau^2}{n^2}\sum_{p=0}^{n-1}\sum_{l=0}^{p}\frac{(2p-l)!}{(p!)(p-l)!}\left[\frac{1}{2}\right]^{2p-l}. \tag{6.21}$$

Using the binomial theorem, we can further simplify this expression, and find

$$\frac{\gamma'}{k_\mathrm{B}\mathcal{T}'}\int_0^\infty C_\nu(\Delta t)\mathrm{d}(\Delta t) = \frac{\tau^2}{n^2}\sum_{m=1}^{n}\sum_{r=0}^{p}\binom{p}{r}\left[\frac{1}{2}\right]^r\left[\frac{1}{2}\right]^{p-r}$$
$$= \frac{\tau^2}{n^2}\sum_{m=1}^{n}\left[\frac{1}{2}+\frac{1}{2}\right]^{2(n-m)} = \frac{\tau^2}{n^2}\sum_{m=1}^{n}1 = \frac{\tau^2}{n}. \tag{6.22}$$

Hence, the weight *vanishes* as $n \to \infty$. Since C_ν is obviously a non-negative function of Δt, this readily implies

$$\lim_{n\to\infty} C_\nu(\Delta t) = 0. \tag{6.23}$$

Limit of X_0 With (6.20) and (6.23), the limit $n \to \infty$ of the projected Eq. (6.2) is indeed *identical* to the original delayed LE (2.43), with

$$X_0(t) \to X(t), \quad X_n(\tau) \to X(t-\tau). \tag{6.24}$$

Limit of Dynamical Equations for X_j So far, we have shown that the colored noise in the dynamical equation for X_0 vanishes in the limit $n \to \infty$, see Eq. (6.23), while the memory kernel approaches a delta peak, Eq. (6.20). By application of the same arguments, one can analogously show that in the projected equation (6.15) for $X_{j>0}$, K_j collapses onto a delta distribution, and ν_j fades away.

Thus, via the Eq. (6.15) all variables $X_{j>0}(t)$ are determined by the values of $X_0(t')$, $t' \in [t-\tau, t]$ alone, as expected. In the unidirectionally coupled ring network, the information is passed from X_0 to all other variables (see Fig. 6.1). The impact of the noise terms $\xi_{j>0}$ on the variables $X_{j>0}$ vanishes in the limit $n \to \infty$, irrespective of the value of $\mathcal{T}'$.

In Eq. (6.24), we have linked X_0 and X_n to the variable X of the delayed LE (2.43). The same can thus be done for all other $X_{j>0}$. We show this in the following and give another perspective on the meaning of the variables $X_{j>0}$ in this limit.

6.1.3 Interpretation of the X_j Variables

We reconsider the dynamical Eqs. (6.1b) for the variables $X_{j>0}$. Using the *difference quotient* to approximate $\dot{X}_j$ yields

$$\frac{n}{\tau}\left[X_{j-1}(t) - X_j(t)\right] \overset{(5.1)}{\approx} \dot{X}_j(t) \approx \frac{X_j(t+\Delta t) - X_j(t)}{\Delta t}. \tag{6.25}$$

As we are interested in the interpretation at $n \to \infty$, it is safe to neglect the noise terms ξ_j here, whose impact on the X_j fades away in this limit, see Eq. (6.15) and below. Equation (6.25) gives rise to the interpretations $\Delta t = \tau/n$ and $X_j(t + \Delta t) \approx X_{j-1}(t)$, and, through iteration, $X_j(t + j\tau/n) \approx X_{j-i}(t)$. In combination with Eq. (6.24), this reveals

$$X_j(t) \to X(t - j\tau/n), \quad \forall j \in \{0, 1, 2, \ldots, n\}. \tag{6.26}$$

The interpretation (6.26) becomes accurate in the limit $n \to \infty$ (at $\Delta t \to 0$ the difference quotient equals the differential quotient). Figure 5.1 (left panel) shows a visualization of the dynamics, which emphasizes the interpretation (6.26), and becomes accurate for $n \to \infty$. Figure 6.1 shows an alternative visualization focusing on the network structure of the coupled Langevin Eqs. (6.1).

6.1.4 Initial Condition

We recall the initial conditions $X_{j\in\{1,\ldots,n\}}(0) \equiv 0$. From the previous consideration, specifically Eq. (6.26), we see that this amounts to a constant zero history function $X(t < 0) \equiv 0$ of the delay LE. Thus, it is safe to let the integral with the memory kernel start at time $t = 0$, instead of $t \to -\infty$. As we are mainly interested in steady states throughout this thesis, we can choose this initial condition without loss of generality. However, for general cases, the transient dynamics is only described correctly by the Markovian set of equations, if the history-function of $X(t)$ is correctly translated to the initial values $X_{j\in\{1,\ldots,n\}}$. Equation (6.15) provides the formula for this translation.

6.1.5 $(n + 1)$-dimensional Markovian Fokker-Planck Equation

On a probabilistic level, the Markovian $(n + 1)$-dimensional system given by the set of coupled LEs (6.1) can be described by the FPE

$$\underbrace{\partial_t \rho_{n+1}}_{\text{(I)}} = \underbrace{-\partial_{x_0}\left[\gamma^{-1} F(x_0, x_n)\rho_{n+1}\right]}_{\text{(II)}} + \underbrace{\frac{k_B \mathcal{T}}{\gamma}\,\partial^2_{x_0}\rho_{n+1}}_{\text{(III)}}$$
$$\underbrace{-\frac{n}{\tau}\sum_{j=1}^{n}\partial_{x_j}\left[(x_{j-1}-x_j)\rho_{n+1}\right]}_{\text{(IV)}} + \underbrace{\frac{k_B \mathcal{T}'}{\gamma}\sum_{j=1}^{n}\partial^2_{x_j}\rho_{n+1}}_{\text{(V)}} \tag{6.27}$$

for the $(n+1)$-point (joint) PDF $\rho_{n+1}(x_0, t; x_1, t; ...; x_n, t) = \langle \delta[x_0 - X_0(t)]\delta[x_1 - X_1(t)]..\delta[x_n - X_n(t)]\rangle$. Equation (6.27) is a self-sufficient FPE on a $(n+1)$-dimensional space, which describes on one subspace the dynamics of the delayed process in a closed form. This description is accurate in the limit $n \to \infty$. Importantly, we will nevertheless always regard n as a finite value when dealing with Eq. (6.27) itself, and only perform the limit at appropriate times (i.e., after marginalization as we explain below).

From a mathematical point of view, the limit $n \to \infty$ of Eq. (6.27) must be treated with special care. Indeed, it not is *per se* clear how to formulate a Fokker-Planck equation on an infinite-dimensional space [5]. Moreover, one drift term in (6.27) is proportional to n itself, making this limit highly non-trivial. Finding a mathematical framework in which the limit $n \to \infty$ of Eq. (6.27) is meaningful and well-defined is subject of ongoing work. Here, we can safely ignore this problem, since we focus on the equations for the marginalized (low-dimensional) PDFs. As shown in the following (Sects. 6.2 and 6.3), they can be derived by keeping in Eq. (6.27) n finite and performing the limit after the marginalization.

Further, we aim to note that Eq. (6.27) again resembles the FPE of a many-particle system, similar to (5.7). However, in sharp contrast to a system of interacting (colloidal) particles, the "interactions" are here *unidirectional* (see Fig. 6.1), which means that the "particle" x_j only feels the "particle" x_{j-1} and *not vice versa*. Hence, the forces between them are *non-reciprocal* (violating Newtons third law) and cannot be described by an interaction potential. This point will be discussed in more detail in Part III of this thesis.

6.2 Derivation of First Member of Fokker-Planck Hierarchy

Here, we will use the Markovian embedding to find a probabilistic description of the delayed system, by a "coarse-graining" of (6.27) in the form of a *marginalization*. To be more specific, we integrate out all variables but the ones relevant for the targeted member of the FP hierarchy. In this way, we obtain an FPE for the marginal PDFs.

We make use of the general relation between joint, $\rho_m(x, y)$, and marginal distributions: $\rho_{m-1}(z) = \int_\Omega \rho_m(y, z)\mathrm{d}y$. Importantly, we take the limit $n \to \infty$ only *after* the marginalization. We further employ again the natural boundary conditions (see

Sect. 3.1.1). For sake of a shorter notation, we will omit the time arguments of the PDFs in lengthy expressions. We begin with the first member of the hierarchy, i.e., the delayed FPE for ρ_1.

Marginalization We start with the FPE (6.27) for the joint PDF, and integrate it over the whole domain of *all* n auxiliary variables, i.e., $\int\int\int_\Omega \mathrm{d}x_1 \mathrm{d}x_2 .. \mathrm{d}x_n$. The terms (I–III) in Eq. (6.27) can be readily marginalized to

$$\begin{aligned} &(\mathrm{I}) \rightarrow \partial_t \rho_1(x_0, t), && (\mathrm{II}) \rightarrow -\partial_{x_0} \int_\Omega [F(x_0, x_n)\rho_2(x_0, t; x_n, t)]\, \mathrm{d}x_n, \\ &(\mathrm{III}) \rightarrow \partial^2_{x_0} \rho_1(x_0, t), && (\mathrm{V}) \rightarrow 0. \end{aligned} \tag{6.28}$$

To show the disappearance of (V), we have employed the natural boundary conditions, which imply $[\partial_{x_j} \rho_2(x_0; x_j)]^{\infty}_{-\infty} \rightarrow 0$. The remaining drift terms (IV) in Eq. (6.27) can be simplified by making use of their linearity w.r.t. the various x_j, and treating the remaining integrals by partial integration, which yields

$$\begin{aligned} (\mathrm{IV}) \rightarrow\ & \frac{n}{\tau}\Bigg\{ n\,\rho_1(x_0) - \sum_{j=2}^{n} \int\!\!\int_\Omega x_j\, \partial_{x_j} \rho_3(x_0; x_{j-1}; x_j) \mathrm{d}x_j \mathrm{d}x_{j-1} + x_0 \underbrace{[\rho_2(x_0; x_1)]^{\omega_2}_{\omega_1}}_{=0} \\ & - \int_\Omega x_1\, \partial_{x_1} \rho_2(x_0; x_1) \mathrm{d}x_1 + \sum_{j=2}^{n} \int_\Omega x_{j-1} \underbrace{\left[\rho_3(x_0; x_{j-1}; x_j)\right]^{\omega_2}_{\omega_1}}_{=0} \mathrm{d}x_{j-1} \Bigg\} \\ =\ & \frac{n}{\tau}\Bigg\{ n\,\rho_1(x_0) + \sum_{j=2}^{n} \underbrace{\left[x_j \rho_3(x_0; x_{j-1}; x_j)\right]^{\omega_2}_{\omega_1}}_{=0} - \sum_{j=2}^{n} \int_\Omega \rho_2(x_0; x_j) \mathrm{d}x_j \\ & - \underbrace{[x_1 \rho_2(x_0; x_1)]^{\omega_2}_{\omega_1}}_{=0} - \rho_1(x_0) \Bigg\} = 0. \end{aligned} \tag{6.29}$$

Again, we have further employed the boundary conditions multiple times. Hence, the contribution of (IV) to the drift term vanishes, and we obtain in total from (6.28), (6.29)

$$\partial_t \rho_1(x_0, t) = -\partial_{x_0} \int_\Omega \frac{1}{\gamma} F(x_0, x_n) \rho_2(x_0; x_n) \mathrm{d}x_n + \frac{k_{\mathrm{B}} \mathcal{T}}{\gamma} \partial^2_{x_0} \rho_1(x_0, t). \tag{6.30}$$

Limiting procedure The marginalized FPE (6.30) still pertains to a finite (but arbitrary) value of $n \in \mathbb{N}$. As a last step, we finally take the limit $n \rightarrow \infty$. To this end, we recall the meaning of the respective stochastic variables in this limit [Eq. (6.24)]. Based on that, we perform the phase space transformation $\{x_0, t\} \rightarrow \{x, t\}$ and

$\{x_n, t\} \to \{x_\tau, t - \tau\}$, and therewith, obtain[3] the delayed FPE for the one-time PDF (3.15).

Next, we demonstrate that our procedure can also be applied to find higher members of the FP hierarchy.

6.3 Derivation of Higher Members via Markovian Embedding

To derive a FPE for the two-time PDF $\rho_2(x, t; x_\tau, t - \tau)$, we marginalize the full $(n + 1)$-dimensional FPE (6.27) w.r.t. all variables but x_0 and x_n (again keeping n finite until after the marginalization). This time, we do not integrate over x_n as this variable represents $X(t - \tau)$ in the limit $n \to \infty$. In a second step, we will again perform a coordinate transformation to take the limit.

Marginalization The integration of the full FPE (6.27) over the variables $x_1, .., x_{n-1}$, is completely analogous to the marginalization presented in Sect. 6.2. In particular, one again performs the same steps, i.e., partial integrations and subsequent application of the boundary conditions. As opposed to the first member, now the term (IV) gives a non-vanishing contribution, namely

$$(\mathrm{IV}) \to \partial_{x_n}\left[\int_\Omega \frac{n}{\tau}(x_n - x_{n-1})\rho_3(x_0, x_{n-1}, x_n)\mathrm{d}x_{n-1}\right], \tag{6.31}$$

which involves a *three*-point PDF. Moreover, now the diffusion term (V) yields a non-zero term, that is

$$(\mathrm{V}) \to \frac{k_\mathrm{B}\mathcal{T}'}{\gamma'}\partial^2_{x_n}\rho_2(x_0, x_n). \tag{6.32}$$

Thus, for the first time, we see a trace of the additional noise terms $\xi_{j>0}$ in the FPE, which yield an additional diffusion term on the level of the two-point PDF in the variable x_n.

In this way, one readily obtains the marginalized FPE

$$\begin{aligned}\partial_t\rho_2(x_0, t; x_n, t) = &- \partial_{x_0}\left[\gamma^{-1}F(x_0, x_n)\rho_2(x_0, t; x_n, t)\right] + \frac{k_\mathrm{B}\mathcal{T}}{\gamma}\partial^2_{x_0}\rho_2(x_0, t; x_n, t)\\ &+ \partial_{x_n}\int_\Omega \frac{n}{\tau}(x_n - x_{n-1})\rho_3(x_0, t; x_{n-1}, t; x_n, t)\mathrm{d}x_{n-1}\\ &+ \frac{k_\mathrm{B}\mathcal{T}'}{\gamma'}\partial^2_{x_n}\rho_2(x_0, x_n).\end{aligned} \tag{6.33}$$

[3] Independently of the choice of $\mathcal{T}'$.

Confirming the consistency of this calculation, we note that integrating (6.33) over x_n yields the first member of the hierarchy (6.30), thus, the exact FPE for ρ_1 of the system with time delay. (By the way, this integration makes the x_n-diffusion term disappear again, independent of $\mathcal{T}'$.) Higher-order equations can be obtained in an analogous manner, yielding the infinite hierarchy of coupled equations, whose $(m+1)$st member for $\rho_m(x_0; x_{n-m+2}, .., x_n)$, with $m \geq 3$, reads

$$\begin{aligned}
\partial_t \rho_m = &- \partial_{x_0}\left[\gamma^{-1} F(x_0, x_n)\rho_m(x_0; x_{n-m+2}, .., x_n)\right] \\
&+ \sum_{l=0}^{m-3} \frac{n}{\tau}\partial_{x_{n-l}}\left[(x_{n-l} - x_{n-l-1})\rho_m(x_0; x_{n-m+2}, .., x_n)\right] \\
&+ \frac{n}{\tau}\partial_{x_{n-m+2}} \int_\Omega (x_{n-m+2} - x_{n-m+1})\rho_{m+1}(x_0; x_{n-m+1}, .., x_n)\mathrm{d}x_{n-m+1} \\
&+ \frac{k_B\mathcal{T}}{\gamma}\partial^2_{x_0}\rho_m(x_0, t; x_{n-m+2}, t, .., x_n, t) \\
&+ \frac{k_B\mathcal{T}'}{\gamma'}\sum_{l=0}^{m-3}\partial^2_{x_{n-1}}\rho_m(x_0, t; x_{n-m+2}, t, .., x_n, t).
\end{aligned} \tag{6.34}$$

In contrast to the FPE (6.30) for ρ_1, all higher members of this FP hierarchy (6.33), (6.34) explicitly contain the linear forces stemming from the auxiliary variable Eqs. (6.1b), and further involve diffusion terms scaled with the temperature $\mathcal{T}'$.

We now inspect in more detail the second member (6.33) in the limit $n \to \infty$, to recover the system with discrete delay. We again need to transform the probability space, hence, the phase space and the PDFs defined on it.

Transformation of phase space, limiting procedure In order to take the limit $n \to \infty$ of Eq. (6.33), we will transform the phase space and the PDFs. We perform the transformation in two steps. First we replace the phase space variable x_{n-1} by $\widetilde{x} = (n/\tau)(x_{n-1} - x_n)$, which will later in the limit $n \to \infty$ receive the meaning of a *velocity*. In a second step, we will transform back to the original variables of the delayed LE (2.43). This transformation will mainly affect the time arguments.

As first step, we change the phase space from $\{x_0, t; x_{n-1}, t; x_n, t\}$ to $\{x_0, t; \widetilde{x}, t; x_n, t\}$. To this end, we start with performing the substitution $x_{n-1} \longrightarrow \widetilde{x} = (n/\tau)(x_{n-1} - x_n)$ to rewrite the integral $\int_\Omega \mathrm{d}x_{n-1}$ as $(\tau/n)\int_{\widetilde{\Omega}} \mathrm{d}\widetilde{x}$. The drift term then becomes

$$\begin{aligned}
&\frac{n}{\tau}\partial_{x_n}\int_\Omega (x_n - x_{n-1})\rho_3(x_0, t; x_{n-1}, t; x_n, t)\mathrm{d}x_{n-1} \\
&= \frac{\tau}{n}\partial_{x_n}\int_{\widetilde{\Omega}} \widetilde{x}\,\rho_3\left(x_0, t; \frac{\tau}{n}\widetilde{x} + x_n, t; x_n, t\right)\mathrm{d}\widetilde{x}.
\end{aligned} \tag{6.35}$$

In the next step, we need to transform the PDF $\rho_3 \to \widetilde{\rho_3}$, where $\widetilde{\rho_3}$ is a normalized (joint) PDF of the new phase space variables $\{x_0, \widetilde{x}, x_n\}$. The relation between both PDFs can be found by comparing the normalizations. In particular,

$$\begin{aligned} 1 &= \iiint_\Omega \rho_3\left(x_0, t; x_{n-1}, t; x_n, t\right) \mathrm{d}x_0 \mathrm{d}x_{n-1} \mathrm{d}x_n \\ &= \frac{\tau}{n} \int_\Omega \int_{\widetilde{\Omega}} \int_\Omega \rho_3\left(x_0, t; \widetilde{x} + x_n, t; x_n, t\right) \mathrm{d}x_0 \mathrm{d}\widetilde{x} \mathrm{d}x_n, \end{aligned} \tag{6.36}$$

and, on the other hand, by definition, $\iiint \widetilde{\rho_3}\left(x_0, t; \widetilde{x}, t; x_n, t\right) \mathrm{d}x_0 \mathrm{d}\widetilde{x} \mathrm{d}x_n = 1$. Combining the constraints due to both normalizations, yields

$$\rho_3\left(x_0, t; \frac{\tau}{n}\widetilde{x} + x_n, t; x_n, t\right) \stackrel{!}{=} \frac{\tau}{n} \widetilde{\rho_3}\left(x_0, t; \widetilde{x}, t; x_n, t\right). \tag{6.37}$$

Therewith, we obtain the drift term

$$\begin{aligned} \frac{n}{\tau} \partial_{x_n} \int_\Omega (x_n - x_{n-1}) \rho_3(x_0; x_{n-1}; x_n) \mathrm{d}x_{n-1} &\longrightarrow \partial_{x_n} \int_{\widetilde{\Omega}} \widetilde{x}\, \widetilde{\rho_3}\left(x_0; \widetilde{x}; x_n\right) \mathrm{d}\widetilde{x} \\ &= \partial_{x_n} \left[\left\langle \widetilde{X}(t) \middle| x_0 = X_0(t), x_n = X_n(t) \right\rangle \rho_2(x_0, t; x_n, t) \right]. \end{aligned} \tag{6.38}$$

In the last line, we have used the conditional average notation, which involves the respective stochastic variables. Now, we recall the meaning of the stochastic variables in the limit $n \to \infty$, i.e., $X_0(t) = X(t)$, $X_n(t) = X(t - \tau)$ from (6.24), and $\dot{X}_n(t) = \frac{n}{\tau}\left[X_{n-1}(t) - X_n(t)\right] + \sqrt{2k_\mathrm{B}\mathcal{T}'/\gamma'}\xi_j$ from (6.1b). Based on this, we perform the transformation

$$\{x_0, t\} \to \{x, t\}, \tag{6.39}$$

$$\{x_n, t\} \to \{x_\tau, t - \tau\}, \tag{6.40}$$

$$\{\widetilde{x}, t\} \to \{\dot{x}_\tau, t - \tau\}. \tag{6.41}$$

A nonzero $\mathcal{T}'$ does not affect this transformation, as we will discuss. Further, we transform

$$\begin{aligned} &\partial_{x_n} \int_{\widetilde{\Omega}} \widetilde{x}\, \widetilde{\rho_3}\left(x_0, t; \widetilde{x}, t; x_n, t\right) \mathrm{d}\widetilde{x} \\ &\quad\longrightarrow \partial_{x_\tau} \left[\left\langle \dot{X}(t - \tau) \middle| x = X(t), x_\tau = X(t - \tau) \right\rangle \rho_2(x, t; x_\tau, t - \tau) \right], \end{aligned} \tag{6.42}$$

with $\dot{x}_\tau \in \widetilde{\Omega} = \mathbb{R}$. We note that the appearance of a *velocity*-like variable is unusual, as we deal with overdamped dynamics. Since $\dot{X}$ has no well-defined corresponding PDF for overdamped dynamics [16], we cannot express the conditional average on the right side of (6.42) in the form an integral.

Due to the coordinate transformation and limiting procedure, we finally obtain from (6.33), which is the FPE for the two-point PDF of the Markovian system (6.1), the FPE for the two-*time* PDF of the original non-Markovian system with time delay

$$\begin{aligned}\partial_t \rho_2(x,t;x_\tau,t-\tau) =& -\partial_x\left[\gamma^{-1}F(x,x_\tau)\rho_2(x,t;x_\tau,t-\tau)\right] + \frac{k_B\mathcal{T}}{\gamma}\partial_x^2\rho_2(x,t;x_\tau,t-\tau)\\&+\partial_{x_\tau}\left[\langle \dot{X}(t-\tau)|x=X(t),x_\tau=X(t-\tau)\rangle\rho_2(x,t;x_\tau,t-\tau)\right]\\&+\frac{k_B\mathcal{T}'}{\gamma'}\partial_{x_\tau}^2\rho_2(x,t;x_\tau,t-\tau). \qquad (6.43)\end{aligned}$$

Before we compare Eq. (6.43) to the respective FPE obtained from Novikov's theorem [Eq. (5.7)] we will discuss the impact of the additional noise terms.

Effect of Additional Noise Terms at $\mathcal{T}'$ Let us consider the effect of additional noise terms. It is easy to see that the transformation $\{\widetilde{x},t\}\to\{\dot{x}_\tau,t-\tau\}$ from (6.41) is exact in the limit $\mathcal{T}'\to 0$. However, because $n\to\infty$, this transformation is as well expected to apply at finite $\mathcal{T}'>0$ (recall that the additional noise term in the LE vanishes as $n\to\infty$ despite $\mathcal{T}'>0$). Thus, from that perspective, $\mathcal{T}'$ may be finite. On the contrary, the resulting FPE (6.43) contains the additional diffusion term $\frac{k_B\mathcal{T}'}{\gamma}\partial_{x_\tau}^2\rho_2(x,t;x_\tau,t-\tau)$, which describes an effect of the additional noise terms on the probabilistic level. This renders a different two-time PDF as at $\mathcal{T}'\to 0$. This is interesting, because the first member of the FP hierarchy was independent of $\mathcal{T}'$. In the following, we will consider the case $\mathcal{T}'=0$.

6.3.1 *Comparison to Equation from Novikov's Theorem*

Comparing the here derived equation (6.43) (at $\mathcal{T}'=0$)

$$\begin{aligned}\partial_t \rho_2(x,t;x_\tau,t-\tau) =& \ \partial_{x_\tau}\left[\langle \dot{X}(t-\tau)|x=X(t),x_\tau=X(t-\tau)\rangle\rho_2(x,t;x_\tau,t-\tau)\right]\\&-\partial_x\left[\gamma^{-1}F(x,x_\tau)\rho_2(x,t;x_\tau,t-\tau)\right]\\&+k_B\mathcal{T}\gamma^{-1}\partial_x^2\rho_2(x,t;x_\tau,t-\tau), \qquad (6.44)\end{aligned}$$

to the FPE (5.7) obtained from Novikov's theorem, reveals that they are in fact quite different. Our approach thus yields an *alternative representation* of the second member. However, as we demonstrate in the Appendix A.4 one can indeed transform (6.44) into (5.7), and *vice versa*. (This further represents a consistency check of our approach.)

The main difference of (6.44) as compared to (5.7) (from Novikov's theorem), is that it involves statistical dependencies between the present and delayed system state and the delayed *velocity*. Instead, Eq. (5.7) implicitly involves the statistical dependencies among the system states at three different times via the three-time PDF $\rho_3(x, t; x_\tau, t-\tau, x_{2\tau}, t-2\tau)$. This indicates that, instead of going back even further *into the past* with every member [by first taking into account $X(t-\tau)$, then $X(t-2\tau)$, then $X(t-3\tau)$ ...], the here derived hierarchy rather collects more and more information about the dynamics at time $t-\tau$ [by taking into account $X(t-\tau)$, then $\dot{X}(t-\tau)$]. This suits to the different pictures of the delayed dynamics provided by the two illustrations in Fig. 5.1.

As both representations describe the exact same process, this demonstrates that to (probabilistically) predict the future (after a time t), either knowledge of the system states at all times $t, t-\tau, t-2\tau, \ldots$ would be needed, or, complete knowledge about the preceding interval $[t-\tau, t]$. On the stochastic level of description, the process which underlies the non-Markovian one, can either be considered as a process of the state vector $\{X(t), X(t-\tau), X(t-2\tau), X(t-3\tau), \ldots\}$, or, equivalently (and actually more natural), of the state vector of all $X(s) \in \mathbb{R}^\infty$, with $s \in [t-\tau, t[$.

A technical advantage of the representation (6.44) is that it does not involve unknown functional derivatives w.r.t. the noise, as opposed to (5.7), and only contains one diffusion term w.r.t. x. In Chap. 8, we will utilize these technical advantages, and use (6.44) as starting point to find an approximate two-time PDF, $\rho_2(x, t; x_\tau, t-\tau)$. To our knowledge, this is the first approach of this kind.

Further Perspective on This Approach As a final remark, we would like to reconsider the Markovian embedding from a conceptual point of view. Non-Markovianity, in general, and specifically the discrete time delay considered here, is a consequence of an incomplete description of a system, i.e., missing degrees of freedom. The delayed force might either stem from a specifically designed external agent, a *feedback controller*, or, more generally, from any part of a more complex system (the "super-system"), which is not explicitly modeled within the considered description, but interacts with the system at hand in form of delayed forces. The Markovian auxiliary system, in which we here embed the delayed process, allows us to mimic the external agent (or other parts of the "super-system") without explicitly modeling it in detail. In this sense, the auxiliary variables can be interpreted as a "substitute" of the hidden degrees of freedom in the system. On the other hand, it is well known that hidden degrees of freedom have a significant impact on the entropy production and other thermodynamic notions, see for example [17, 18]. Hence, it is interesting to study the consequences of our treatment of delayed systems from a stochastic-thermodynamical (and information-theoretical) perspective. This motivates the investigations presented in Part III, Chap. 10.

6.4 Side Note: Discrete Versus Distributed Delay

The Markovian embedding from Sect. 6.1 has revealed that the discrete delay, white noise LE can be considered as the $n \to \infty$ limit of a LE (6.2) with distributed delay and colored noise. In particular, the dynamics of the first degree of freedom, X_0, in the $(n+1)$-dimensional network (6.1) is then identical to the time delayed process (with delta-correlated memory kernel). Keeping n finite amounts to approximating the delayed force $F_d[X(t), X(t-\tau)]$ in the LE with a force $F_d[X(t), \int K(t-t')X_0(t')dt' + \nu(t)]$ that involves a Gamma-distributed memory kernel (6.3) with finite width and colored noise 6.4 (if $\mathcal{T}' > 0$). For $n > 1$, the delay had similar features as the discrete delay case, i.e., a pronounced maximum around τ, and a vanishing kernel $K(t-t')$ at $t' = t$. On the contrary, the case $n = 1$, this yields a LE with an exponentially decaying delay [shown in Fig. 6.2a]. To get an impression of the system with distributed delay, let us here compare the trajectories, focusing on $n = 1$, i.e., the most distinct case. In a second step, we briefly consider the corresponding PDFs.

Trajectories of the Delayed Ornstein-Uhlenbeck Process For the comparison between discrete and distributed delay, we consider the simplest possible example, in particular, the linear delay equation

$$\gamma\dot{X} = -\int_{-\infty}^{t} K(t-t')X(t')dt' + \sqrt{2D_0}\xi(t). \tag{6.45}$$

Discrete Delay We start with the discrete delay, which is recovered at $K(t-t') = \delta(|t-t'| - \tau)$, yielding

$$\gamma\dot{X} = -(\gamma/\tau)X(t-\tau) + \sqrt{2D_0}\xi(t). \tag{6.46}$$

This is a special case of the linear delay LE considered in Sects. 2.4.5 and 5.2.2 (with $\alpha = 0$, $\beta = \gamma/\tau$). In regards to the colloidal particle in the trap with optical feedback, this corresponds to the case $k = -1, a = 1$, (i.e., positive feedback).

Figure 6.3 depicts some example steady-state trajectories and the associated position-autocorrelation function C of a Brownian particle subject to a linear feedback force, whose motion is given by (6.46).

As expected from the considerations in Sects. 2.4.5 and 5.2.2, the correlation function C shows damped oscillation, with negative regions $C(t') < 0$, i.e., anticorrelations. We have discussed this point in a similar model in Sect. 2.4.5. Here, the period is roughly 4τ. We further recall that oscillatory solutions of first-order scalar differential equations are indeed a characteristic of delay differential equations (see Sect. 2.4.4).

6.4.1 Distributed Delay

Let us now consider the *exponentially distributed delay*, specifically

$$\gamma\dot{X} = -(\gamma/\tau)\int_{-\infty}^{t} e^{-(t-t')/\tau} X(t')\mathrm{d}t' + \sqrt{2D_0}\,\xi(t), \tag{6.47}$$

which corresponds to the case $n = 1$ of the LE (6.2). To better see the impact of the distributed delay, we do not consider colored noise here ($\mathcal{T}' = 0$). Figure 6.4 shows some trajectories and the position-autocorrelations. Remarkably, the trajectories and correlations are found to be quite similar to the discrete delay in Fig. 6.3, despite the very different type of memory. To make both cases more comparable, the first moment of the memory kernel [given by the delay time in (6.46)] is the same. One likewise notices damped oscillatory behavior with a similar oscillation period and anticorrelations. However, the oscillatory component seems a bit less pronounced, and the decay is faster. Furthermore, the initial value $C(0) = \langle X^2\rangle$, which gives the second moment of the steady-state PDF, is smaller, i.e., the density is *more located* in the case of distributed delay.

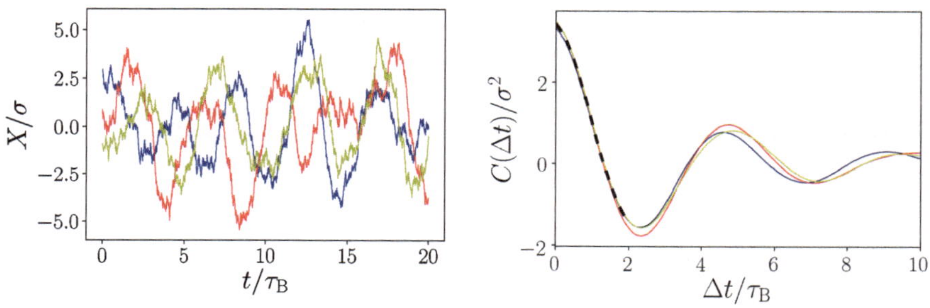

Fig. 6.3 *Left panel:* Three example trajectories of an Brownian particle subject to a linear time-delayed force $F_\mathrm{d} \propto -X(t-\tau)$ in the steady state. The LE is (6.46) at $\tau = \tau_\mathrm{B}$. *Right panel:* Corresponding position-autocorrelation function from the three single numerically generated trajectories, and additionally plotted by a dashed black line from the analytical results (5.23), (2.53) for $t \in [0, 2\tau]$, which we will explain in detail in Sect. 2.4.5. The delay force applies positive feedback to the colloid

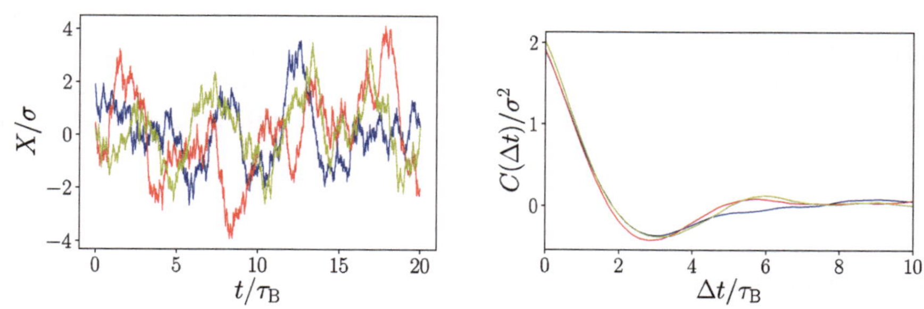

Fig. 6.4 *Left panel:* Three example trajectories of a Brownian particle subject to a linear feedback force involving *distributed* delay with an exponential memory kernel at $\tau = \tau_\mathrm{B}$, in the steady state. The LE reads (6.47). *Right panel:* Corresponding temporal position-autocorrelation function from the three single numerically generated trajectories

We will consider this type of memory in more detail in Part III of this thesis from a thermodynamic point of view. Let us now briefly consider the probability densities for discrete and distributed delay.

6.4.2 Probability Densities in the Presence of Discrete and Distributed Delay

Above, we have compared the trajectories of LEs with discrete and distributed delay, corresponding to the cases $n = 1$ and $n \to \infty$. Let us now switch to the probabilistic perspective. We are particularly interested in the idea that keeping n finite readily implies an approximation for the PDF, which is the main objective of the remainder of Part II.

We recall that the dynamics is described as the first degree of freedom of the Markovian set of Langevin equations (6.1). This system can also be described by a self-sufficient set of $n + 1$ Fokker-Planck equations (6.27). By keeping n finite, Eq. (6.27) is a well-defined, closed FPE, which can be readily used to calculate the $(n + 1)$-time PDF by means of standard techniques for Markovian FPEs (at least numerically). By marginalization of the resulting PDFs, one would obtain an approximate PDF of the delayed process. On the level of the LE, approximating the delay process by keeping n finite is in fact a common method [19–22]. However, a systematic investigation of this approximation (including an error estimate), has, to the best of our knowledge, not been performed yet.

Such an approximation has the advantage that one readily has a corresponding LE. In this sense, one can immediately "see" the consequences of this approximation on the stochastic level. This is very appealing from a conceptional point of view. However, while being closed, the resulting FPE (6.27) is still a set of multiple coupled partial differential equations, whose analytical solution is a very involved problem on its own. This is especially true for systems with nonlinear forces F.

As we are mostly focus on analytical approximations for the PDFs, we do not further proceed with this finite n idea in the present context. Let us still consider some exemplary, numerical results to get an idea of the impact of n on the probability densities. Figure 6.5 depicts some numerical results from this "approximation", i.e., the steady-state one-time PDF for different finite values of n, in the doublewell potential. These results indicate that the probability density converges, but it seems to converge relatively slowly. Our preliminary numerical investigations further suggest that the convergence is faster, if $\mathcal{T}' = \mathcal{T}$, i.e., if the temperature of the auxiliary variables matches the temperature of X. This is an interesting observation, which might be worth to further elucidate in future research.

As we have already mentioned, it is not obvious how to solve the set of $(n + 1)$ FPEs (6.27). It is noteworthy that this is because the coupling terms between the individual equations (e.g., the coupling between X_i and X_{i+1}) resemble non-reciprocal interactions, which we have introduced in Sect. 4.7.3. For this reason,

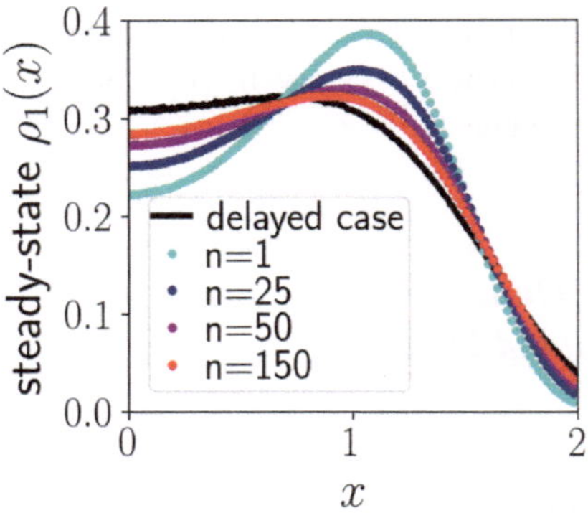

Fig. 6.5 Comparison of the one-time PDFs for different values of n in the doublewell potential with linear delay force, from numerical simulations. At $n = 1$ (cyan symbols), the memory kernel (6.3) is exponentially decaying, while it is non-monotonic for all $n > 1$ with a maximum around τ. The black line plots the PDF of the system with discrete delay, which corresponds to the limit $n \to \infty$. The parameters are set to $V_0 = 1.5\,k_B\mathcal{T}$, $k = k_B\mathcal{T}/\sigma^2$, $\tau = \tau_B$, $\mathcal{T}' = 0.5\mathcal{T}$. Thanks to Simon Hermann for the numerical data

there is for example no Boltzmann-like steady-state solution. We will come back to this observation in Chap. 10 and discuss its implications from a thermodynamic perspective in great detail. In Chap. 10, we will also consider the impact of changing n on thermodynamic and dynamical properties of the system, and reconsider the "discrete delay" limit $n \to \infty$.

In the following, we will discuss alternative approximation schemes that yield an approximate first member of the FP hierarchy, and even introduce an approach that operates on the level of the second member of the FP hierarchy. The focus is on approaches that can be treated analytically.

References

1. of Nature CM (1997) Global biodiversity. Bd. 7–8 (The Museum. https://books.google.de/books?id=O5TzAAAAMAAJ
2. Tichelli A, Rovó A (2013) A clever person solves a problem; a wise person avoids it. Biol Blood Marrow Transplant 19:1004–1005
3. Longtin A (2010) Complex time-delay systems: theory and applications. Springer, Berlin Heidelberg
4. Smith HL (2011) An introduction to delay differential equations with applications to the life sciences, vol 57. Springer, New York
5. Niculescu S-I, Gu K (2012) Advances in time-delay systems, vol 38. Springer Science & Business Media
6. Siegle P, Goychuk I, Talkner P, Hänggi P (2010) Markovian embedding of non-Markovian superdiffusion. Phys Rev E 81
7. Siegle P, Goychuk I, Hänggi P (2011) Markovian embedding of fractional superdiffusion. EPL 93:20002

8. Bao J-D, Hänggi P, Zhuo Y-Z (2005) Non-Markovian Brownian dynamics and nonergodicity. Phys Rev E 72
9. Villamaina D, Baldassarri A, Puglisi A, Vulpiani A (2009) The fluctuation-dissipation relation: how does one compare correlation functions and responses? J Stat Mech Theory Exp 2009:P07024
10. Puglisi A, Villamaina D (2009) Irreversible effects of memory. EPL 88:30004
11. Crisanti A, Puglisi A, Villamaina D (2012) Nonequilibrium and information: the role of cross correlations. Phys Rev E 85
12. Loos SAM, Klapp SHL (2019) Fokker-Planck equations for time-delayed systems via Markovian embedding. J Stat Phys 177:95–118
13. Novikov EA (1965) Functionals and the random-force method in turbulence theory. Sov Phys JETP 20:1290–1294
14. Grabert H, Graham R, Green MS (1980) Fluctuations and nonlinear irreversible processes. ii. Phys Rev A 21:2136
15. Risken H (1984) The Fokker-Planck equation. Springer, Berlin
16. Gardiner CW (2002) Handbook of stochastic methods, 2 edn. Springer, Berlin
17. Mehl J, Lander B, Bechinger C, Blickle V, Seifert U (2012) Role of hidden slow degrees of freedom in the fluctuation theorem. Phys Rev Lett 108
18. Kawaguchi K, Nakayama Y (2013) Fluctuation theorem for hidden entropy production. Phys Rev E 88
19. Rosinberg M, Tarjus G, Munakata T (2018) Influence of time delay on information exchanges between coupled linear stochastic systems. Phys Rev E 98
20. De Vries B, Príncipe JC (1991) A theory for neural networks with time delays. Advances in neural information processing systems. Morgan Kaufmann, San Mateo, CA, pp 162–168
21. Rateitschak K, Wolkenhauer O (2007) Intracellular delay limits cyclic changes in gene expression. Math Biosci 205:163–179
22. Karmeshu Gupta V, Kadambari K (2011) Neuronal model with distributed delay: analysis and simulation study for gamma distribution memory kernel. Biol Cybern 104:369–383

Chapter 7
Force-Linearization Closure

In this Chapter, we turn to the problem of finding (approximate) probabilistic solutions of systems with delay. In particular, we discuss an approximation scheme for the steady-state one-time PDF, which we have introduced in Ref. [1], called *Force-linearization closure* (FLC).

Starting point The FLC applies to systems described by the delayed Langevin equation (2.43), recall

$$\dot{X}(t) = \gamma^{-1} F\,[X(t), X(t-\tau)] + \sqrt{2D_0}\,\xi(t).$$

Consistent with the example systems considered throughout this thesis, we assume that the total deterministic force F can be written as

$$F(x, x_\tau) = F_\mathrm{s}(x) + F_\mathrm{d}(x_\tau, x), \tag{7.1}$$

where the force $F_\mathrm{s}(x) = -\partial_x V_\mathrm{s}(x)$ stems from a static potential V_s, and F_d is a delayed force involving the system state at times t and $t-\tau$.

We further recall the corresponding FPE (3.15) for the one-time PDF ρ_1 [discussed in Sect. 3.2], which we here recast into the form

$$\partial_t \rho_1(x,t) = -\,\gamma^{-1}\partial_x\big[\widehat{F}(x,t)\rho_1(x,t)\big] + D_0\partial_{xx}\rho_1(x,t), \tag{7.2a}$$

with

$$\widehat{F}(x,t) = F_\mathrm{s}(x) + \int_\Omega F_\mathrm{d}(x, x_\tau)\rho_\mathrm{c}(x_\tau, t-\tau|x,t)\mathrm{d}x_\tau. \tag{7.2b}$$

The FPE (7.2) is not self-sufficient as it involves ρ_c (which in turn depends on ρ_2). As we have discussed in Sect. 3.2, in fact, an infinite hierarchy of equations emerges, whose $(n+1)$st member depends on the n-time probability density $\rho_n(x,t;\,...;\,x_{n\tau}, t-n\tau)$.

© The Author(s), under exclusive license to Springer Nature Switzerland AG 2021

S. A. M. Loos, *Stochastic Systems with Time Delay*, Springer Theses,
https://doi.org/10.1007/978-3-030-80771-9_7

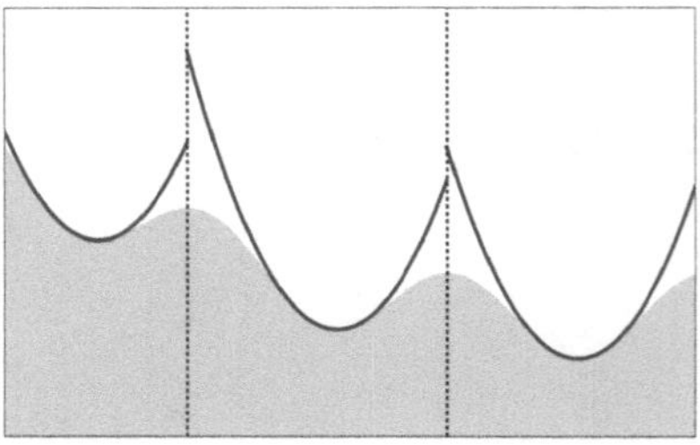

Fig. 7.1 Illustration of the main idea behind the Force-linearization closure, that is, piecewise approximating the static potential by parabolic functions in the higher members of the Fokker-Planck hierarchy. To this end, an effective static potential V_{STAT} is defined, see Eq. (7.3), and the spatial domain is chopped between its maxima

By finding suitable approximations for ρ_c, Eqs. (7.2) can be closed and the hierarchy truncated. The FLC represents such a closure scheme.

Main idea The FLC bases on the fact that the exact steady-state solution of the FPE hierarchy is known for systems where all deterministic forces are linear in x and x_τ [see Eqs. (5.19), (5.24)]. The main idea is to *linearize the deterministic forces* in all members of the infinite FPE hierarchy apart from the first one, in this way obtaining a closed approximate FPE for the one-time PDF ρ_1. Importantly, this linearization is carried out separately on suitable subdomains, as sketched in Fig. 7.1, which is crucial in situations where the potential landscape has multiple minima. In this way, we reduce the problem to a solvable, linear situation, keeping just enough aspects of the nonlinear original system to get an interesting result.

Having this main idea in mind, one would already expect—without further reasoning—that the FLC renders a good approximation, whenever the PDF is strongly centered around to the potential minima. This is, for instance, the case, when the potential barriers are high compared to thermal fluctuations. Indeed, our results presented at the end of this Chapter, confirm these expectations.

7.1 Details of the Approximation

Here we present the details of the FLC, which is a multi-step procedure. To understand the results discussed in Sects. 7.3 and 7.4, it is, however, not necessary to be aware of all the mathematical details, and the Reader who is mainly interested in the general idea and the results may safely skip this Part and proceed immediately with Sect. 7.2.

We won't make further physical assumptions in the present Section. However, as will become clear towards the end of the Section, the expressions severely simplify for systems with natural boundary conditions and linear delay forces, which is the case in our example applications. The following text is closely related to Ref. [1].

7.1.1 Linearization of the Deterministic Forces

Slitting the space into subdomains We start by considering the history-dependent energy landscape corresponding to the deterministic force $F(x, x_\tau) = F_s(x) + F_d(x, x_\tau)$ in the LE (7.1). As a first step, we introduce a simple estimate for the total steady-state energy landscape. In particular, we assume at this point the system to be at rest, i.e., $X(t - \tau) \equiv X(t)$, such that the total (static) potential is given by

$$V_{\text{STAT}}(x) = -\int^x F(x', x_\tau = x')\text{d}x'. \tag{7.3}$$

We number each minimum $x_{0,i}$ of V_{STAT} with an integer $i \in \mathcal{I}$. To formulate the FLC, we split the spatial domain Ω into non-overlapping intervals Ω_i, such that each local maximum of V_{STAT} (as well as both boundaries of Ω) represents a bound of an interval, and every Ω_i contains exactly one minimum $x_{0,i}$. By this procedure, we obtain a splitting of Ω into subdomains Ω_i, whose union is again the entire spatial domain: $\Omega = \bigcap_{i \in \mathcal{I}} \Omega_i$.

Linearization We return to the actual deterministic force $F = F_s + F_d$, and perform for each i a Taylor expansion of F in both spatial variables x and x_τ. More specifically, we expand around the deviations from the enclosed minimum, i.e., $\Delta x_i = x - x_{0,i}$ and $\Delta x_{\tau,i} = x_\tau - x_{0,i}$. Neglecting all terms of quadratic orders $\mathcal{O}(\Delta x_i^2)$, $\mathcal{O}(\Delta x_i \Delta x_{\tau,i})$, $\mathcal{O}(\Delta x_{\tau,i}^2)$ or higher, we obtain

$$F^{\text{lin},i}(x, x_\tau) = -\alpha_i \Delta x_i - \beta_i \Delta x_{\tau,i}, \tag{7.4}$$

where $-\alpha_i$ and $-\beta_i$ are the first-order derivatives of F with respect to x and x_τ, respectively, evaluated at the minimum $x_{0,i}$. [Note that the constant terms always vanish as we expand around the minima of V_{STAT} (7.3), thus, $F(x_{0,i}, x_{0,i}) = 0$.]

This procedure yields an approximation of the steady-state energy landscape composed of a sequence of quadratic polynomials, each subdomain Ω_i reaching from one local maximum to the following one (or to a bound of the entire spatial domain). An example is sketched in Fig. 7.1.

7.1.2 Analytical Probabilistic Solution for Linearized Forces

We now turn back to the full system which involves the original nonlinear deterministic forces and thermal noise, and apply the linearization procedure of Sect. 7.1.1 to the deterministic forces in the higher members of the Fokker-Planck hierarchy. The following steps are performed separately for each subdomain Ω_i. We first apply the linearization $F \approx F^{\text{lin},i}$ [from Eq. (7.4)] to the force terms in all members of the infinite FPE hierarchy starting from the second, i.e., the equation for the two-time PDF $\rho_{2,\text{ss}}(x, t; x_\tau, t - \tau)$.

Then, we utilize the linear solutions for the PDF, which we have discussed in Sect. 5.2.2. Assuming natural boundary conditions at every subdomain bound in the higher members, the FPEs for all $\rho_{\mathrm{n,ss}}$ can be solved by multivariate Gaussian distributions [2]. We have given the explicit results for $\rho^{\mathrm{lin}}_{1,\mathrm{ss}}$, $\rho^{\mathrm{lin}}_{2,\mathrm{ss}}$ and $\rho^{\mathrm{lin}}_{3,\mathrm{ss}}$ before in Eq. (5.19) and below. For each subdomain, the corresponding conditional PDF[1] reads

$$\rho^{\mathrm{lin},i}_{\mathrm{c,ss}}(x_\tau|x;\tau) = \sqrt{\frac{2K_i}{\pi}(d_{1,i})^2\left[1-(d_{2,i})^2\right]}\, e^{\frac{\Delta x_i{}^2}{2K_i}}\, e^{d_{1,i}\left[2d_{2,i}\,\Delta x_i\,\Delta x^i_\tau - \Delta x_i{}^2 - (\Delta x^i_\tau)^2\right]} \tag{7.5}$$

with the coefficients

$$\begin{aligned}
\Omega_i \equiv\ & \omega(\alpha^i,\beta^i) &&= \sqrt{\alpha_i^2-\beta_i^2}/\gamma \quad \in \mathbb{C}, && (7.6\mathrm{a})\\
K_i \equiv\ & K(\alpha_i,\beta_i,\tau) &&= D_0\frac{\gamma+(\beta_i/\omega_i)\sinh(\tau\omega_i)}{\alpha_i+\beta_i\cosh(\tau\omega_i)}, && (7.6\mathrm{b})\\
d_{1,i} \equiv\ & d_1(\alpha_i,\beta_i,\tau) &&= \frac{\beta_i^2 K_i/2}{(\beta_i K_i)^2-(D_0\gamma-\alpha_i K_i)^2}, && (7.6\mathrm{c})\\
d_{2,i} \equiv\ & d_2(\alpha_i,\beta_i,\tau) &&= (D_0\gamma-\alpha_i K_i)/(\beta_i K_i), && (7.6\mathrm{d})
\end{aligned}$$

where $\Delta x_i = x - x_{0,i}$, with $x_{0,i}$ being is the enclosed minimum of V_{STAT} from Eq. (7.3). d_2 scales the "correlation" between Δx and Δx_τ as one can see from Eq. (7.5), and is the most important parameter [as we will see in Eq. (7.12) below]. Further, K is the variance of the respective one-time PDF $\rho_1(\Delta x)$, while ω scales how sensitive the variance K changes with τ. More details and the stability conditions of the solutions are given in Sect. 5.2.2.

With Eq. (7.5), the first member of the approximate FPE hierarchy accordingly reads

$$\gamma D_0\,\partial_{xx}\rho^{\mathrm{FLC}}_{1,\mathrm{ss}}(x) = \partial_x\left[\widehat{F}^{\mathrm{FLC}}(x)\,\rho^{\mathrm{FLC}}_{1,\mathrm{ss}}(x)\right], \tag{7.7a}$$

with

$$\widehat{F}^{\mathrm{FLC}}(x) = \int_\Omega F(x,x_\tau)\rho^{\mathrm{lin},i}_{\mathrm{c,ss}}(x_\tau|x;\tau)\mathrm{d}x_\tau,\ \forall x\in\Omega_i, \tag{7.7b}$$

where, importantly, the deterministic force F is *not* linearized. Equations (7.5)–(7.7b) [together with the linearization rule (7.4)] form a closed set of equations. $\rho^{\mathrm{FLC}}_{1,\mathrm{ss}}$ from (7.7) represents the approximate steady-state one-time PDF we aimed for. Thus, the force linearization applied to the second and all higher members yielded a closure of the FPE hierarchy. This is the main idea of our approach.

Since $\widehat{F}^{\mathrm{FLC}}$ can be calculated separately [from Eq. (7.5)], the FLC formally converts the delayed, not self-sufficient equation for $\rho_{1,\mathrm{ss}}$ itself into a closed, *quasi-*

[1] Recall that $\rho_2(x;x_\tau) = \rho_{\mathrm{c}}(x_\tau|x)\rho_1(x)$.

Markovian FPE [Eq. (7.7a)]. The fact that the delay and the non-Markovianity are still somewhat present, becomes apparent by considering the special case of a linear force F. Then, our approximate FPE (7.7) is *identical* to the exact delayed FPE. This is a crucial difference as compared to earlier approaches known from the literature (discussed below, see Sect. 7.2) effectively yield Markovian descriptions.

Furthermore, for the sake of consistency, one can easily check that the usual Markovian FPE is recovered from Eq. (7.7) in the absence of delay (where $\widehat{F}^{\mathrm{FLC}} \to F$).

7.1.3 Vanishing Steady-State Probability Current

The key equations of the FLC (7.5)–(7.7b) are valid for steady states of, in principle, arbitrary systems which can be meaningfully linearized according to the procedure described in Sect. 7.1.1. A particularly simple situation arises when the probability current given by $J = [\widehat{F}^{\mathrm{FLC}}(x) - \gamma D_0 \partial_x]\rho_{1,\mathrm{ss}}^{\mathrm{FLC}}(x)$ vanishes.[2] If $J = 0$, the formal solution of Eq. (7.7a) takes the simple Boltzmann-like form [compare with Eq. (3.13)]

$$\rho_{1,\mathrm{ss}}^{\mathrm{FLC}}(x) = Z^{-1} e^{-V_{\mathrm{eff}}^{\mathrm{FLC}}(x)/(k_{\mathrm{B}}\mathcal{T})} \tag{7.8}$$

with the *effective static potential*

$$V_{\mathrm{eff}}^{\mathrm{FLC}}(x) = -\int_{\hat{x}}^{x} \mathrm{d}x' \widehat{F}^{\mathrm{FLC}}(x'), \tag{7.9}$$

where $\widehat{F}^{\mathrm{FLC}}$ is given in Eq. (7.7b), and $\hat{x} \in \Omega$ is arbitrary, but fixed. Here and in the following, we denote the normalization constant by Z.

We stress that the assumption $J = 0$ simplifies the analysis, but it is not a necessary condition. Also for non-zero currents, Eq. (7.7a) can be treated using standard techniques for (Markovian) Fokker-Planck equations [3, 4].

7.1.4 Specification to Linear Delay Force

In some situations, the expressions further simplify severely. As a particularly important example in the present context, represent systems with vanishing probability current and linear delay force (2.45). The nonlinear models introduced in Sect. 2.5 fall into this class. Later, in Sects. 7.3 and 7.4, we will consider these nonlinear models as exemplary applications of the FLC.

The general expression for $\rho_{1,\mathrm{ss}}^{\mathrm{FLC}}$ from solution (7.8) is obtained as follows. The delay force is already linear and yields $\beta_i = -k, \forall i \in \mathcal{I}$, where β_i and k are the

---

[2] This is, for example, automatically the case, when a system obeys natural boundary conditions.

coefficients appearing in Eqs. (7.4) and (2.45), respectively. After performing several Gaussian integrals, Eq. (7.7b) yields

$$\widehat{F}^{\mathrm{FLC}} = -\partial_x V_{\mathrm{s}}(x) - k\Delta x_i + kxd_{2,i}\sqrt{2K_i d_{1,i}}\,\sqrt{1-d_{2,i}{}^2}\,e^{[d_{1,i}d_{2,i}{}^2 - d_{1,i} + 1/(2K_i)]\Delta x_i{}^2}. \tag{7.10}$$

$\widehat{F}^{\mathrm{FLC}}$ can be further simplified by using the identity

$$d_{1,i}d_{2,i}{}^2 - d_{1,i} + 1/(2K_i) = 0, \tag{7.11}$$

where the quantities $d_{1,i}$ and $d_{2,i}$ are given in Eqs. (7.6c) and (7.6d). In this way, we obtain the piecewise defined effective potential

$$V_{\mathrm{eff}}^{\mathrm{FLC}}(x) = V_{\mathrm{s}}(x) + k/2\big(1 - |d_{2,i}|\big)\Delta x_i{}^2, \tag{7.12}$$

for $x \in \Omega_i$. Together with (7.8), this yields the approximate PDF $\rho_{1,\mathrm{ss}}^{\mathrm{FLC}} = Z^{-1} e^{V_{\mathrm{eff}}^{\mathrm{FLC}}/(k_{\mathrm{B}}\mathcal{T})}$.

In the case of linear delay forces, the estimate for the steady state energy landscape further coincides with the static potential $V_{\mathrm{STAT}} = V_{\mathrm{s}}$ [since $F_{\mathrm{d}}(x, x) = 0$], such that $x_{0,i}$ and the bounds of Ω_i are readily determined by the extrema of V_{s}. Thus, for a given V_{s}, one only needs to calculate α^i by linearizing F_{s}, and therewith the coefficient $d_{2,i}$ [Eq. (7.6d)].

7.2 Comparison to Earlier Approaches

As we have mentioned in Sect. 3.2, alternative schemes to obtain approximate one-time PDFs are known from the literature, particularly the small delay expansion and the perturbation theory (see Sect. 3.2.1). Let us show the resulting approximations for the one-time PDFs in the present case. Later, we will compare them with the FLC.

7.2.1 Small Delay Expansion

The small delay approximation on the level of the LE [5, 6] is essentially a Taylor expansion up to linear order of the total deterministic forces in powers of τ. It renders a time-local LE that approximately describes the dynamics, where τ appears as a parameter. We recall that this approach is, by construction, appropriate for very short delay times only. At the same time, this is a regime where the dynamics is qualitatively very similar to a Markovian one. Thus, we expect that many effects induced by the memory may not captured by this approximation.

For the case of linear delay force $F_\mathrm{d} = k(x_\tau - x)$ (which we will consider in Sects. 7.3 and 7.4), the small delay expansion gives rise to a Boltzmann-like one-time PDF, analogously to the FLC (7.8), with

$$\rho_{1,\mathrm{ss}}^{\mathrm{approx}}(x) = Z^{-1} e^{-V_{\mathrm{eff}}^{\mathrm{approx}}(x)/(k_\mathrm{B}\mathcal{T})}, \tag{7.13}$$

and the effective potential $V_{\mathrm{eff}}^{\mathrm{approx}} \equiv V_{\mathrm{eff}}^{\mathrm{small}\tau}$ given by [7]

$$V_{\mathrm{eff}}^{\mathrm{small}\tau}(x) = V_\mathrm{s}(x)/(1 - k\tau). \tag{7.14}$$

Thus, the small delay expansion renders an effective potential which is proportional to the static potential V_s. This further implies that also the approximate PDF has the same functional form as in the case $F_\mathrm{d} \equiv 0$ (given if k or $\tau = 0$). We will come back to this observation in short, and consider it from the perspective of effective temperatures [see Sect. 7.2.3 below].

7.2.2 *Perturbation Theory*

Another approximation scheme introduced in earlier literature involves a first-order perturbation-theoretical ansatz for the density on level of the FPE. Here, the delay force F_d is regarded as a *small perturbation* to the non-delayed, Markovian dynamics, i.e., $|F_\mathrm{d}| \ll |F_\mathrm{s}|$. The resulting first-order equation has the same form as Eq. (7.2a), but the $\widehat{F}$-term [given in Eq. (7.2b)] contains the conditional probability density with respect to the unperturbed, i.e., non-delayed system with $F_\mathrm{d} \equiv 0$, which we will refer to as $\rho_c^{F_d\equiv 0}$. The latter follows a closed *Markovian* FPE, and consequentially, the combination of the resulting two FPEs is self-sufficient.

Thus, the FLC and the PT both render closed Fokker-Planck equations, where the conditional PDF $\rho_c(x_\tau|x)$ in the drift term is an approximate one. While in the FLC, it is approximated with the conditional PDF of the *linearized, delayed* system; the PT involves the one corresponding to the (nonlinear) *non-delayed* system, $\rho_\mathrm{c}^{F_d\equiv 0}$.

From a practical perspective, the PT is particularly appealing for systems where an analytical expression for $\rho_\mathrm{c}^{F_d\equiv 0}$ is available. However, such solutions are known, in fact, only for very few nonlinear static forces F_s. For this reason, the application of the PT approach usually requires additional approximations.

Short time propagator

As suggested in Ref. [8], one can combine the perturbation-theoretical ansatz with the *short time propagator* (s.t.)

$$\rho_\mathrm{c}(x_\tau|x;\tau) = \frac{1}{\sqrt{4\pi D_0 \tau}} \exp\left[-\frac{\{x_\tau - x - F_\mathrm{s}(x)\tau/\gamma\}^2}{4D_0\tau}\right], \tag{7.15}$$

which is derived and discussed in Ref. [3] on p. 73 and prior. The usage of the short time propagator implies a first-order approximation in τ [3, 7].

For the case of a linear delay force $F_\mathrm{d} = k(x_\tau - x)$, the resulting one-time PDF is a Boltzmann distribution, like Eqs. (7.8), (7.13), with the effective potential

$$V_\mathrm{eff}^\mathrm{PT}(x) = (1 + k\tau)V_\mathrm{s}(x). \tag{7.16}$$

Interestingly, when the short time propagator is used for $\rho_\mathrm{c,ss}^{F_d\equiv 0}$, the PT approach yields a qualitatively very similar expression as the small delay expansion [compare Eqs. (7.14) and (7.16)]. According to both approximations, the only effect of the delay force, is a constant (yet different) factor within the exponent of $\rho_\mathrm{1,ss}^\mathrm{approx}$. In fact, the small delay expansion and the PT with short time propagator even become equivalent for small τk, since then these factors coincide. This can be seen from a Taylor expansion around $\tau k = 0$, which yields $1/(1 - k\tau) = (1 + k\tau) + \mathcal{O}([\tau k]^2)$.

Ornstein-Uhlenbeck approximation Alternatively to the short-time propagator, we propose to approximate $\rho_\mathrm{c,ss}^{F_d\equiv 0}$ by the conditional PDF from the corresponding unperturbed ($F_d \equiv 0$), linearized system, i.e., the corresponding Ornstein-Uhlenbeck (OU) process, given in Ref. [3] (p. 100). We have introduced this idea in [1]. The resulting effective potential from the PT approach reads

$$V_\mathrm{eff}^\mathrm{PT}(x) = V_\mathrm{s}(x) + (k/2)\Big[1 - e^{-(\alpha^i - k)\tau/\gamma}\Big](\Delta x_i)^2,\ \forall x \in \Omega_i. \tag{7.17}$$

As opposed to the result obtained with the short-time propagator [Eq. (7.16)], the effective potential in Eq. (7.17) has a different functional form than V_s. More specifically, the delay force now effectively adds to the static potential a fixed quadratic potential around the minima of V_s. Since the linear delay force [Eq. (2.46)] indeed stems from a quadratic confining potential (with history-dependent center), this effective potential appears to be more realistic than Eq. (7.16). However, the history-dependency of the position of the "trap" imposed by F_d is, of course, not captured by this Markovian approximation.

As one can see from Eq. (7.17), this PT-OU approach also requires a multi-step procedure, where one first needs to define suitable subdomains, just like in the FLC. Further, one might intuitively expect this approach to be cruder than the FLC, as it amounts to neglecting the delay as well as the nonlinearity of the forces, in the higher members of the hierarchy. Indeed, this approach renders a poorer approximation compared to the FLC, as we will show in the following. Only in the regime of very small delay force magnitudes $|F_\mathrm{d}| \ll |F_\mathrm{s}|$, the results are comparable.

7.2.3 *Effective Temperatures*

At this point, we briefly resume the discussion from Sect. 5.2.3, about of the notion of effective temperatures. In Sect. 5.2.3, we have discussed purely linear delay equations only.

Among the expressions above, the physically relevant quantities are the PDFs. Our notation and wording indicate the standpoint that the one-time PDF describes the diffusion of a particle in an "effective potential" at the (original) temperature $\mathcal{T}$. Thus, the feedback force alters the energy landscape but not the temperature. However, considering the results of the small delay expansion or the PT with short-time propagator [given in Eqs. (7.13), (7.14, 7.16)], the Boltzmann-distributed PDFs could just as well be interpreted as describing a particle in the (original) potential V_s at the *effective temperature* $\mathcal{T}(1 - k\tau)$, or $\mathcal{T}/(1 + k\tau)$, respectively. This offers a convenient view on the problem, and the concept of effective temperatures is indeed quite common in other contexts. According to both approximations, the effective temperature can either be reduced or increased due to the delay force, depending on the sign of k. We note that the effective temperature can also become negative for large magnitudes of $|k\tau|$, where the approximations break down.

However, this viewpoint loses meaning for the other approaches, where the effective potentials have a different functional form than V_s. This includes the FLC [see Eq. (7.12)] and the other version of the PT, with Ornstein-Uhlenbeck approximation (7.17). The fact that these other approaches yield indeed more accurate PDFs (as we show below), is an argument against the notion of an effective temperature in the present context.

Furthermore, as we have shown in Sect. 5.2.3, the concept of effective temperature is also not appropriate when one considers higher-time PDFs. In Sect. 5.2.3, we have explicitly shown this for the two-time PDF of the linear system.

7.3 Application to the Periodic Potential

We apply the different approximations to the periodic washboard potential (7.18) introduced in Sect. 2.5.2, which is given by a cosine function

$$V_s(x) = -(V_0/2)\, \cos(x/\sigma), \tag{7.18}$$

with barriers of height V_0. We further consider a linear delay force (2.45), i.e.,

$$F_d(x, x_\tau) = -k\,(x - x_\tau).$$

If $k \geq 0$, the force is always directed towards the delayed position counteracting a deviation between the instantaneous and delayed particle position, i.e., negative feedback, and can be associated with the delayed confining potential $V_d(x, x_\tau) =$

$(k/2)(x - x_\tau)^2$. We recall that such quadratic feedback potentials are commonly used to model optical traps, see Sect. 2.4.1. Thus, we will focus here on $k > 0$ meaning nonnegative feedback strengths.[3]

Approximations The approximations from the small delay expansion and the PT are readily given by the Boltzmann-distributions with effective potentials (7.14) and (7.16, 7.17), respectively. To apply the FLC, a more elaborate procedure is required. However, due to the linearity of the feedback (2.45), we can now make use of the results from Sect. 7.1.4, where we have already carried out the main steps.

The only remaining ingredient is splitting the domain by the procedure described in 7.1.1. As $F_\mathrm{d}(x, x) = 0\,\forall x \in \Omega$, V_STAT is identical to V_s. Hence, $x_{0,i}$ and the interval bounds of Ω_i are given by the extrema of V_s. Thus, we only need the minima of the washboard potential, which are located at $x_{0,i} = 2\pi\sigma i$, $i \in \mathbb{Z}$. We split the domain $\mathbb{R}$ in subdomains $\Omega_i = [(2i - 1)\pi\sigma, (2i + 1)\pi\sigma]$, such that each interval contains one minima $x_{0,i}$, with $i \in \mathcal{I} = \mathbb{Z}$. Now, linearizing the static potential yields $\alpha^i = V_0/(2\sigma^2) + k$ (and $\beta^i = -k$) for all $i \in \mathcal{I}$. Thus, on all subdomains Ω_i, the density is given by the Boltzmann distribution (7.8) with effective potential (7.12), where

$$d_{2,i} = \frac{\gamma\omega \cosh(\tau\omega) - [V_0/(2\sigma^2) + k] \sinh(\tau\omega)}{\gamma\omega - k \sinh(\tau\omega)} \tag{7.19}$$

with $\gamma\omega\sigma = \sqrt{(V_0/2\sigma)^2 + kV_0}$ and $\Delta x_i = x - x_{0,i}$.

7.3.1 *Discussion of Results*

Figure 7.2 (left panel) displays the steady-state one-time PDF, which is a periodic function in this case. In particular, it shows one peak of the PDF sitting in one potential well. To be more precise, we here consider the "reduced PDF", i.e., the normalization is one within each well, see Sect. 2.5.2. Besides the different approximation results, also data generated by BD simulations of the delayed LE (2.43) are shown. One clearly sees that the FLC renders the best approximation of the PDF. Similar convincing results are found for all other considered values of τ and k in the tested range $\tau/\tau_\mathrm{B} \in [10^{-2}, 10]$ and $k/(\sigma^{-2}k_\mathrm{B}\mathcal{T}) \in [0, 32]$. To compare approximation and simulation results more systematically, we calculate the moments μ_n of the (reduced) PDFs within one potential well (for x values between two maxima of V_s). Due to the spatial inversion symmetry of the total potential around the enclosed minimum, all odd moments automatically vanish. We hence consider the second central moment $\mu_2 = \langle (X - \mu_1)^2 \rangle_\mathrm{ss}$ (with $\mu_1 = \langle X \rangle_\mathrm{ss}$ and $X/\sigma \in [-\pi, \pi]$). Figure 7.2 (right panel) shows μ_2 as a function of the delay time τ for two exemplary values of delay force strength k. We find that in the presence of the delay force, the variance of the distribution is reduced. Thus, the thermal fluctuations are suppressed by the delay

[3] In later Chapters, we will also consider negative values of k.

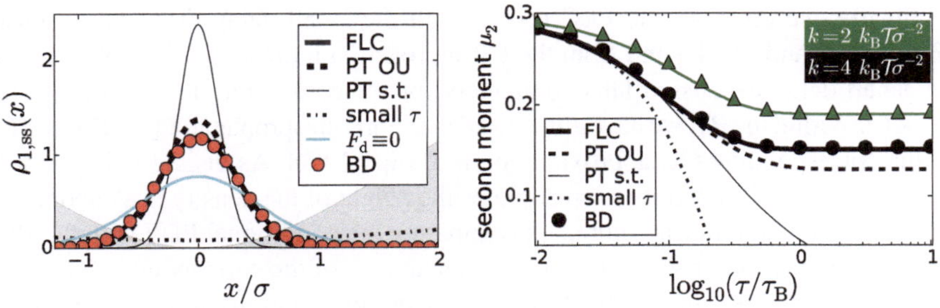

Fig. 7.2 Steady-state PDF $\rho_{1,\mathrm{ss}}$ within one well of the periodic washboard potential with linear delay force from several approximations. Symbols: numerical data (BD), thick solid lines: force-linearization closure (FLC), dashed lines: perturbation theory with Ornstein-Uhlenbeck approximation (PT OU), thin lines: PT with short time propagator (PT s.t.), dashed-dotted line: small delay expansion (small τ). *Left panel:* The PDF itself, at delay force strength $k=8\,k_\mathrm{B}\mathcal{T}/\sigma^2$ and delay time $\tau=\tau_\mathrm{B}$. The light blue line additionally shows the PDF $\rho_{1,\mathrm{ss}}(x)\propto\exp[-V_\mathrm{s}(x)/(k_\mathrm{B}\mathcal{T})]$ for the case without delay force ($F_\mathrm{d}\equiv0$). *Right panel:* The second moment μ_2 of the PDF. The green color refers to $k=2\,k_\mathrm{B}\mathcal{T}/\sigma^2$, while the black color refer to $k=4\,k_\mathrm{B}\mathcal{T}/\sigma^2$. The lines for PT and small τ are only displayed for the second parameter case for sake of clarity. The barrier height is set to $V_0=8\,k_\mathrm{B}\mathcal{T}$. These results are also published in [1]

force. This is not surprising, since the delay force arises from a quadratic potential which "traps" the particle. Within the considered parameter regime, the variance is seen to decrease with increasing τ, until at about $\tau_\mathrm{sat}^\mathrm{W}\approx0.5\,\tau_\mathrm{B}$, where the variance appears to saturate. (Due to strongly increasing simulation times, we have, however, not investigated the behavior for much larger τ than shown here, but we will discuss our expectations below.)

On the level of the PDFs and the FPE, the saturation of μ_2 at large τ indicates that the delay-averaged drift $\widehat{F}$, and hence $\rho_{\mathrm{c,ss}}(x_\tau|x;\tau)$, are essentially constant for $\tau>\tau_\mathrm{sat}^\mathrm{W}$, in the considered range of delay times. We recall that we have already observed such a saturation of the PDFs for the linear delayed system, see Fig. 5.4 and below. The fact that the conditional probability does not change when the time difference τ is further increased, suggests that a relaxation mechanism within the valley takes place, where the history of the stochastic process becomes "washed out" on the level of ensemble averaged quantities. In this context it is interesting to note that, for the case without delay force ($k=0$), the relaxation time within a potential well can roughly be estimated to be around $\tau_\mathrm{ir}\approx V_\mathrm{s}/\gamma=0.25\,\tau_\mathrm{B}$, see Sect. 2.6. The saturation of μ_2, thus, sets in *after* the density relaxation within a potential well. This provides a "hand having" argument to understand the observed saturation.

We, however, expect a different behavior when τ has an even much larger value (than in Fig. 7.2); in particular, when τ gets into the range of the mean escape times. Since in this parameter regime, the particle has a high probability to escape the well, transport between the potential valleys becomes important. We thus expect again a τ-dependency of $\rho_{\mathrm{c,ss}}$ and therewith of μ_2. We will discuss the escape processes later in Sect. 7.5. This behavior is not captured by the FLC which yields a saturated PDF for all $\tau\to\infty$.

Finally, we compare the FLC results with the small delay (Taylor) expansion [Eqs. (7.13) and (7.14)] and from the PT approach in more detail. It is obvious that the small delay expansion fails completely for the considered delay times (unless $\tau \to 0$). Within the PT, we have either used the short time propagator [see Eq. (7.16)], or the corresponding OU approximation [see Eq. (7.17)]. As visible in both panels of Fig. 7.2, the PT generally overestimates the height of the density peak and underestimates its width. Furthermore, approximating the conditional PDF with the short time propagator yields worse results than using $\rho_{c,ss}$ of the corresponding Ornstein-Uhlenbeck process. This observation matches our expectations, as the latter scheme seems to be the more sophisticated and accurate one. When the OU approximation is used, the PT at least reproduces the quantitative behavior of the function $\mu_2(\tau)$, see Fig. 7.2 (right panel).

Markovian limits If $\tau \to 0$ or $k \to 0$, the delay force vanishes and the system is Markovian. Thus, one expects the system to reach thermal equilibrium where the one-time PDF is a Boltzmann-distribution $\rho_1 \propto e^{-V_s/(k_B \mathcal{T})}$ (depicted as light blue line in the left panel of Fig. 7.2). Indeed, all approximations considered here correctly describe the limits, as can be seen from Fig. 7.2, right panel [and as is also evident from Eqs. (7.12), (7.14, 7.16)].

Conclusion Among the tested approaches, the FLC generates the best approximation of the steady-state one-time PDF in the periodic potential, especially in the regime of large τ or large k. On the one hand, the small delay expansion involves a truncated Taylor expansion in τ, making it plausible that this approach fails for large delay times. On the other hand, in this regime the "perturbation", i.e., $|F_d|$ is not small compared to the force applied by the static potential, such that the basic assumption of the perturbation theory is not fulfilled. It is noteworthy that the PT-OU nevertheless yields quite reasonable results. The FLC is a reliable approximation for the intrawell dynamics even for large τk. This is because it renders an approximate, but still *non-Markovian*, description. We will discuss below, under which conditions the FLC breaks down.

7.4 Application to the Bistable Potential

As a second example, we consider the doublewell potential (2.54) introduced in Sect. 2.5, namely

$$V_s(x) = V_0\,[(x/\sigma)^4 - 2\,(x/\sigma)^2],$$

with a barrier of height V_0 at $x/\sigma = 0$, and two minima at $x_{0,i} = i\sigma$ with $i \in \mathcal{I} = \{-1, 1\}$. We further reconsider the linear delay force (2.45). The application of the different approximations works analogously to the washboard potential described in the last section.

More specifically, for the FLC, we again have to employ the procedure described in Sect. 7.1.1, yielding $\alpha^i = 8\,V_0/\sigma^2 + k$ for both minima, and $\beta^i = -k$. For $x/\sigma \in$

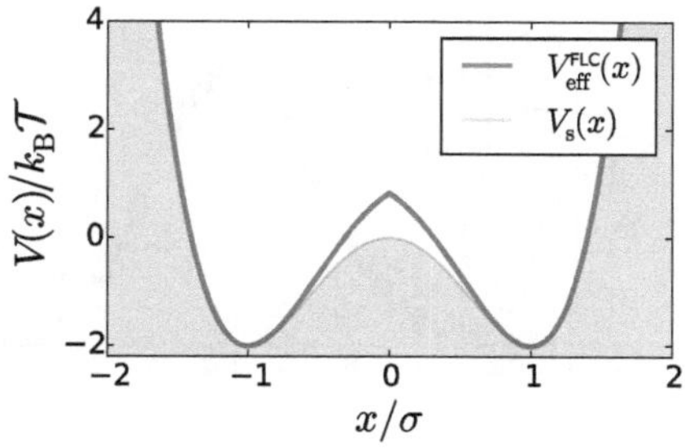

Fig. 7.3 *Gray shaded:* Bistable potential V_s (2.54), *magenta line:* the corresponding effective potential $V_{\rm eff}^{\rm FLC}$ from the FLC given in Eq. (7.12). The parameters are $V_0 = 8k_BT$, $k = 8\,k_BT/\sigma^2$ and $\tau = \tau_B$

$[-\infty, 0]$ and $x_{0,i=-1} = -\sigma$, or $x/\sigma \in [0, \infty]$ and $x_{0,i=1} = \sigma$, the effective potential $V_{\rm eff}^{\rm FLC}$ is given by Eq. (7.12) with

$$d_{2,i} = \frac{\gamma\omega\cosh(\tau\omega) - [(8V_0/\sigma^2) + k]\sinh(\tau\omega)}{\gamma\omega - k\sinh(\tau\omega)} \tag{7.20}$$

and $\gamma\omega\sigma = \sqrt{(8V_0/\sigma)^2 + kV_0}$. The resulting effective potential is displayed in Fig. 7.3.

7.4.1 Discussion of Results

As demonstrated in Fig. 7.4, the FLC again yields a very good approximation of the PDF, which is a bimodal distribution in this case. Contrary to Fig. 7.2, we consider for this example the third moment (because the second moment is here found to be barely affected by the delay force). More specifically, we consider the skew, i.e., the third central moment normalized with the standard deviation, of one peak of the bimodal PDF, which is defined as $\mu_3^{x>0} = \int_{x>0}[(x-\mu_1)/\sqrt{\mu_2}]^3 \rho_{1,\rm ss}(x)\mathrm{d}x$. The comparison with the numerical data in Fig. 7.4 reveals that the FLC approach renders reasonably good predictions of $\mu_3^{x>0}$. We moreover find convincing results for all other tested parameters within the studied ranges of $\tau/\tau_B \in [10^{-2}, 5]$ and $k/(\sigma^{-2}k_BT) \in [0, 18]$.[4]

However, the FLC is not as accurate as it is in the case of the periodic potential (compare with Fig. 7.2). One reason might be the fact that the linearized potential "seen" by the higher-order PDFs is, by construction, symmetric with respect to the enclosed minimum (within each subdomain). In this sense, the doublewell potential, which is asymmetric around each minimum, is not as well approximated as the symmetric washboard potential. This manifests in a larger magnitude of the higher order Taylor terms which are neglected within the linearization procedure. We also

---

[4] This range spans from rather small delay forces that merely affect the shape of the energy landscape, to k values so high that the total potential has no second minimum due to the superimposed delayed potential. For all these qualitatively different cases, the agreement between FLC and numerical results is convincing.

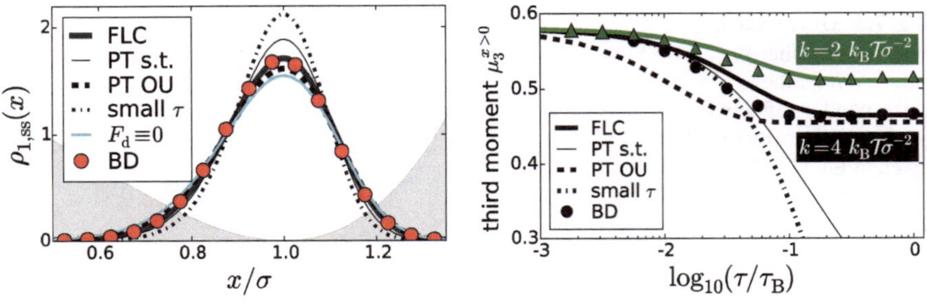

Fig. 7.4 *Left:* One peak of the bimodal PDF $\rho_{1,ss}$ in the doublewell potential. The delay force amplitude and delay time are set to $k = 18\,k_B\mathcal{T}/\sigma^2$ and $\tau = 0.1\,\tau_B$, respectively. *Right:* Third moment $\mu_3^{x>0}$ of the PDF within the right well of the doublewell potential, i.e., of $\rho_{1,ss}(x > 0)$. The color code and parameters are as those in Fig. 7.2. These results are also published in [1]

note that the third moment is *per se* very sensitive against inaccuracies, since it involves cubic terms in the relative position. The deviations between the FLC and the exact result are hence expected to be larger.

Regarding the impact of increasing k and τ, we see from Fig. 7.4 (right panel) that the skew $\mu_3^{x>0}$ becomes smaller, which means that the probability distribution becomes more symmetric. This is not surprising since the pure delay potential is quadratic (i.e., symmetric) in the system state variable. Moreover, similar to the second moment in the washboard potential (see Fig. 7.2), $\mu_3^{x>0}$ approaches a saturation value within the considered parameter range. The corresponding delay time is about one order of magnitude smaller than in the washboard case: $\tau_{sat} \approx 0.1\tau_B$. This value of τ_{sat} is, as in the case of the washboard, significantly larger than that for the intrawell relaxation time τ_{ir}, which can be roughly estimated to be around $\tau_{ir} \approx 1/64\tau_B$ for the present case (see Sect. 2.6). Thus, the saturation of μ_3 can be explained by the same arguments as in the case of the washboard potential (Sect. 7.3).

Figure 7.4 also show the results according to the PT [Eqs. (7.16) and (7.17)], and from small delay expansion, see Eq. (7.14). Very similar to the case of the washboard, only the PT-OU is capable of reproducing the quantitative behavior (especially for large τ), and the FLC clearly provides the best approximation of the one-time PDF.

Conclusion The FLC yields reasonable predictions for the one-time PDFs in quite large parameter regimes, and in both considered examples. Also the PT-OU approach is quite successful. In contrast, the PT with short time propagator is only appropriate for very small values of the delay time, where the system indeed behaves more or less like a Markovian one. The same thing can be said about the small delay expansion.

At the same time, the FLC and the PT-OU are more complicated to apply. They are not simply given by a single equation, but require a procedure with multiple steps (determination of V_{STAT}, finding the subdomains, then determining the parameters of the estimate). Thus, the more accurate approximations come at the "cost" of the more involved formalism. Among these two, the FLC is even superior as compared to the PT-OU, clearly yielding the best results.

A disadvantage of the FLC is, however, that it does not give an explicit estimate of the involved total error in the PDF. As opposed to this, the PT, for example, is a *first order* estimation in the "perturbation". Such a statement is not available for the FLC. But, due to the required secondary approximation in the PT approach, i.e., the Ornstein-Uhlenbeck approximation or the short-time propagator, this "error-estimation", is indeed also not readily available for the PT.

We finally add a comment concerning the limitations of these statements. First, we observe a bad performance of the FPE-based approaches, i.e., the PT and FLC, if k is negative, corresponding to a delay force that pushes the particle away from its delayed position (i.e., positive feedback). Interestingly, then, there even exists a parameter regime, where the small delay expansion outperforms the other approaches (for small delay times and negative k). An example will be given in the next Chapter (see Figs. 8.5, 8.6). Second, all of these approaches break down in parameter regimes, where the particle is very likely to escape its potential valley and delay induces interwell oscillations. The effectively Markovian descriptions given by the small delay expansion and the PT clearly fail as the delay force and delay time have large amplitudes; but also the FLC is expected to fail, because in these regimes the interplay of nonlinearities and memory matters. We will turn to the interwell jump dynamics in the next Section.

Thermodynamic application As we will explicitly show in Chap. 9, these approximations put as in the position to analytically calculate thermodynamic quantities in systems with discrete delay.

7.5 Estimation of Escape Times

7.5.1 Interwell Dynamics

So far, we have focused on the intrawell dynamics, i.e., the fluctuations of the particle position X within one potential valley. However, an important feature of the studied systems is that the particle can also overcome the potential barriers and escape its valley, yielding *interwell dynamics* (see Fig. 7.5). The energy for such a jump is provided by the random thermal fluctuations. Due to the Gaussianity of the noise, an escape is, in fact, expected to happen eventually, no matter how high the barriers are compared to the thermal energy $k_B\mathcal{T}$ (which scales with the temperature $\mathcal{T}$). However, the expected *time* that passes until the jump occurs will, of course, depend on this very ratio $k_B\mathcal{T}/V_0$. Moreover, in the non-Markovian systems considered here (i.e., when $k\tau \neq 0$), the energy landscape is actually history-dependent, which may induce more intriguing jump dynamics than the one found in a Markovian system and dependency of the waiting times upon k and τ.

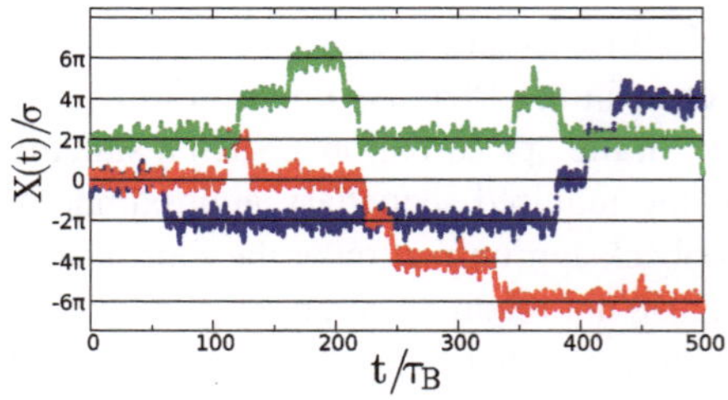

Fig. 7.5 Exemplary random trajectories in the washboard potential with $V_0 = 4\,k_B\mathcal{T}$ in the absence of delay force ($k = 0$), from BD simulations. The horizontal lines are at the positions of the potential minima. One can see from these trajectories that the dynamics is characterized, on the one hand, by intrawell fluctuations around the (current) potential minimum and, on the other hand, by random (interwell) jumps between the potential wells

Markovian case For the Markovian case ($k\tau = 0$), the total time between two jumps Δt_{jump}, is described by the famous Kramers theory [9, 10], which we have briefly discussed in Sect. 2.6.1. This theory[5] predicts exponentially decaying waiting time probabilities, with a mean escape time given by Arrhenius formula [given by Eq. (2.60)]. We indeed find such an exponential decay for the considered systems, as displayed for an exemplary parameter setting in Fig. 7.6 (blue histograms).

Non-Markovian case Fig. 7.6 further shows exponentially decaying distributions in non-Markovian cases where the delay force is present. Thus, the functional form is here identical to the Markovian case. However, the delay force yields a slower decay, corresponding to the particle staying longer in one well. As expected, this effect is enhanced by increasing k, i.e., by stiffening the feedback trap. The fact that the delay force does not change the functional form of the distributions motivates us to investigate the question: "*Is it possible to estimate the mean escape times in the non-Markovian models by similar means like in the Markovian case?*" To tackle this problem, we will utilize the FLC approach, which brings us into the position to apply the Kramers theory.[6] The results of this idea are presented in the next paragraph.

As we will later show, the similarity of the waiting time distributions between Markovian and non-Markovian case is limited to the regime of short delay times. At the end of this Chapter, we will discuss situations, where the non-Markovian effects yield qualitatively different jump dynamics. In this regime, the Kramers-FLC approach must break down, since one of the main assumptions of Kramers theory is

[5] Which involves several approximations, e.g., that thermal energy is low compared to barrier height, such that the jumps occur rather seldom.

[6] By construction, escape times involve two or more locations, and multiple instants in time, and are therefore naturally connected to PDFs ρ_n with $n > 1$. Within the FLC, the higher order PDFs $\rho^{\mathrm{FLC}}_{(n>2),\mathrm{ss}}$ are multivariate Gaussian densities belonging to quadratic static potentials which have only one valley and, thus, no escape process. Therefore, there is no direct way to calculate the mean escape time on the basis of the FLC approach alone without Kramers theory.

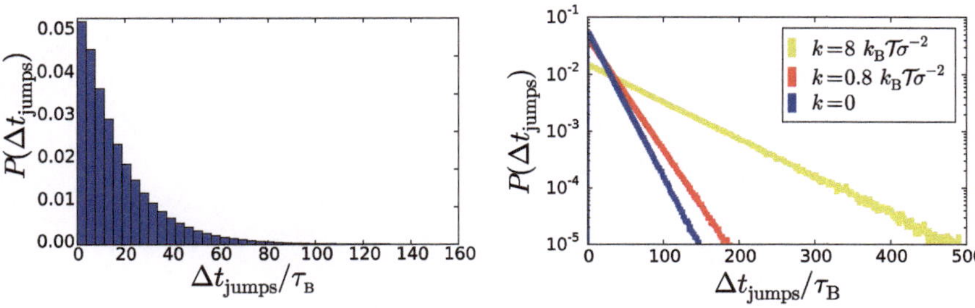

Fig. 7.6 Waiting time distribution in the doublewell potential, with ($k > 0$) and without ($k = 0$) linear delay force, for a short time delay $\tau = 0.1\tau_B$. The distributions are numerically (BD) generated (normalized) histograms of the times between sequential jumps Δt_{jumps}. *Left panel:* k=0. *Right panel:* logarithmic plots with $k = 0$ shown again as blue histogram (we here only display the envelope for sake of clarity), red is the histogram for $k = 0.8k_B\mathcal{T}/\sigma^2$, yellow is $k = 8k_B\mathcal{T}/\sigma^2$. All distributions are found to decay exponentially. The barrier height is set to $V_0 = 4\,k_B\mathcal{T}$

violated, i.e., that the jumps should be rather seldom. More importantly in the present context, here also the FLC is expected to fail as we have mentioned before. This is because the dynamics is dominated by the interplay of memory and the nonlinearities, which is neglected in this approximation.

7.5.2 The Kramers-FLC Estimate

The escape time τ_K denotes the average time X spends in the vicinity of a potential minimum, until it leaves the valley by jumping to an adjacent one. We recall that the Kramers theory, which we have discussed in Sect. 2.6.1, was developed for Markovian systems characterized by a *static* potential landscape U, gives an approximation for τ_K. When the potential barriers ΔU are large compared to the thermal energy, i.e., $\gamma D_0 \ll \Delta U$, the Kramers theory provides an estimation for the escape rate r_K by the Arrhenius formula given in Eq. (2.60); recall $r_K = \sqrt{U''(x_{\min})|U''(x_{\max})|}/(2\pi\gamma)e^{-\Delta U/\gamma D_0}$. This estimate for r_K relies on a quasi-static approximation for ρ_1[7] For the washboard potential, the resulting mean escape time can be approximated by $2\tau_K = 1/r_K$ due to the two symmetric ways to leave a valley (see Ref. [10]). For the doublewell potential with the unique direction to exit each valley, $\tau_K = 1/r_K$.

Main idea For non-Markovian systems, the energy landscape is history-dependent, and there is no direct route to estimate the mean escape times by Kramers theory. However, in the FLC approach, the first member of the FPE hierarchy Eq. (7.7a) is formally identical to a Markovian (non-delayed) FPE for $\rho_{1,ss}$ with a static effective

[7] And does not involve ρ_2, or any higher order correlation between different times and spaces.

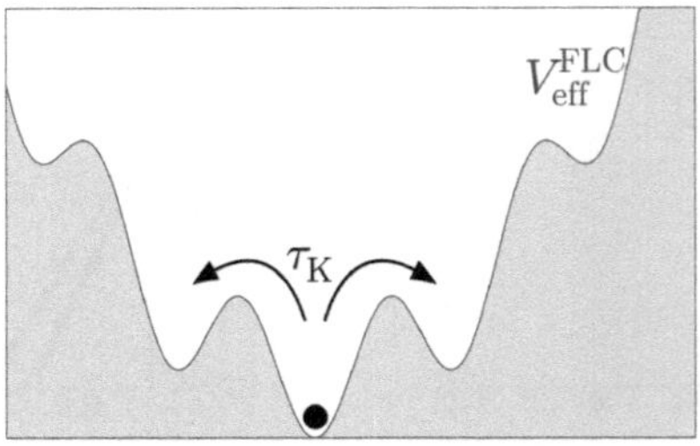

Fig. 7.7 Schematic illustration of the idea behind the Kramers-FLC estimate of the mean escape times τ_K in the washboard potential with linear delay force. Here, we pretend to particle to move in static (effective) potential, while it actually moves within a history-dependent energy landscape

potential $V_{\text{eff}}^{\text{FLC}}$ given by Eq. (7.9). Thus, we can in fact directly apply the Arrhenius formula (2.60) with $U \equiv V_{\text{eff}}^{\text{FLC}}$. For the linear delay force in our examples, one just needs to substitute $\Delta U = V_0 + (k/2)(x_{\text{max}} - x_{\text{min}})$ and $U''(x_{\text{ex}}) = V_s''(x_{\text{ex}}) + k(1 - |d_{2,i}|)$. Taken altogether, our estimate of τ_K involves the following approximations: first, the FLC to obtain $V_{\text{eff}}^{\text{FLC}}$, second, the fact that we apply the Arrhenius formula to a non-Markovian system, third, all underlying simplifications made within the Kramers theory. The main idea behind this approach is illustrated in Fig. 7.7.

We note that one could employ the same strategy to get approximated escape times from the perturbation theory. We have indeed also tested this idea for the PT-OU, obtaining results very comparable to those from the FLC presented in the following.

7.5.3 Comparison with Numerical Results

We test the Kramers-FLC approximation by applying it to the example systems, comparing the predicted escape times to numerically obtained ones from BD simulation of the full model (2.43).

Periodic system Figure 7.8 displays exemplary results for τ_K in the (delayed) washboard potential. For $F_d \equiv 0$ (which corresponds to $\tau k = 0$), the numerical data and the Kramers-FLC estimate generally coincide. For $\tau k > 0$, the mean escape time τ_K generally increases by increasing k or τ. As one can see from Fig. 7.8, the FLC estimates are roughly in agreement with the actual escape rates. This is rather remarkable, due to the crudeness of applying the Kramers theory to this non-Markovian system. However, at large values of τ, they have a qualitatively different behavior: while the escape times resulting from the Kramers-FLC estimate saturate, the BD results continue to increase with τ. This difference becomes particularly prominent for large k, i.e., large delay force strengths.

We think that the saturation of the approximate results (and the discrepancy) can be explained as follows. As discussed in Sects. 7.3, 7.4, the conditional probability saturates at a finite value τ, which is related to the intrawell relaxation time τ_{ir}.

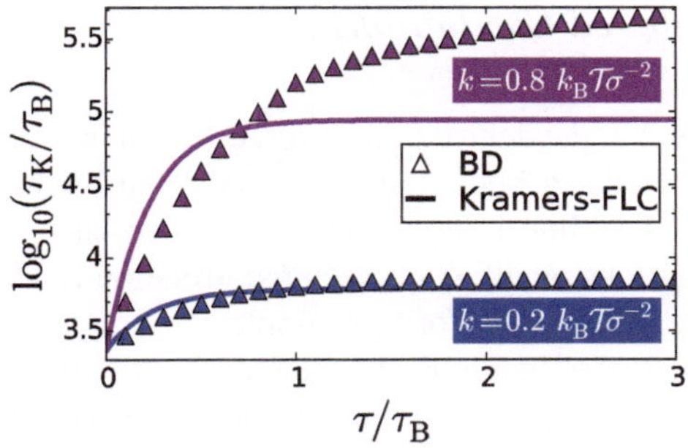

Fig. 7.8 Mean escape times τ_K in the washboard potential with $V_0 = 8\,k_B\mathcal{T}$, as a function of the delay time τ (in units of the Brownian timescale τ_B), in a logarithmic plot. Blue: delay force strength $k=0.2\,k_B\mathcal{T}/\sigma^2$, Purple: $k=0.8\,k_B\mathcal{T}/\sigma^2$, symbols: (BD) simulation results, lines: Kramers-FLC estimate as described in Sect. 7.5.2. These results are also published in [1]

Accordingly, also $V_{\text{eff}}^{\text{FLC}}$, and therewith τ_K from the Kramers-FLC approximation, saturate at this delay time. However, in the true non-Markovian system, the escape times do not only depend on $\rho_{1,\text{ss}}$, but also on higher n-time probabilities. Therefore, they do not necessarily saturate together with $\rho_{1,\text{ss}}$. Indeed, our numerical investigations reveal that τ_K does not saturate at all. Moreover, for larger τ, another, not yet taken into account timescale becomes increasingly important that is the jump duration time (see Sect. 2.6). If the delay time τ is in the range of the jump duration times, the temporal changes of the energy landscape and the dynamics of the system state variable X occur on similar timescales. The interplay between both motions then leads to new (interwell) dynamical behavior, which is not captured by the approximations. One example for such new dynamical behavior are the quasi-regular *delay-induced oscillations* of X between two valleys, which we discussed in Sect. 2.7 and reconsider below in Sect. 7.5.4. It hence becomes less justified to treat the delayed system as a quasi-Markovian one, whose stochastic (non-delayed) dynamics evolves in a static (effective) potential. Therefore, neither the FLC, PT, small τ, nor the application of Kramers theory are appropriate.

Bistable system Finally, we briefly look at the espace times in the doublewell potential. We find that the FLC estimated escape times again reach a saturation value (for the same reason as in the washboard potential explained above). However, here the BD results show a peak, and *decrease* for larger τ. For $k = 2k_B\mathcal{T}/\sigma^2$, the peak is at about $\tau \sim 0.2\tau_B$, for $k = 4k_B\mathcal{T}/\sigma^2$ it is at $\tau \sim 0.3\tau_B$, while the peak is around $\tau \sim 0.5\tau_B$ for $k = 8k_B\mathcal{T}/\sigma^2$. Interestingly, these delay times correspond quite accurately to the mean *jump duration times* within the respective potentials. We have determined the latter numerically, see Sect. 2.6 for more details. We find that in the regime of the peak delay-induced spontaneous oscillations occur.

7.5.4 Delay-Induced Oscillations

As discussed in Sect. 2.7, the delayed linear force F_d can induce more or less regular jumps between the wells in both static potentials, the washboard as well as the doublewell, if the delay time is large enough. As we have mentioned there, the delay-induced oscillations manifest in the autocorrelation function $C(\Delta t)$, yielding hight peaks at $C(\tau)$, see Fig. 7.11 for an example.

Here we briefly consider their impact on the waiting time distributions. Let us first think about the expected effect on the mean value of the distributions. In the present case, the delay-induced oscillations have a mean period of about τ (in both static potentials) as $k > 0$. Because the mean escape times are usually much larger than τ, the occurrence of oscillations should reduce τ_K significantly. This is indeed visible in Fig. 7.9. Further, in addition to changing the mean value, these jumps manifest in a significantly altered functional form of the waiting time distributions (compare Figs. 7.10 and 7.6). In the presence of the oscillations, most waiting times lie within an interval $\Delta t_{\text{jump}} \in [0, \tau]$ around a single, yet broad peak at about $\tau/2$. The width of the peak illustrates the stochastic character of the oscillations, which start at random times, occasionally pause and randomly set in again.

With the frequent occurrence of delay-induced oscillations, the dynamics of the model deviates so strongly from a Markovian one that it is no surprise that the Kramers-FLC estimate completely breaks down.

One can understand the breakdown of all approaches so far from a very general perspective, when one thinks about the probability densities. In the parameter regime where jumps are very likely, the "memory of a jump", which is expressed in the non-Markovian ρ_2 (and higher PDFs), is what dominates the process. For example, it can trigger subsequent jumps, thus, oscillations. Since the higher PDFs correspond to the ones of Markovian systems in all approaches, the essence of the dynamics is not captured.

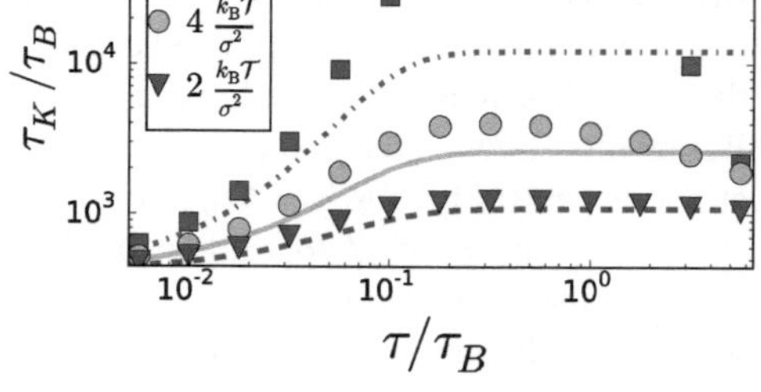

Fig. 7.9 Escape times τ_K versus delay time τ in the doublewell potential with barrier height $V_0 = 8k_B\mathcal{T}$. Blue: $k = 2k_B\mathcal{T}/\sigma^2$, yellow: $k = 4k_B\mathcal{T}/\sigma^2$, magenta: $k = 8k_B\mathcal{T}/\sigma^2$. Lines plot the FLC-Kramers estimates, symbols show the Brownian Dynamics simulations result

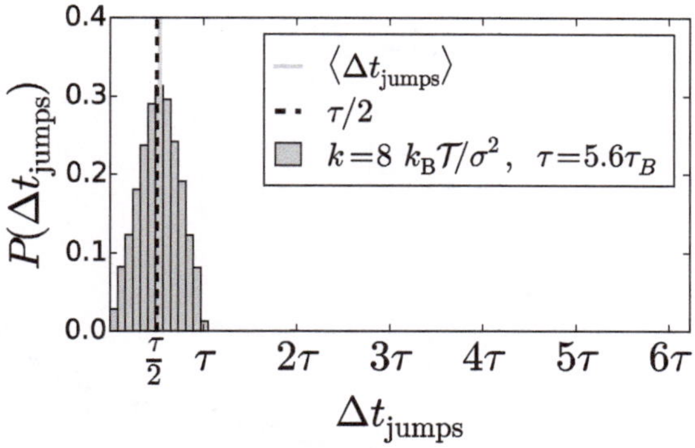

Fig. 7.10 Waiting time distribution in the delayed doublewell potential with $k = 8\, k_B \mathcal{T}/\sigma^2$ and $\tau = 5.6\tau_B$. The vertical purple line marks the mean waiting time $\langle \Delta t_{\text{jumps}} \rangle$, while the dashed black line marks $\tau/2$. The barrier height is set to $V_0 = 8\, k_B \mathcal{T}$. Due to the delay-induced jumps $P(\Delta_{\text{jump}})$ is strongly altered; compare to Fig. 7.6

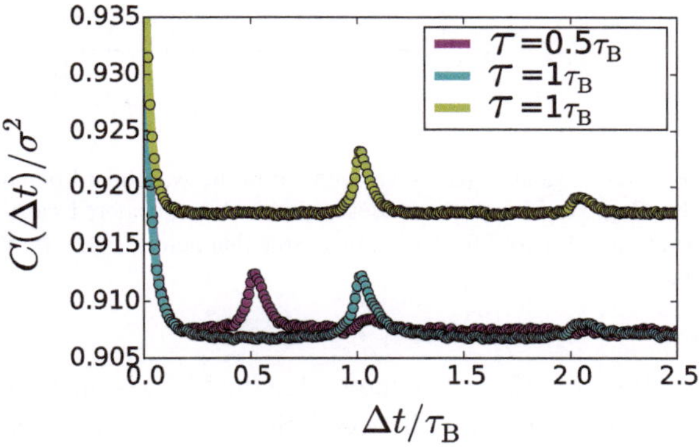

Fig. 7.11 Position-autocorrelation function in the doublewell with delay force. Cyan and magenta lines and symbols: $k = 8k_B\mathcal{T}/\sigma^2$, yellow: $k = 12k_B\mathcal{T}/\sigma^2$. The delay time is $\tau = \tau_B$ (yellow and cyan), or $\tau = 0.5\tau_B$ (magenta)

7.5.5 Side Note: Normal Diffusion Despite Non-Markovianity

As we have mentioned in Chap. 2, a common way to classify diffusion processes is via the mean square displacement $\langle [X(t) - X(0)]^2 \rangle$. In particular, when in the long-time limit, the latter grows linearly with time, the process is denoted normal diffusion, while nonlinear behavior is called anomalous diffusion. Let us briefly consider, in this regard, the washboard potential with delay force. Please note that the reason for not also considering the bistable potential is that the latter is confining.

In the presence of time delay, it is not immediately clear, which behavior to expect. However, the finite memory time (the memory kernel does not reach arbitrarily long into the past, but decays exponentially fast), one might expect normal diffusion at very large times. Indeed, Fig. 7.12 shows preliminary numerical results for a particle within the washboard potential with linear delay force, indicating a linear growth of the MSD at large times. Note that this behavior seems to be irrespective of the occurring oscillations, as we have considered values of τ, where the delay-induced interwell oscillations are very prominent, e.g. $\tau = 4\tau_B$ shown by a green curve in Fig. 7.12, and values where they are not observed, e.g. $\tau = 2\tau_B$, blue curve. In the present case, the delay-induced oscillations are found to occur with high probability for about $\tau > 3\tau_B$.

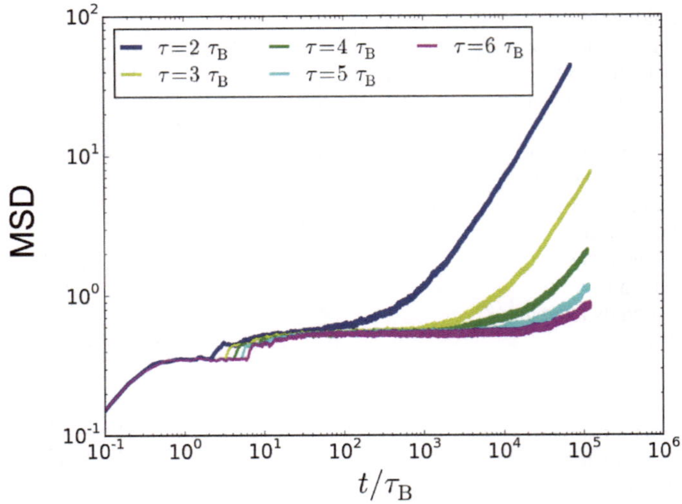

Fig. 7.12 Mean square displacement in the washboard potential with barrier height $V_0 = 1k_B\mathcal{T}$, $k = 2k_B\mathcal{T}/\sigma^2$, at different delay times. All considered cases seem to approach linear behavior (normal diffusion) for long times. Note that here $X(t') = 0,\ \forall t' \in [-\tau, 0]$. The MSD is given in units of σ^2

Figure 7.12 further shows that the MSD is linear on short times (corresponding to the diffusion within the well before the particle "sees" the barriers), and saturates at intermediate times. This *plateau* roughly begins at the intrawell relaxation time and ends at the typical escape time out of the well (see Sect. 2.6 for a description of the timescales).

Please note that the kink at $t = \tau$ is a consequence of the initial condition, which was kept fixed along $[-\tau, 0]$, specifically $X(-\tau < t < 0) \equiv 0$. This yields an abrupt change in the diffusive behavior at $X(\tau)$, where the delayed positions suddenly start to be realization-dependent.

References

1. Loos SAM, Klapp SHL (2017) Force-linearization closure for non-Markovian Langevin systems with time delay. Phys Rev E 96
2. Küchler U, Mensch B (1992) Langevins stochastic differential equation extended by a time-delayed term. Stoch Stoch Rep 40:23–42
3. Risken H (1984) The Fokker-Planck equation. Springer
4. Gardiner CW (2002) Handbook of stochastic methods, 2 edn. Springer, Berlin
5. Guillouzic S, L'Heureux I, Longtin A (1999) Small delay approximation of stochastic delay differential equations. Phys Rev E 59:3970
6. Guillouzic S, Heureux IL, Longtin A (2000) Rate processes in a delayed, stochastically driven, and overdamped system. Phys Rev E 61, 4906

7. Frank TD (2005) Delay Fokker-Planck equations, perturbation theory, and data analysis for nonlinear stochastic systems with time delays. Phys Rev E 71
8. Frank TD (2005) Delay Fokker-Planck equations, Novikov's theorem, and Boltzmann distributions as small delay approximations. Phys Rev E 72
9. Kramers HA (1940) Brownian motion in a field of force and the diffusion model of chemical reactions. Physica 7:284–304
10. Hänggi P, Talkner P, Borkovec M (1990) Reaction-rate theory: fifty years after Kramers. Rev Mod Phys 62:251

Chapter 8
Approximation for the Two-time Probability density

So far, we have discussed different approximations of the one-time probability density function, stemming from approaches based on the first member of the Fokker-Planck hierarchy. In particular, we have discussed the perturbation theory and the force-linearization closure (Chap. 7). While these approaches render quite accurate descriptions of the intrawell dynamics, they do not yield any information about the *two*-time PDF. The latter is, however, a very important quantity for the probabilistic description of the delayed dynamics. Many important features are only captured by two-time PDFs (or higher PDFs), which reflects the memory in this non-Markovian processes. As a matter of fact, only by studying them, non-Markovian steady-states can be qualitatively distinguished at all from thermal equilibria.[1] Moreover, we have seen that the FLC and PT estimates break down when escape events are likely, thus, when the interwell dynamics dominate. This is for example the case when the delay time is in the range of (or larger than) the jump duration times. Under such conditions, the interplay between the dynamics of the system state variable X and the time-dependent energy landscape causes oscillatory motion, whose description is clearly beyond the rather crude FLC-Kramers approximation described in Chap. 7. For this reason, an extension of the theory towards higher members of the FPE hierarchy would be worthwhile.

However, despite the clear benefits of a probabilistic treatment and the apparent limits of a study focusing on the first member of the FP hierarchy alone, no general approximation schemes for the two-time PDF or temporal position-correlation functions have been proposed in earlier literature (one exception being the two-state reduction employed in [1, 2], which we will discuss in Chap. 9).

[1]As an illustrative example, consider the following. Let $\rho_1^{\mathrm{nM}}(x)$ be the non-Markovian one-time PDF of an arbitrary nonequilibrium steady state. This PDF can in general not be distinguished from a PDF $\rho_1^{\mathrm{M}}(x)$ of a Markovian (equilibrium) process in a (fictive) potential defined by $U_{\mathrm{ficitve}}(x) \equiv k_{\mathrm{B}}\mathcal{T} \ln \rho_1^{\mathrm{nM}}(x)$, which would equally yield a distribution $\rho_1^{\mathrm{M}} = Z^{-1} e^{-U_{\mathrm{fictive}}/k_{\mathrm{B}}\mathcal{T}} = \rho_1^{\mathrm{nM}}$. Furthermore, the probability current vanishes, just like in an equilibrium systems.

© The Author(s), under exclusive license to Springer Nature Switzerland AG 2021

S. A. M. Loos, *Stochastic Systems with Time Delay*, Springer Theses,
https://doi.org/10.1007/978-3-030-80771-9_8

In this spirit, we propose an approximation for the steady-state two-time PDF in the regime of large delay times, based on the new equation for the two-time PDF which we have derived in Chap. 6. We have introduced this approach in Ref. [3].

After reviewing the main idea, we will apply this approximation to our bistable example system with linear delay force. We will focus on the regime where the system exhibits the delay-induced particle oscillations. As the delay-induced oscillations manifest in the two-time PDF, this system is ideal to test our approximation, and, at the same time, study non-Markovian effects on the level of the two-time PDF.

Main idea We reconsider the previously derived Eq. (6.44), recall

$$\begin{aligned}\partial_t \rho_2(x,t;x_\tau,t-\tau) = &-\partial_x\left[\gamma^{-1}F(x,x_\tau)\rho_2(x,t;x_\tau,t-\tau)\right] + D_0\partial_x^2\rho_2(x,t;x_\tau,t-\tau)\\ &+\partial_{x_\tau}\left[\langle\dot{X}(t-\tau)|x=X(t),x_\tau=X(t-\tau)\rangle\rho_2(x,t;x_\tau,t-\tau)\right].\end{aligned}$$

By neglecting statistical dependence between the (instantaneous and future) system state and the displacement, we obtain a closed equation that can be solved analytically in the steady state (where $\partial_t\rho = 0$). As we will show in Sect. 8.1, it yields a surprisingly good approximation of the non-Markovian dynamics.

Technical details As a first step, we assume that the statistical dependence between the system state $X(t)$ and the delayed displacement $\mathrm{d}X(t-\tau)$ can be neglected, as well as the dependence between system state and instantaneous displacement. This amounts to the ansatz

$$\langle\dot{X}(t-\tau)|X(t),X(t-\tau)\rangle \approx \langle\dot{X}(t-\tau)\rangle. \tag{8.1}$$

With this simplification, Eq. (6.44), simplifies to

$$\begin{aligned}-\langle\dot{X}\rangle\,\partial_{x_\tau}\rho_2(x,t;x_\tau,t-\tau) = &-\partial_x\left[\gamma^{-1}F(x,x_\tau)\rho_2(x,t;x_\tau,t-\tau)\right]\\ &+D_0\,\partial_x^2\rho_2(x,t;x_\tau,t-\tau),\end{aligned} \tag{8.2}$$

which involves the *particle current* $\langle\dot{X}\rangle$.

In order to find an ansatz for the particle current, one can proceed by utilizing the framework [4] (developed for Markovian systems), which we have mentioned in Sect. 3.1.3. In particular, we can employ the relation $\langle\dot{X}\rangle = \int_\Omega J(x,t)\mathrm{d}x$ with the probability current J. Recall that the latter is given by the FPE $\partial_t\rho_1 = -\partial_x J$. For the considered delayed systems, it reads

$$J(x,t) = \int_\Omega \gamma^{-1}F(x,x_\tau)\rho_2(x,x_\tau)\mathrm{d}x_\tau - D_0\partial_x\rho_1(x). \tag{8.3}$$

In combination, these simplifications give rise to the approximate, closed FPE (8.2) with $\langle\dot{X}\rangle = \iint_\Omega F(x,x_\tau)\rho_2(x,x_\tau)\mathrm{d}x_\tau\mathrm{d}x$.

Vanishing probability current When $\langle\dot{X}\rangle = 0$ in the steady-state, the approximate Eq. (8.2) automatically simplifies, without making further assumptions. This is indeed true for a common type of system, in particular, for any system that obeys a symmetry w.r.t. coordinate inversions $x \rightarrow -x$, like both of our example systems. In these cases (8.2) has the formal steady-state solution

$$\rho_2(x,t;x_\tau,t-\tau) = Z^{-1}h(x_\tau)e^{(\gamma D_0)^{-1}\int F(x,x_\tau)\,\mathrm{d}x}. \tag{8.4}$$

Z is again the normalization constant, and h is a function of x_τ which is not determined by (8.2) alone. However, h is determined by the steady-state constraint $\partial_t\rho_1 = 0$, which implies $\rho_1(x) = \rho_1(x_\tau) \Rightarrow \int_\Omega \rho_2(x,x_\tau)\mathrm{d}x = \int_\Omega \rho_2(x,x_\tau)\mathrm{d}x_\tau$. After specifying the delay force, which we do next, we can give a concrete example.

Linear delay force We again consider systems with *linear* delay force kx_τ, and (inversion-symmetric) nonlinear, nondelayed part of the force $F_s - kx$, (the total force being $F(x,x_\tau) = F_s(x) - kx + kx_\tau$), as in our two example systems (see Chap. 2). Then, we immediately have $h = e^{\int F_s(x_\tau)\mathrm{d}x_\tau/(\gamma D_0)}$, yielding the approximate two-time PDF

$$\rho_2^{\mathrm{approx}}(x,t;x_\tau,t-\tau) = Z^{-1}\exp\frac{\int[F_s(x)-kx]\,\mathrm{d}x + kxx_\tau + \int[F_s(x_\tau)-kx_\tau]\,\mathrm{d}x_\tau}{\gamma D_0}, \tag{8.5}$$

which fulfills both, the approximate FPE (8.2) and the steady-state constraint.

A peculiar feature of (8.5) is that it does not explicitly involve τ. However, we have already seen before that the PDFs indeed saturate for large τ values above the intrawell relaxation time (see Sect. 5.2.2 for linear, and in Sects. 7.4 and 7.3 for the nonlinear example systems). The approximation discussed here is indeed designed to capture the regime of large τ values (above τ_{ir}). We suspect that we can understand the independence of (8.5) on τ in this sense. We further recall that the approximation (8.5) is *per se* expected to break down for small values of τ, because then the statistical dependence between $\dot{X}(t-\tau)$ and $X(t)$ is non-negligible.

8.1 Application to the Bistable Delayed System

We test the approximation (8.5) by applying it to our bistable example system, which we have already considered in Sect. 7.4 in the context of the approximations for the one-time PDF. In particular, we again consider the LE (2.55), recall

$$\gamma\dot{X}(t) = (4V_0/\sigma)[-(X(t)/\sigma)^3 + X(t)\sigma] - k[X(t) - X(t-\tau)] + \sqrt{2D_0}\,\xi(t),$$

which involves the doublewell potential $V = V_0[(x/\sigma)^4 - 2(x/\sigma)^2]$ from Eq. (2.54), with minima at $x \pm \sigma$ introduced in Sect. 2.5 and the linear delayed force (2.45), stemming from an optical tweezers potential.

From Eq. (8.5), readily follows the approximate steady-state two-time PDF

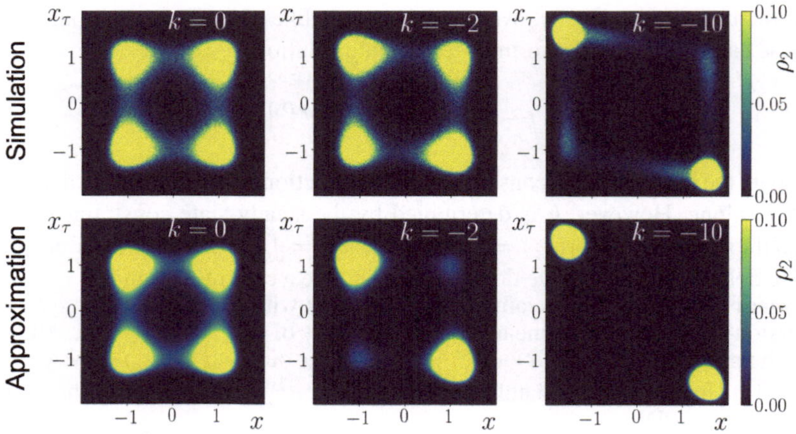

Fig. 8.1 Two-time PDF in the doublewell potential with a linear delay force of increasing (negative) magnitude k (positive feedback). *Upper panels*: numerical results from LE (2.55), *lower panels*: approximated ρ_2 from Eq. (8.6). Other parameters $V_0 = 3.5\gamma D_0$, $\tau = 10\tau_B$. The feedback strength k is given in units of $(\gamma D_0/\sigma^2)$, while x is given in units of σ. These results are also published in [3]

$$\rho_2(x, t; x_\tau, t-\tau) = Z^{-1} \exp \frac{-\frac{V_0}{\sigma^2}(\frac{x^4}{\sigma^2} - 2x^2) - \frac{k}{2}x^2 - \frac{V_0}{\sigma^2}(\frac{x_\tau^4}{\sigma^2} - 2x_\tau^2) - \frac{k}{2}x_\tau^2 + kxx_\tau}{\gamma D_0}. \tag{8.6}$$

As we have discussed already in Sects. 2.7 and 7.5.4, the bistable system (2.55) exhibits delay-induced oscillations between the two potential minima, when the delay time is larger than the typical jump duration time, and when $|k|$ has the same order of magnitude as V_0. Because these jumps manifest in the two-time PDF, as will be explained below, we consider a parameter regime, where they are pronounced (in particular, we fix $V_0 = 3.5\gamma D_0$ and $\tau = 10\tau_B$) and only vary k, particularly focusing on whether the approximation can capture the delay-induced jumps. We test the results from a quantitative point of view by comparing with direct BD simulations[2] of Eq. (2.55).

Markovian limit First, we consider the system in the absence of the delay force ($k = 0$), where the dynamics is Markovian and the system reaches thermal equilibrium. We expect the probability to find X to be the highest around the potential minima $x/\sigma = \pm 1$. As thermal jumps are very likely on the timescale of τ, the chance of finding the system in one well should be quasi independent of the question whether it was on the same or on the other side τ ago, meaning that the probabilities to find $X(t)$ and $X(t-\tau)$ are independent from each other. We hence expect the

[2] With > 2000 realizations, $\Delta t = 10^{-4}$, and > 10^6 time steps cut off to reach steady-state conditions.

joint two-time PDF[3] $\rho_2(x, x_\tau)$ to have pronounced peaks of identical heights around $\{x, x_\tau\}/\sigma \approx \{\pm 1, \pm 1\}$ and $\{\pm 1, \mp 1\}$. This expectation is confirmed by numerical simulations, and is correctly described by our approximation (8.6), as plotted in Fig. 8.1 (left plots, upper and lower panel, respectively). We now consider the impact of $k \neq 0$ on the long time behavior, where the system is in a NESS.

Delay-induced oscillations For negative $k < 0$, i.e, when the linear feedback force is directed away from the delayed system state (positive feedback), the induced oscillations are known to have a mean period of roughly 2τ (see Sect. 2.7). This means that the jumps *reduce* the joint probability to find the system in the same well at times $t - \tau$ and t. We hence expect the delay-induced oscillations to lower the two-time PDF around $\{x, x_\tau\}/\sigma \approx \{\pm 1, \pm 1\}$, and increase it around $\{\pm 1, \mp 1\}$. This effect is indeed seen in both, the numerically obtained PDF and the approximation from (8.6), see Fig. 8.1. Interestingly, by changing the sign of k, the mean period of the oscillations can be switched from 2τ to τ (see Sect. 2.7). Thus, we expect the delay-induced oscillations to have, in fact, the opposite effect for positive k, i.e., enhance the PDF at $\{\pm 1, \pm 1\}$, and reduce it around $\{\pm 1, \mp 1\}$. Again, this behavior is confirmed by simulations, and as well captured nicely by the proposed approximation, see Fig. 8.2, where we have considered negative feedback, $k > 0$.

The comparison between numerical results and the approximate ρ_2 from (8.6) in Figs. 8.1 and 8.2 includes a broad range of k values. Remarkably, we find that the proposed approximation yields very good results in all cases considered. In particular, it captures the main characteristics of the two-time PDF described above, for negative and for positive k. Even quantitatively, the agreement is very reasonable. We have further considered higher barrier heights ($V_0 = 5\,\gamma D_0$), smaller delays ($\tau = 5\,\tau_B$), as well as higher feedback strength (up to $|k| = 20\,\gamma D_0/\sigma^2$) and found as well very good agreement (not shown here). On the contrary, the approximation breaks down if the delay time is very short, see for example Fig. 8.3, where $\tau = 0.1\,\tau_B$ (left panel). We suspect that in this regime, the statistical dependence between $\dot{X}(t - \tau)$ and $X(t)$ is not negligible (hence, the assumption (8.1) is not justified).

Further interesting features of the two-time PDF We note that besides the features discussed above, the numerically obtained two-time PDFs in Figs. 8.1 and 8.2 show very interesting further details. An example is the broken symmetry $\rho_2(y, t; y', t - \tau) \neq \rho_2(y', t; y, t - \tau)$, which is not captured by the approximation (8.6). We have not further studied these features, but it indeed would be interesting to see how these details are connected to dynamical and thermodynamic properties of the non-equilibrium steady-state. (Equilibrium states are always symmetric.) This might be an interesting perspective for future research.

[3] Due to the inversion symmetry of the external potentials, we further always expect that $\rho_2(x, x_\tau)$ is symmetric w.r.t. $\{x, x_\tau\} \leftrightarrow -\{x, x_\tau\}$.

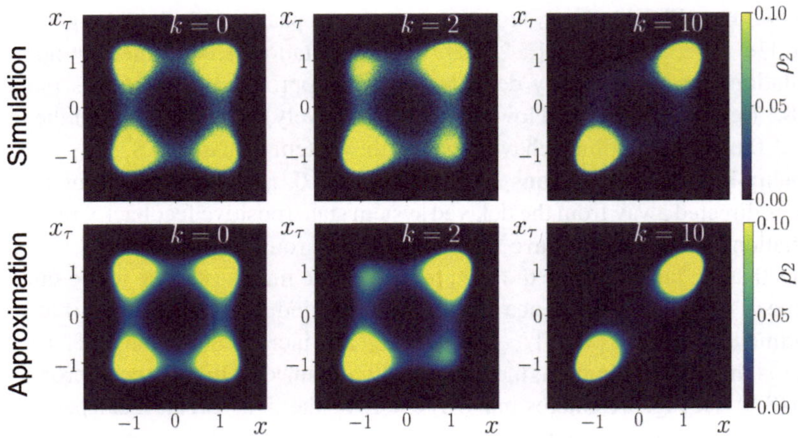

Fig. 8.2 Two-time PDF in the doublewell potential with linear delay force of increasing (positive) magnitude k (negative feedback). *Upper panels*: numerical results from LE (2.55), *lower panels*: approximated ρ_2 from Eq. (8.6). Other parameters $V_0 = 3.5\gamma D_0$, $\tau = 10\tau_B$. The feedback strength k is given in units of $(\gamma D_0/\sigma^2)$. These results are also published in [3]

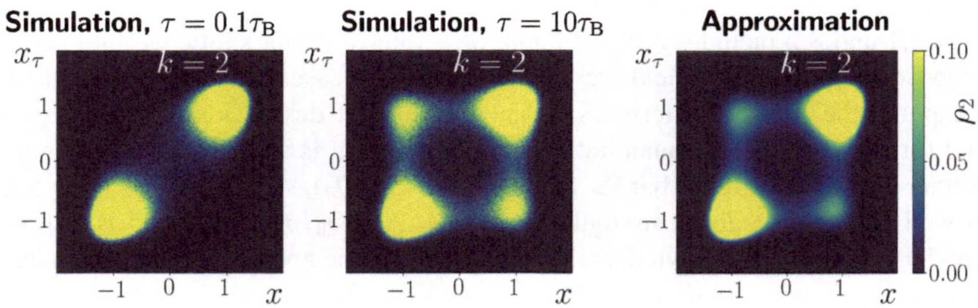

Fig. 8.3 Two-time PDF in the doublewell potential with linear delay force $kX(t-\tau)$, with positive feedback strength $k = 2(\gamma D_0/\sigma^2)$, and different time delays τ. *Left:* Simulation results with $\tau = 0.1\tau_B$, *Center:* Simulation results with $\tau = 10\tau_B$, *Right:* Approximation. Other parameters $V_0 = 3.5\gamma D_0$. The approximation matches better to the results at large values of τ, as expected

8.1.1 Comparison with Approaches for One-time Probability Density

The accuracy of the approximate ρ_2 described above is indeed remarkable, given that the underlying assumption appears to be rather crude. Further, the approximation result has a simple closed form (8.6). This indicates that an approximation scheme on the level of the second member of the FP hierarchy is somewhat superior as compared to an approach on the level of the first member. To further support this argument, we compare the prediction of approximation (8.6) to the approaches for the one-time PDF which we have discussed in the previous Chapter (see Chap. 7).

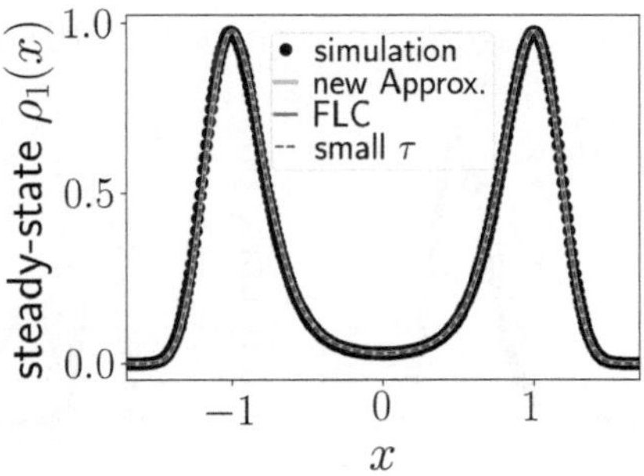

Fig. 8.4 Consistency check: The small delay expansion (small τ), the force-linearization closure (FLC), and the new approximation (8.6), all reproduce the correct equilibrium one-time PDF ρ_1 in the Markovian limit $k \to 0$. The potential barrier height is $V_0 = 3.5\,\gamma D_0$, x is in units of σ, ρ_1 is in units of $1/\sigma$

To this end, we marginalize ρ_2 from Eq. (8.6), i.e., $\rho_1(x) = \int_\Omega \rho_2(x, x_\tau)\mathrm{d}x_\tau$ and compare the resulting approximate one-time PDF with the corresponding PDF from the small delay expansion, and the force-linearization closure (FLC), given in Sect. 7.4. As we have discussed in detail in Sects. 7.4, and 7.5, the FLC and the small τ, are both not justified in the regime where the delay-induced oscillations dominate the dynamics. We nevertheless use them to compare with, due to the lack of other approximation schemes for this nonlinear system.

Consistency check To make sure the consistency of the results, we briefly consider the Markovian limit $k \to 0$, where the system is expected to reach thermal equilibrium. Indeed, all approximations are found to correctly reproduce the Boltzmann-distribution $\rho_1 \propto e^{-V_s/k_B\mathcal{T}}$, as displayed in Fig. 8.4.

Large τ regime Let us return to the case with feedback. In the regime of large delay times, the novel approximation yields the best results for ρ_1, for all considered values of the feedback strength k, as confirmed by Fig. 8.5. In this regime, delay-induced oscillations dominate, such that the underlying assumption of the FLC is violated. The small delay expansion, on the other hand, is *per se* not justified, due to the large delay time $\tau = 10\,\tau_B$.

Short delay times By construction, the approximate two-time PDF (8.6) is not reliable when the delay time is very short, because the underlying assumption is then violated. For the moderate $|k|/V_0$ and small τ values considered in Fig. 8.6 (a), the corresponding one-time PDF is found to still give a reasonable approximation. In this regime, also the other approaches from Chap. 7, e.g., the FLC, and the small delay expansion are justified (as well as the perturbation theory, which is not considered here. The two-time PDF approximation performs worse than to the FLC, but still better then the small delay expansion. Interestingly, for the large, negative k/V_0, the situation is reversed and only the small delay expansion yields a somewhat reasonable approximation, see Fig. 8.6b.

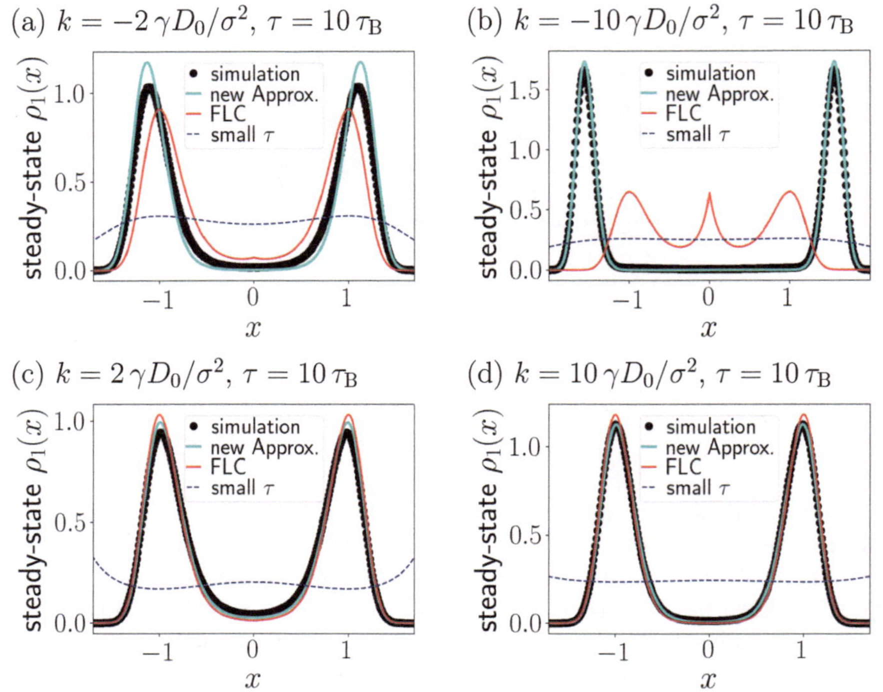

Fig. 8.5 Comparison of different approximations for the one-time PDF in the doublewell potential with linear delay force, at $V_0 = 3.5\,\gamma\,D_0, \tau = 10\,\tau_B$, and at various values of k: **a, b** positive feedback with **a** $k = -2\gamma\,D_0/\sigma^2$, **b** $k = -10\gamma\,D_0/\sigma^2$. **c, d** negative feedback, with **c** $k = 2\gamma\,D_0/\sigma^2$, **d** $k = 10\gamma\,D_0/\sigma^2$. *blue dashed lines:* small delay expansion (which is only justified for very short τ), *cyan lines:* the new approximation $\int_\Omega \rho_2(x, x_\tau)\mathrm{d}x_\tau$ with ρ_2 from Eq. (8.6), *red lines:* Force-linearization closure (FLC) discussed in Chap. 7. *black disk*: Brownian dynamics simulations. These figures stem from [3]

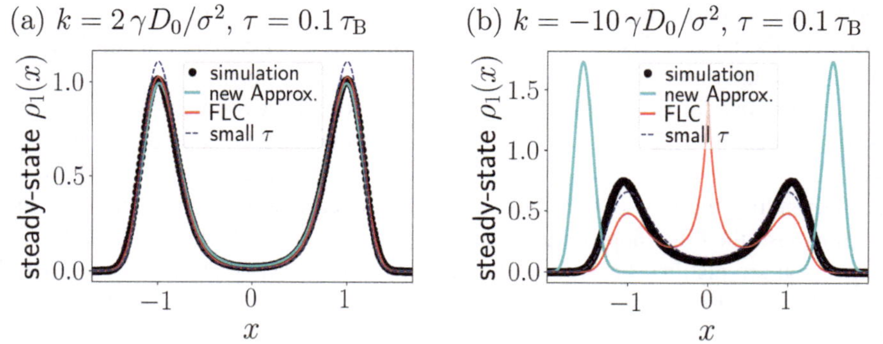

Fig. 8.6 Comparison of different approximations for the one-time PDF in the doublewell potential with linear delay force at $V_0 = 3.5\,\gamma\,D_0$, for two exemplary values of k: **a** negative feedback $k = 2\gamma\,D_0/\sigma^2$, **b** positive feedback $k = -10\gamma\,D_0/\sigma^2$. In contrast to Fig. 8.5 where $\tau = 10\,\tau_B$, here a *shorter* delay time of $\tau = 0.1\,\tau_B$ is considered. These figures stem from [3]

8.2 Concluding Remarks

By neglecting statistical dependencies between the system state and the displacement, we have obtained a closed equation for the two-time PDF that can be solved analytically. To test the approximation, we have applied it to our bistable example system involving a doublewell potential and a linear delay force, which we have also used in Chap. 7 to test the approaches with base on the first FP hierarchy member. Here we have focused on the regime where the bistable model exhibits delay-induced particle oscillations between the two minima.

The described results demonstrate that the proposed scheme yields a good approximation of the actual[4] PDFs. This is remarkable given the apparent crudeness of the underlying assumptions. In particular, the approximation captures the main characteristics of the two-time PDF, in both situations, where the delay force is directed towards, and where it points away from the delayed position. To the best of our knowledge, this is the first approximation for the two-time probability density of continuous stochastic systems with time delay.

Projecting the obtained two-time PDF on one time argument, we can further compare the results to the approximations for the one-time PDF from Chap. 7. We find that the new approximation outperforms the other approaches in the regime of large delay times, demonstrating that a framework on the level of the second FP member, is somewhat superior compared to approaches on the first member.

In the following, we turn to the thermodynamic description of systems with time delay.

References

1. Tsimring LS, Pikovsky A (2001) Noise-induced dynamics in bistable systems with delay. Phys Rev Lett 87
2. Loos SAM, Klapp SHL (2019) Heat flow due to time-delayed feedback. Sci Rep 9:2491
3. Loos SAM, Klapp SHL (2019) Fokker-Planck equations for time-delayed systems via Markovian embedding. J Stat Phys 177:95–118
4. Reimann P (2002) Brownian motors noisy transport far from equilibrium. Phys Rep 361

[4] Which are obtained with numerical simulations of the full model.

Part III
Thermodynamic Notions for Systems with Time Delay

Wenn sich die hier zu behandelnde [Brownsche] Bewegung samt den für sie zu erwartenden Gesetzmäßigkeiten wirklich beobachten läßt, so ist die klassische Thermodynamik schon für mikroskopisch unterscheidbare Räume nicht mehr als genau gültig anzusehen

—Albert Einstein, 1905 [176]

After having discussed the probabilistic description in Part II, we now turn to another cornerstone of statistical physics: thermodynamics. As we deal with systems that are inherently far from equilibrium, a promising candidate for a suitable theoretical framework is stochastic thermodynamics. In Chap. 3, we have introduced the key concepts of this theory, and formulated the main research questions addressed in this thesis.

In the following, we will first apply stochastic thermodynamics to investigate the mean heat rate and medium entropy production for nonlinear systems with time delay. In particular, we will clarify whether energy extraction from the single heat bath is possible, address the heat ow in the limit of vanishing delay times, and discuss the consequence of the overdamped limit on thermodynamic properties. We aim to go as far as possible with analytical arguments, but also present numerical results.

Furthermore, we will discuss the notion of total entropy production, and calculate information ows in systems subject to time-delayed feedback. To this end, we will step away from the case of discrete delay, and consider systems with distributed delay and colored noise. To obtain an entropic description, we will employ a strategy related to the Markovian embedding introduced in Part II.

Chapter 9
The Heat Flow Induced by a Discrete Delay

Finding thermodynamic notions for non-Markovian systems is a major problem. As a first step in this direction, we consider the *heat rate* $\dot{Q} = \langle \delta q/\mathrm{d}t \rangle_{\mathrm{ss}}$ of nonlinear systems with discrete delay. This quantity can be calculated via well-established concepts from stochastic thermodynamics, in particular, from the framework of stochastic energetics [1]. The heat flow and related medium entropy production are key thermodynamic quantities, which already provide important physical insight. They further constitute a nontrivial part of the total entropy production. Based on this consideration, we can already address several fundamental questions. First, we aim to clarify: *Can the delayed feedback induce a net energy flow from the bath to the colloid?* Here we recall the connection to the "Maxwell demon"-type of devices mentioned in Chap. 4. Next we want to find out: *Is the overdamped limit problematic in regard to the thermodynamic description?* Indeed, in the dynamical and probabilistic treatment, this limit was not problematic at all (see Chap. 7). However, from a thermodynamic perspective the situation might be different. Specifically, the overdamped limit, which amounts to a neglect of the *inertia term* in the LE,[1] implies that the kinetic energy is not described. In turn, the kinetic energy would yield additional contribution in the energy balance (see Chap. 4), thus, possibly affecting the calculated heat flow. Last, we also ask ourselves whether *thermodynamic quantities reflect delay-induced dynamical behavior?* As an example for the latter, we will again consider the delay-induced oscillations, which we have mentioned in Chaps. 7, and 8. Indeed, a previous, numerical study [2] has indicated that the medium entropy production has a maximum around coherence resonance. However, in this study only a single parameter regime was considered, no analytical arguments are used to support the finding, and no physical explanation is provided. Therefore it seems worthwhile to revisit this observation and investigate the heat and medium entropy production around coherence resonance more thoroughly.

We have published in Ref. [3] the main results presented in the following.

[1] We recall that lists of all abbreviations and symbols are given on pages VII and VIII, respectively.

© The Author(s), under exclusive license to Springer Nature Switzerland AG 2021
S. A. M. Loos, *Stochastic Systems with Time Delay*, Springer Theses,
https://doi.org/10.1007/978-3-030-80771-9_9

Previous results for linear systems For linear delay systems, it has already been shown that in the long-time limit a nonequilibrium steady state with finite heat production [4] is approached (in the absence of time-dependent forces), as the delay pushes the system out of equilibrium. However, this previous result fails to correctly describe the limit of vanishing delay time, and the question whether energy extraction from the single heat bath is possible, was not answered. Another issue is that the approach in [4] is limited to linear systems only.

9.1 Main Idea

For nonlinear cases, the linear response function method employed in [4] cannot be applied. In the following, we present an alternative strategy to get analytical results. Focusing on polynomial energy landscapes brings us into the position to find some analytical statements for nonlinear, time-delayed systems. In particular, by using both, the FPE and LE, we can derive several identities linking different noise-position correlations as well as position-autocorrelations (with time difference 0 or τ) to positional moments. By means of these identities, we can then derive an equation for the mean heat rate that solely depends on positional moments. In combination with the approaches presented in Part II, this equation can be utilized to obtain closed approximate expressions for the heat, which become exact in the Markovian limits.

9.1.1 Polynomial Energy Landscapes

In this Chapter, we assume that the conservative forces in the delay LE (2.43) can be expressed as a *polynomial*, as it is the case for our bistable example, and further holds for a wide class of nonlinear potentials (even non-polynomial ones due to Taylor expansions). This assumption simplifies our analytical treatment, as will become clear below. The feedback control force is again chosen to be linear (as in the case of optical tweezers feedback, see Sect. 2.4.1). Hence, we here consider the overdamped LE

$$\gamma \mathrm{d}X(t) = -\sum_{i=1}^{m} a_i X(t)^i \mathrm{d}t - bX(t-\tau)\mathrm{d}t + \gamma\sqrt{2D_0}\,\xi(t)\,\mathrm{d}t, \tag{9.1}$$

with $a_i, b \in \mathbb{R}$. The non-delayed part of the deterministic force $-\sum_{i=1}^{m} a_i X(t)^i$ might stem from a static potential $V_s(x) = \sum_{i=1}^{m} (a_i/i+1)x^{i+1}$, and is hence a conservative force.[2]

[2] Note that we here include the non-delayed part $-kX(t)$ of the "optical tweezers" force $-k[X(t) - X(t-\tau)]$ into the a_1-term of the polynomial sum. In this case, $b = -k$.

Following the stochastic energetics framework by Sekimoto described in Chap. 4 [specifically, in Sect. 4.2, see Eqs. (4.1–4.3)], the fluctuating work δw done by the *nonconservative* delayed feedback force to the particle, the change of its internal energy $\mathrm{d}u$, which is given by the potential energy change, and the heat δq flowing to the reservoir during the infinitesimal time $\mathrm{d}t$ are given by [4]

$$\delta w(t) = -bX(t-\tau) \circ \mathrm{d}X(t), \tag{9.2}$$

$$\mathrm{d}u(t) = V_s'[X(t)] \circ \mathrm{d}X(t), \tag{9.3}$$

$$\delta q(t) = \gamma\left[\mathrm{d}X(t)/\mathrm{d}t - \sqrt{2D_0}\,\xi(t)\right] \circ \mathrm{d}X(t) = \delta w(t) - \mathrm{d}u(t). \tag{9.4}$$

Plugging the LE (9.1) into Eqs. (9.2, 9.3) results in the following expressions for the NESS ensemble averages[3]

$$\dot{W} \equiv \left\langle \frac{\delta w}{\mathrm{d}t} \right\rangle_{\mathrm{ss}} = \frac{b}{\gamma}\left\{ \sum_{i=1}^{m} a_i C_i(\tau) + b\left\langle X^2 \right\rangle_{\mathrm{ss}} - \gamma\sqrt{2D_0}\,\langle X(t-\tau)\xi(t)\rangle_{\mathrm{ss}} \right\}, \tag{9.5}$$

$$\dot{U} \equiv \left\langle \frac{\mathrm{d}u}{\mathrm{d}t} \right\rangle_{\mathrm{ss}} = \frac{1}{\gamma}\sum_{i=1}^{m}\left\{ a_i\gamma\sqrt{2D_0}\left\langle X^i\xi\right\rangle_{\mathrm{ss}} - a_i b\, C_i(\tau) - a_i^2\left\langle X^{2i}\right\rangle_{\mathrm{ss}} - \sum_{j>i}^{m} 2a_i a_j \left\langle X^{i+j}\right\rangle_{\mathrm{ss}} \right\}, \tag{9.6}$$

$$\dot{Q} \equiv \left\langle \frac{\delta q}{\mathrm{d}t} \right\rangle_{\mathrm{ss}} = \dot{W} - \dot{U}. \tag{9.7}$$

Because of the polynomial from of the static potential, the mean values of the energy flows linearly depend on various positional and position–noise cross-correlations. Further, due to the time delay, the thermodynamic quantities depend on the temporal position-autocorrelation functions at time difference τ,

$$C_i(\tau) = \left\langle X(t)^i X(t-\tau) \right\rangle_{\mathrm{ss}}, \tag{9.8}$$

mirroring the non-Markovian nature of (9.1).

9.2 Mean Heat Rate & Medium Entropy Production

In the following, we derive exact expressions for (9.6, 9.5) which only involve positional moments at one time. To this end, we employ the FPE and the LE. Specifically, by projecting the FPE onto positional moments and subsequently inserting the LE, we obtain relations between different correlation functions, as was previously sug-

[3] Recall that capital letters denote ensemble averages.

gested in [5]. In a second step, these expressions can then be evaluated for example on the basis of known approximation schemes, which we have discussed in Part II of this thesis.

Derivation of Relations Between Correlation Functions First we derive useful relations between temporal position-autocorrelation functions. To this end, we project the FPE (3.15) onto its moments by multiplying the equation with x^{n+1}, with an arbitrary nonnegative integer n, and integrating over the spatial domain $x \in (-\infty, \infty)$. After several partial integrations, (where the boundary terms vanish due to the natural boundary conditions), yield

$$\gamma \frac{\mathrm{d}}{\mathrm{d}t} \frac{\langle X(t)^{n+1} \rangle}{(n+1)} = \gamma D_0\, n \left\langle X(t)^{n-1} \right\rangle - \sum_{i=1}^{m} a_i \left\langle X(t)^{n+i} \right\rangle - b \left\langle X(t)^n X(t-\tau) \right\rangle . \quad (9.9)$$

In a NESS, the left side vanishes and (9.9) reduces to a relation between different moments and the temporal position-autocorrelation function [5]

$$b\, C_n(\tau) = \gamma D_0\, n \left\langle X^{n-1} \right\rangle_{\mathrm{ss}} - \sum_{i=1}^{m} a_i \left\langle X^{n+i} \right\rangle_{\mathrm{ss}} , \quad (9.10)$$

valid $\forall n \geq 1$. We will use this relation to substitute all the $C_i(\tau)$-terms in the mean work rate and internal energy from (9.6, 9.5).

Further, by plugging in the LE (9.1), one can deduce directly

$$\frac{\mathrm{d}}{\mathrm{d}t} \frac{\langle X(t)^{n+1} \rangle}{(n+1)} = \left\langle X(t)^n \frac{\mathrm{d}X(t)}{\mathrm{d}t} \right\rangle \overset{\mathrm{LE}}{=} -\sum_{i=1}^{m} \frac{a_i}{\gamma} \left\langle X(t)^{i+n} \right\rangle + \sqrt{2D_0} \left\langle X(t)^n \xi(t) \right\rangle$$
$$- \frac{b}{\gamma} \left\langle X(t)^n X(t-\tau) \right\rangle . \quad (9.11)$$

A comparison of Eqs. (9.11) and (9.9) readily provides for all $n \geq 1$ the surprisingly simple and generic relation

$$\left\langle X(t)^n \xi(t) \right\rangle = n \sqrt{D_0/2} \left\langle X(t)^{n-1} \right\rangle , \quad (9.12)$$

which, in fact, generally holds for any (nonlinear) force. This can be shown analogously, see Ref. [5]. Using Eq. (9.12), the instantaneous noise–position cross-correlations in (9.6) can be replaced by positional moments.

Finally, to evaluate $\langle X(t-\tau)\xi(t) \rangle$ at $\tau \neq 0$, we combine Eq. (9.12) with the causality argument (see Sects. 2.3.6 and 4.4.1). Because no physical quantity shall be influenced by future noise, statistical independence follows, and $\left\langle X(t)\xi(t') \right\rangle = \langle X(t) \rangle \left\langle \xi(t') \right\rangle \equiv 0$ must hold $\forall t' > t$. Hence

$$\langle X(t-\tau)\xi(t)\rangle = \sqrt{\frac{D_0}{2}}\delta_\tau = \begin{cases} \sqrt{D_0/2}, & \tau = 0 \\ 0, & \tau > 0 \end{cases}. \tag{9.13}$$

With these relations, we can now consider the mean rates of the energy flows. Substituting Eqs. (9.10) and (9.12) at $n = 1, 2, \ldots, m$, into Eq. (9.6) yields

$$\dot{U} = 0. \tag{9.14}$$

This is expected, since the net internal energy should be a conserved quantity in a NESS [6]. Thus $\dot{W} \equiv \dot{Q}$. Substituting (9.10) at $n = 1, 2, \ldots, m$ into Eq. (9.5) and using (9.13) further yields

$$\begin{aligned} \dot{Q} = \dot{W} &= \sum_{i=1}^{m}\left\{ i a_i D_0 \left\langle X^{i-1}\right\rangle_{\rm ss} - \sum_{j=1}^{m} \frac{a_i a_j}{\gamma}\left\langle X^{i+j}\right\rangle_{\rm ss} \right\} + \frac{b^2}{\gamma}\left\langle X^2\right\rangle_{\rm ss} - b D_0 \delta_\tau \\ &= \dot{S}_{\rm m}(\gamma D_0/k_{\rm B}). \end{aligned} \tag{9.15}$$

Please note that this simplification step is possible due to the polynomial form of the deterministic forces.

Equation (9.15) is an exact expression only involving positional moments, i.e., *one*-time ensemble averages over X^n. $\dot{Q}$ and $\dot{S}_{\rm m}$ can therefore be computed directly from the steady-state one-time PDF, and hence, on the basis of the various approximations discussed in Part II.

The δ_τ-term suggests discontinuous behavior of $\dot{Q}$ at $\tau \to 0$. But in order to study this limit properly, one also has to investigate the behavior of the PDF to clarify whether the moments behave continuously. Obviously, this can be readily done for linear systems, where the exact solutions are known. In fact, we can as well find exact statements for nonlinear systems in the limits. We will discuss both in the following, starting with the linear case. As in the limit $\tau \to 0$ Markovianity is recovered, one expects that the system equilibrates (because of the absence of external driving), and therefore $\dot{Q} = \dot{S}_{\rm m} = 0$. The same is expected when $b \to 0$, which we also will discuss below.

9.2.1 Linear Systems

Let us first consider the purely linear systems with LE (2.51), where the force has the form

$$F = -\alpha x - \beta x_\tau. \tag{9.16}$$

Thus, comparing to notation here and there, $a_1 = \alpha,\ b = \beta$. Recall that for this case, the exact steady-state PDFs are known, as we have discussed in Sect. 5.2. Therefore, we can immediately write down a closed expression for the mean heat rate from Eq. (9.15), which reads

$$\frac{\dot{Q}}{D_0} = \frac{\gamma \dot{S}_\mathrm{m}}{k_\mathrm{B}} = \alpha - \beta\delta_\tau + (\alpha^2 - \beta^2)\frac{1 + \beta \sinh\left(\sqrt{\alpha^2 - \beta^2}\,\tau/\gamma\right) \Big/ \sqrt{|\alpha^2 - \beta^2|}}{\alpha + \beta \cosh\left(\sqrt{\alpha^2 - \beta^2}\,\tau/\gamma\right)}. \tag{9.17}$$

Equation (9.17) is consistent with Ref. [4], where only the linear case is considered (and a different approach is used via linear response functions), apart from the *additional* δ_τ-term. Consequentially, only Eq. (9.17) correctly predicts $\dot{S}_\mathrm{m}(\tau \to 0) = 0$. We will discuss this point in detail below.

Let us now take a closer look to the mean heat rate $\dot{Q}$ and the medium entropy production $\dot{S}_\mathrm{m}$ (which are proportional to each other) from Eq. (9.17). Figure 9.1 depicts the results for two exemplary parameter settings. One immediately sees that the heat rate is a nonzero quantity for $|\beta|\tau > 0$, proving the true non-equilibrium nature of this steady state. Here, $\dot{Q} > 0$ implying heating of the surrounding fluid. However, according to (9.17), the heat rate can also be negative for certain β, τ values. An example is given in Fig. 9.3 (dashed line[4]). At this point, we can answer one of our initially posed research questions about the thermodynamic behavior induced by time-delayed feedback (stated in the Introduction and in Sect. 4.6.1), that is, whether a delay force can be used to extract energy from the surrounding heat bath (like a Szilárd engine). Indeed, a linear delay force is capable of performing this task, already in this fully linear system.

Moreover, according to Eq. (9.17), $\dot{Q}$ continuously approaches zero as $\beta \to 0$, and nullifies at $\beta = 0$, i.e., when the steady state is the thermal equilibrium. On the contrary, the limit $\tau \to 0$ has a different characteristic. Remarkably, the mean heat rate has an apparent offset at $\tau = 0$, where it *discontinuously* drops to zero. This can be seen in Fig. 9.1, as well as in the following Figs. 9.3 and 9.4 (black dashed lines), where $\dot{Q}$ is plotted as a function of τ (for two different parameter settings). In fact, for some parameters, $|\dot{Q}|$ even *grows*, when starting from large τ and decreasing τ towards zero, (until a saturation value is reached giving the apparent offset).

The discontinuity and the growth are odd features of the heat flow. We will investigate them in detail in the following. But before, let us clarify the limit behavior in *nonlinear* cases, because the situation turns out to be analogous there.

9.2.2 *Markovian Limits in Nonlinear Systems*

Now we turn to the nonlinear cases, where we have no general analytic solutions. By utilizing the results presented in Part II, we can nevertheless obtain some strict statements.

[4] The other results shown in this Figure will be discussed below.

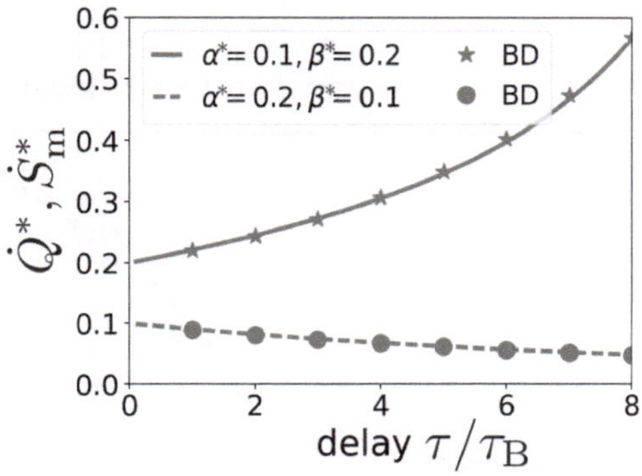

Fig. 9.1 Mean heat rate $\dot{Q}^* = \dot{Q}\,(\tau_B/(k_B T))$ and medium entropy production rate $\dot{S}_m^* = \dot{S}_m\,(\tau_B/k_B)$ plotted against the delay time τ/τ_B, in the linear system with time delay at different $\alpha^* = \alpha\,\tau_B/\gamma$ and $\beta^* = \beta\,\tau_B/\gamma$. *Solid line and stars:* $\alpha^* = 0.1$, $\beta^* = 0.2$, *dashed line and disks:* $\alpha^* = 0.2$, $\beta^* = 0.1$. The lines show the exact solution (9.17), the symbols stem from simulations (BD). The parameters are identical to Fig. 3 of Ref. [4], demonstrating the consistency of the results at $\tau > 0$

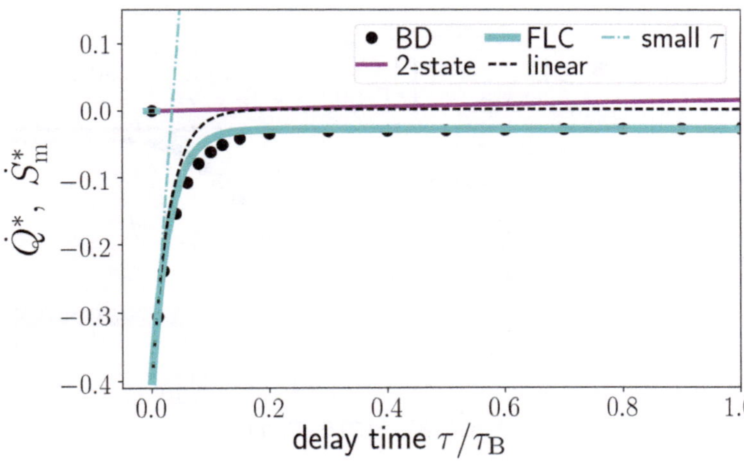

Fig. 9.2 Mean heat rate $\dot{Q}^* = \dot{Q}\,(\tau_B/(k_B T))$ and medium entropy production rate $\dot{S}_m^* = \dot{S}_m\,\sigma^2/(k_B D_0)$ versus delay time (scaled with the Brownian time $\tau_B = \sigma^2/D_0$), in the delayed bistable system at $V_0 = 4k_B T$, $k = 0.4k_B T$. The results stem from simulations (BD), and from several approximations: force-linearization closure (FLC), small delay small expansion (small τ), and the 2-state reduction. The dashed line shows $\dot{Q}^*$ and $\dot{S}_m^*$ of the corresponding linearized system at $\alpha = -(8V_0 + k)/\sigma$ and $\beta = -k/\sigma$ from Eq. (9.17). The same results are again shown in Fig. 9.3 below, in a logarithmic plot

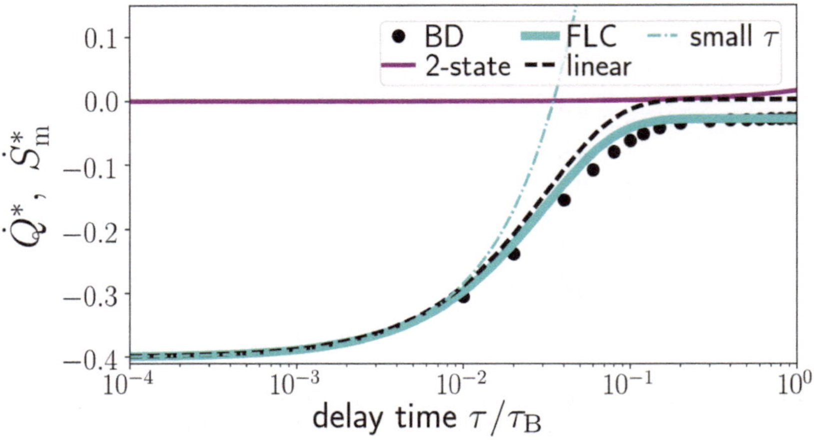

Fig. 9.3 Mean heat rate $\dot{Q}^* = \dot{Q}\,(\tau_{\rm B}/(k_{\rm B}\mathcal{T}))$ and medium entropy production rate $\dot{S}^*_{\rm m} = \dot{S}_{\rm m}\,\sigma^2/(k_{\rm B}D_0)$ versus delay time as in Fig. 9.2. Note the logarithmic x-axis here, which makes it easier to see the behavior at very small τ values

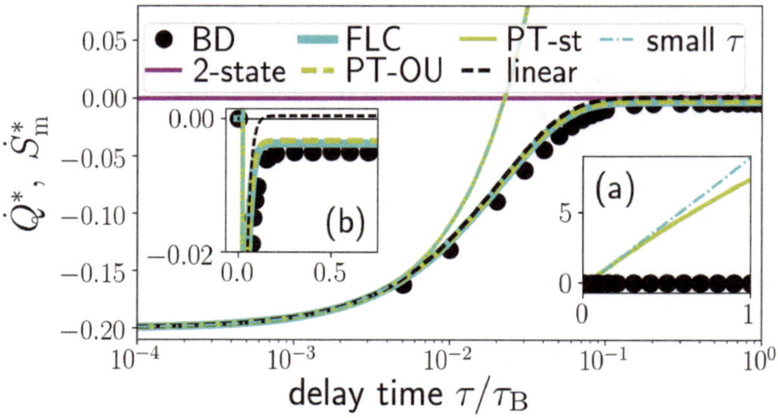

Fig. 9.4 Mean heat rate $\dot{Q}^* = \dot{Q}\,\tau_{\rm B}/(k_{\rm B}\mathcal{T})$ and medium entropy production (EP) rate $\dot{S}^*_{\rm m} = \dot{S}_{\rm m}\,\sigma^2/(k_{\rm B}D_0)$ in the bistable system with different parameters similar to Figs. 9.2 and 9.3, but here at $V_0 = 6k_{\rm B}\mathcal{T}$, $k = 0.2k_{\rm B}\mathcal{T}$. This plot is further complemented with the results from the perturbation theory, once with short time propagator (PT-st), one with Ornstein-Uhlenbeck approximation (PT-OU), shown as dashed and solid yellow lines, respectively. **a** zoom out, **b** magnification. These results are also published in Ref. [3]

Main Idea For nonlinear systems, no exact solutions for the one-time PDF in the presence of delay are known. However, we can nevertheless analytically address the question how the PDFs behave in the Markovian limits, since we have approximative PDFs which become *exact*. This will give us exact results from Eq. (9.15) in the limits. In particular, the small delay expansion (which we have introduced in Sect. 7.2.1), is obtained by a first-order Taylor expansion around $\tau = 0$ in the LE. By construction, it becomes exact[5] in the limit $\tau \to 0$. Further, the perturbation theory (see Sect. 7.2.2), where the delay force is treated as a small perturbation to the Markovian dynamics, becomes exact in the limit $b \to 0$ (where this perturbation vanishes). We refer the interested Reader to Chap. 7, where detailed descriptions of the approximation schemes and their limitations are provided.

9.2.2.1 Limit of Vanishing Feedback Strength

We first consider the Markovian limit of vanishing feedback strength, i.e., $b \to 0$. By construction, the one-time steady-state PDFs from the PT (and also from the FLC) become exact at $b \to 0$. They converge in a *continuous* manner to the equilibrium (Boltzmann) distribution $\rho_{\mathrm{ss}}^{b=0}$, as can be easily seen from the PT-st approximation from Eq. (7.16). This reads

$$\begin{aligned} \rho_{\mathrm{ss}}^{\mathrm{PT}}(x) &= Z \exp\left\{-\frac{[1-(b\tau/\gamma)]\left[V_{\mathrm{s}}(x)+(b/2)x^2\right]}{\gamma D_0}\right\} \\ &\overset{b\to 0}{\longrightarrow} \rho_{\mathrm{ss}}^{b=0}(x) = Z^{-1}\exp\left[-\frac{V_{\mathrm{s}}(x)}{\gamma D_0}\right], \end{aligned} \tag{9.18}$$

with the static potential giving the polynomial force $V_{\mathrm{s}}'(x) = \sum_{i=1}^{m} a_i x^i$ and normalization constant Z. After performing a partial integration step and plugging in the natural boundary conditions, the $(i-1)^{st}$ positional moment can therewith be expressed as

$$\begin{aligned} \left\langle X^{i-1}\right\rangle_{\mathrm{ss}} &= \int_{-\infty}^{\infty} Z^{-1}\exp\left[-\frac{V_{\mathrm{s}}(x)}{\gamma D_0}\right] x^{i-1}\mathrm{d}x \\ &= \left[\rho_{\mathrm{ss}}^{b=0}(x)\,\frac{x^i}{i}\right]_{-\infty}^{\infty} - \int_{-\infty}^{\infty}\frac{F_{\mathrm{con}}(x)}{\gamma D_0} Z\exp\left[-\frac{V_{\mathrm{s}}(x)}{\gamma D_0}\right]\frac{x^i}{i}\mathrm{d}x \\ &= \sum_{j=1}^{m}\frac{a_j}{i\gamma D_0}\int_{-\infty}^{\infty} x^j \rho_{\mathrm{ss}}^{b=0}(x) x^i \mathrm{d}x = \sum_{j=1}^{m}\frac{a_j}{i\gamma D_0}\left\langle X^{j+i}\right\rangle_{\mathrm{ss}}, \quad \forall i \geq 1, \end{aligned} \tag{9.19}$$

[5] As we have noted in Sect. 7.2.1, this is in fact a delicate point, due to the question whether the Taylor expansion of the stochastic quantity has the same properties as in deterministic cases.

from which immediately follows

$$\Rightarrow \sum_{i=1}^{m} i\, a_i D_0 \left\langle X^{i-1} \right\rangle_{\rm ss} = \sum_{i=1}^{m} \sum_{j=1}^{m} \frac{a_j a_i}{\gamma} \left\langle X^{j+i} \right\rangle_{\rm ss}. \tag{9.20}$$

Plugging the last expression into Eq. (9.15), we finally conclude that $\lim_{b\to 0} \dot{Q} = 0$ and $\lim_{b\to 0} \dot{S}_{\rm m} = 0$ in a *continuous* manner. This is in agreement with the linear case, see Eq. (9.17).

9.2.3 Limit of Vanishing Delay Time

The other relevant limit is that of vanishing delay time, $\tau \to 0$. The one-time PDF [7] from the small delay expansion [also see Eq. (7.14)]

$$\rho_{\rm ss}^{ST}(x) = Z^{-1} \exp\left[-\frac{V_{\rm s}(x) + (b/2)x^2}{(\gamma + b\tau)D_0} \right] \overset{\tau\to 0}{\longrightarrow} \rho_{\rm ss}^{\tau=0}(x) = Z \exp\left[-\frac{V_{\rm s}(x) + (b/2)x^2}{\gamma D_0} \right] \tag{9.21}$$

clearly becomes exact for $\tau \to 0$, where it *continuously* converges to the Boltzmann distribution $\rho_{\rm ss}^{\tau=0}$. Performing steps analogously to Eq. (9.20), we obtain for all $i \geq 1$

$$\begin{aligned} \left\langle X^{i-1} \right\rangle_{\rm ss} &= \left[Z \rho_{\rm ss}^{\tau=0}(x)\, (x^i/i) \right]_{-\infty}^{\infty} - \int_{-\infty}^{\infty} \frac{F_{\rm con}(x) - bx}{i\gamma D_0} x^i \rho_{\rm ss}^{\tau=0}(x) {\rm d}x \\ &= \sum_{j=1}^{m} \frac{a_j}{i\gamma D_0} \left\langle X^{j+i} \right\rangle_{\rm ss} + \frac{b}{i\gamma D_0} \left\langle X^{i+1} \right\rangle_{\rm ss}, \end{aligned} \tag{9.22}$$

as $\tau \to 0$, which results in the identities

$$\sum_{i=1}^{m} i\, a_i D_0 \left\langle X^{i-1} \right\rangle_{\rm ss} = \sum_{i=1}^{m} \sum_{j=1}^{m} \frac{a_j a_i}{\gamma} \left\langle X^{j+i} \right\rangle_{\rm ss} + \sum_{i=1}^{m} \frac{b a_i}{\gamma} \left\langle X^{i+1} \right\rangle_{\rm ss}, \tag{9.23}$$

$$\left\langle X^0 \right\rangle_{\rm ss} \equiv 1 = \frac{b}{\gamma D_0} \left\langle X^2 \right\rangle_{\rm ss} + \sum_{j=1}^{m} \frac{a_j}{\gamma D_0} \left\langle X^{j+1} \right\rangle_{\rm ss}. \tag{9.24}$$

Plugging both into Eq. (9.15) in an iterative manner, yields the *discontinuous* limit

$$\lim_{\tau\to 0} \dot{S}_{\rm m} = \lim_{\tau\to 0} \frac{k_{\rm B} b}{\gamma}(1 - \delta_\tau) = \begin{cases} 0, & \tau = 0 \\ k_{\rm B} b/\gamma, & \tau > 0 \end{cases}. \tag{9.25}$$

This ubiquitous jump-discontinuity of $\dot{Q}$ and $\dot{S}_m$ at $\tau \to 0$ is a central theoretical result. Remarkably, it is independent of the details of the potential landscape. This apparent offset indicates an abrupt qualitative change of the thermodynamics when non-Markovianity sets in. It arises due to the discontinuity of the noise–position cross correlations (9.13) at the onset of causal relationship. Explicit examples are shown Figs. 9.3, 9.4, which we will describe in Sect. 9.3.

We recall that the apparent offset has already been observed and discussed in the context of linear systems [4, 8–10]. In [4] it has been considered as a consequence of inconsistent usage of Itô and Stratonovich calculus. However, as shown here, it is not a mathematical error but also arises within consistently applied Stratonovich calculus. The underlying reason is the interplay of white noise and delay below the short (ballistic) relaxation timescale [11], which becomes relevant as $\tau \to 0$. Let us further differentiate between physical effects which should be measurable in experiments, and consequences of the model. To better understand the impact of the overdamped limit, we now consider the full Langevin equation.

9.2.4 *Influence of Inertia Term*

To further elucidate the behavior at $\tau \to 0$, we also briefly consider the limit in the *underdamped* case, where the LE

$$m\ddot{X}(t) = -\gamma\dot{X}(t) + F_{\text{con}}[X(t)] - bX(t-\tau) + \gamma\sqrt{2D_0}\,\xi(t), \qquad (9.26)$$

involves the inertial term with mass m, yielding ballistic motion below the velocity-relaxation timescale, m/γ. For this system, the heat can, in principle, be calculated using the same expression as in the overdamped case [Eqs. (9.2, 9.4)]. For linear systems, it has been shown by [8] via a Green's functions method, that $\dot{Q}$ smoothly decays to zero with τ in the underdamped case. For nonlinear systems, this strategy cannot be applied and the calculations are much more involved.

To obtain a statement for nonlinear systems, we should consider the correlation $\langle X(t-\tau)\xi(t)\rangle$ appearing in Eq. (9.5), which, in our framework, "causes" the discontinuity from a mathematical point of view [compare Eq. (9.13, 9.25)].

9.2.4.1 Derivation of Noise–position (9.32) and Noise–velocity Cross-Correlations for Underdamped Dynamics

We aim to calculate the correlations $\langle X(t-\tau)\xi(t)\rangle$ and $\langle \dot{X}(t-\tau)\xi(t)\rangle$ for underdamped motion with time-delayed feedback. To this end, we first again use the causality argument (i.e., the noise cannot influence the past position or velocity, see Sect. 2.3.6). This implies that both correlations must vanish at $\tau > 0$. Second, we rewrite the correlations at $\tau = 0$ as follows, starting from a formal integration over the LE (9.26)

$$m\dot{X}(t) = -\gamma X(t) + \sum_i a_i \int_0^t X(s)^i \mathrm{d}s - b \int_0^t X(s-\tau)\mathrm{d}s + \gamma\sqrt{2D_0} \int_0^t \xi(s)\mathrm{d}s + C, \quad (9.27)$$

with constant C accounting for initial conditions. From this equation, we readily find

$$\begin{aligned} m\langle \dot{X}(t)\xi(t)\rangle =& -\gamma \langle X(t)\xi(t)\rangle + \sum_i a_i \int_0^t \left\langle X(s)^i \xi(t)\right\rangle \mathrm{d}s - b \int_0^t \langle X(s)\xi(t)\rangle \,\mathrm{d}s \\ &+ \gamma\sqrt{2D_0} \int_0^t \langle \xi(s)\xi(t)\rangle \,\mathrm{d}s \\ =& -\gamma \langle X(t)\xi(t)\rangle + \sum_i a_i \int_0^t \left\langle X(t)^i \xi(t)\right\rangle \delta_{s,t}\mathrm{d}s - b \int_0^t \langle X(t)\xi(t)\rangle \,\delta_{s,t}\mathrm{d}s \\ &+ \gamma\sqrt{D_0/2}. \end{aligned} \quad (9.28)$$

In the last step, we have simplified the integrals of types $\int_0^t \left\langle X(s)^i \xi(t)\right\rangle \mathrm{d}s$ by using the causality argument which implies that the integrands are zero for all $s < t$. Third, we use the identity $X(t) = \int_0^t \dot{X}(s)\mathrm{d}s + X(0)$, and find

$$\langle X(t)\xi(t)\rangle = \int_0^t \left\langle \dot{X}(s)\xi(t)\right\rangle \mathrm{d}s = \int_0^t \langle \dot{X}(t)\xi(t)\rangle \delta_{s,t}\, \mathrm{d}s. \quad (9.29)$$

Fourth, via a proof by contradiction, we establish that Eqs. (9.28, 9.29) only allows for *finite* solutions. Starting with the assumption $\langle \dot{X}(t)\xi(t)\rangle \to \infty$, Eq. (9.29) implies a finite value of $\langle X(t)\xi(t)\rangle$. This, then again yields a finite value of $\langle \dot{X}(t)\xi(t)\rangle$ from (9.28), thus, a contradiction. Using this finiteness, we finally obtain for $\tau \geq 0$

$$\langle X(t-\tau)\xi(t)\rangle = 0, \quad (9.30)$$

$$\langle \dot{X}(t-\tau)\xi(t)\rangle = (\gamma/m)\sqrt{D_0/2}\,\delta_\tau. \quad (9.31)$$

Thus, we find that in contrast to the overdamped case, the position–noise cross-correlation is continuous at $\tau = 0$,

$$\langle X(t-\tau)\xi(t)\rangle = 0, \ \forall \tau \geq 0 \quad (9.32)$$

(not only for linear systems), indicating that $\dot{Q}$ behaves continuously. Therefore, the discontinuity is absent in the underdamped description. This is in agreement with previous results for underdamped systems, e.g., see Fig. 1 in [8].

9.2.5 *Discussion of the Behavior for Small Delay Times*

Here we discuss three interesting observations that follow from the previous considerations: The discontinuous limit, the sign of the heat rate and the growth of its absolute value when τ is reduced.

Meaning of discontinuous limit from a broader perspective

Above we have shown that the heat flow is discontinuous at $\tau \to 0$ in the overdamped limit, while it behaves continuously in the underdamped equation. The discontinuity is thus a consequence of the overdamped limit [recall (9.32) versus (9.13)].

From the viewpoint of theoretical frameworks of stochastic modeling, this is very interesting, because the overdamped limit is an extremely common and well-established assumption, which normally does not yield unphysical behavior, especially not on the large timescales. Here, we consider the long-time behavior $t \to \infty$, and find that it nevertheless breaks down. The reason is that we—although focusing on steady-state behavior—indeed implicitly address the ballistic timescale, when we let τ go to zero.

We would like to remind the Reader that in models of feedback-controlled systems, the assumption of (vanishingly) small τ, is very common. While it often renders nice results for the dynamical quantities, the thermodynamic considerations reveal that it is indeed somewhat problematic. This is an example for a general observation we make: the thermodynamic quantities are more "sensitive" then the typically considered dynamical quantities.[6]

9.2.5.1 Reversed Heat Flow

Turning back to our analytical formula for the limit in the overdamped case, i.e., Eq. (9.25), we further find that at small, positive τ, $\dot{S}_{\mathrm{m}}$ and $\dot{Q}$ have negative values. This is true if $b < 0$, which is always the case in an optical tweezers setup (the optical trap is attractive, i.e., pulls the particle towards the delayed position). The negative sign of $\dot{Q}$ indicates a steady heat flow from the bath to the particle. This corresponds to energy *extraction* from the heat bath, or in other words, feedback cooling of the surrounding fluid. This is a delay-induced phenomenon, which would be impossible in the Markovian counterpart of this system due to the second law, and is one of the goals of "Maxwell demon"-types of devices. We will explain this point in more detail in the next Chapter (see Sect. 10.4.2). The fact that $\dot{S}_{\mathrm{m}} < 0$ (the bath constantly loses entropy) underlines that further entropic terms must contribute to the (nonnegative) total entropy production (as discussed for underdamped dynamics in [8–10]). In the regime of reversed heat flow, the amount of medium entropy loss

[6] The underlying mathematical reason is that the thermodynamic quantities naturally involve higher order correlations.

thus provides a lower bound to the *entropic cost* of the feedback. We will explicitly discuss the entropy balance, as well as the information flow which is a needed to generate such a negative heat flow, in Chap. 10.

9.2.5.2 Growth upon Reduction of Delay Time

The (negative) heat flow is quite pronounced in the regime of small τ, and *increases* in absolute value, when reducing τ (compare with Fig. 9.3). Notably, this is also true for underdamped linear systems, where $\dot{Q}$ only decays for τ *below* the velocity-relaxation timescale (compare with Fig. 1 in [8]). Analytical evaluation of the heat rate for a nonlinear example system, which we will provide below, reveals the same behavior.

Before proceeding with this concrete examples, we here propose an explanation of the phenomenon that $|\dot{Q}|$ increases with τ. When τ is so small that

$$X(t-\tau) \approx X(t) - \tau\dot{X}(t-\tau),$$

the delayed feedback force $F_{\mathrm{d}} = -bX(t-\tau)$ gets a contribution proportional to the strongly fluctuating *velocity*. This changes the induced steady-state heat flow $\dot{Q} = \langle F_{\mathrm{d}} \circ \dot{X}(t)\rangle$ significantly. For $b < 0$, $F_{\mathrm{d}} \sim \tau b\dot{X}(t-\tau)$ has a *friction*-like contribution. As is well known for underdamped systems with velocity-dependent (delayed [8] or non-delayed [12, 13]) feedback, such a control yields medium entropy reduction due to *entropy pumping*. Please note that a velocity-dependent feedback $F_{\mathrm{d}} \sim b\dot{X}(t)$ drives a system out of equilibrium [12, 13] even *without* delay, contrary to a (non-delayed) position-dependent control. Entropy pumping is for example utilized to realize "molecular refrigerators", see [8, 12, 13].

The reason is that the additional friction-like force reduces the thermal fluctuations of the particle, inducing an energy transfer from the bath to the particle, hence, a heat flow. Reversely, the particle fluctuations are enhanced for $b > 0$, yielding a positive heat flow. This means, an additional "entropy pumping" contribution to the heat flow arises for small τ, explaining the enlarged heat flow at small delay times. Please note that this effect is independent of the question whether the overdamped limit is used, or not.

When τ is even smaller and gets below the velocity-relaxation time m/γ, both systems behave differently. For underdamped dynamics, the velocities are correlated, implying that the delay-induced heat flow $\dot{Q} \sim \tau\, b\langle\dot{X}(t-\tau)\dot{X}(t)\rangle \sim \tau\, e^{-\gamma\tau/m}$ eventually decays smoothly to zero (resulting in a maximum of $|\dot{Q}|$ around $\tau = m/\gamma$, see Fig. 1 in [8]). For overdamped dynamics, the velocities are, in contrast, uncorrelated. This is because the white noise directly acts on it $\dot{X} \sim \xi$, giving rise to the nontrivial limit $\tau\,\delta(\tau)$ of type "$0 \times \infty$". This yields a finite value at $\tau \to 0$, as we know from Eq. (9.25).

Hence, in an experimental setup, a position-dependent feedback is expected to induce heat flow, unless the (typically unavoidable) delay is well-below the velocity-relaxation timescale.

9.3 Application to the Bistable Potential

Now, we reconsider the bistable example system involving the doublewell potential with minima at $\pm\sigma$ and potential barrier height V_0, and the delayed optical trap potential $(k/2)[(x/\sigma) - (x_\tau/\sigma)]^2$, which we have already discussed in previous Chapters, see e.g. Sects. 2.5, 7.4, 7.5.2, and 8.1.

The potentials yield a polynomial deterministic force $F = -a_3x^3 - a_1x - bx_\tau$ with coefficients

$$b = -k\sigma^{-2} \tag{9.33}$$

$$a_1 = (-4V_0 + k)\sigma^{-2} \tag{9.34}$$

$$a_3 = 4V_0\sigma^{-4}. \tag{9.35}$$

For this example system, the mean heat rate from Eq. (9.15) involves the even moments up to $6th$ order.

As in Part II, we will consider both: the nontrivial intrawell dynamics within the asymmetric potential wells, and the noise- and delay-induced positional oscillations between the wells. Again, we focus on the regime where the deterministic counterpart of the system has no delay-induced oscillations, i.e., $0 < k < 4V_0$ (see Sect. 2.7). Thus, the oscillations are delay- and noise-induced.

We will also use BD simulations (see Appendix A.1), to validate the analytical results. As a first step, we check the consistency between simulations and theory, and, at the same time, confirm the analytical expression (9.15). Computing the NESS heat rate, both, from the stochastic definition along each simulated trajectory (9.4) and from the numerically obtained PDF (9.15), we indeed find perfect agreement.

In the following, we compare the BD results to those obtained from our analytical approach, i.e., Eq. (9.15) combined with established approximations. Since the approximations are known to perform best when the particle is likely to stay around a potential minimum (see Chap. 7), this approach seems appropriate in the low thermal energy regime ($\gamma D_0 \ll V_0$). At the end, we will introduce a complementary approach for larger noise levels.

9.3.1 *Low Thermal Energy—Intrawell Dynamics*

Figure 9.3 shows $\dot{Q} = \dot{W}$ and $\dot{S}_{\mathrm{m}}$, as functions of the delay time τ, for an exemplary parameter setting in the low noise regime. The simulation results confirm the discontinuous τ-limit with apparent offset given by Eq. (9.25), and the predicted reversed heat flow. Similar results as in Fig. 9.3 but for a different parameter setting are also displayed in Fig. 9.4, which additionally shows the PT approximations. The PT-OU predictions are quantitatively very similar to the ones obtained by the FLC, while the PT-st yields very similar results to the small delay expansion. We will therefore not mention the PT approximations explicitly in the following, but rather discuss only the FLC and the small delay expansion.

While the small delay expansion fails outside the Markovian limit, our analytical approach with FLC makes quantitatively correct predictions for all delay times considered in Fig. 9.3. The thermodynamic quantities grow with τ, until they approach constant values.

9.3.1.1 Saturation

Interestingly, this saturation occurs around the intrawell relaxation timescale, which can roughly be estimated by $\tau_{\rm ir} \approx 1/8\, k_{\rm B}\mathcal{T}/V_0$ in the present case (see Sect. 2.6). Note that this estimation relies on arguments for Markovian systems and is hence rather a crude estimate. The underlying reason for this saturation is the same as of the PDF's saturation, which we have discussed in Sect. 7.4. Figure 9.3 also displays the results from the corresponding linear system, $F = -\alpha x - \beta x_\tau$. The parameters are $\alpha = -(8V_0 + k)/\sigma$ and $\beta = -k/\sigma$, as obtained by a second order Taylor expansion of $V_{\rm s} + V_{\rm d}$ around a (deterministically) stable fix point $(x, x_\tau) = \pm(\sigma, \sigma)$. Interestingly, $\dot{Q}$ is not only equivalent to the nonlinear case for very small τ [as already expected from (9.25)], but also saturates on the same timescale.

9.3.1.2 Reversed Heat Flow and Break down of Approximations

The agreement between BD and FLC persists at larger barrier heights V_0 and at smaller k. However, for larger $k/(\gamma D_0)$, also the FLC approach breaks down, as can be seen in Fig. 9.5. The inset shows that the FLC nevertheless captures the qualitative behavior (but quantitatively, it is really far off). Interestingly, upon increase of k (i.e., the laser intensity in case of optical tweezers [14]), the direction of the mean heat flow can be changed. Thus, in the nonlinear case, one may switch from feedback cooling to heating of the surrounding fluid, and both can be induced by employing *negative feedback*. We recall that this is the type of feedback realizable with an optical trap, where always $k > 0$. This contrasts the linear system (dashed line in Fig. 9.5). There, we could not find a parameter regime where the direction of the heat flow can be revered upon increase of positive k. Instead, the reversal seems to occur when k changes its sign, or upon increase of τ.

The current reversal implies the existence of a certain finite delay force strength k, where the mean heat flow is zero, despite the presence of time-delayed driving. For example, in Fig. 9.5, the BD simulations indicate vanishing heat around $k \approx 1.7\gamma D_0$. Upon increase of τ, the linear system also shows a point of zero heat rate, despite the presence of the delay force, see Fig. 9.4b for an example. Recall that $\delta W = 0$ further implies $\delta W = 0$, while $\mathrm{d}U = 0$ is readily given in the steady state. Thus, *all* energy (ex)changes are zero, as well as the medium EP. Based on this consideration, we cannot see any trace of nonequilibrium here, but we do not think that the system is in thermal equilibrium.[7] We will clarify this point in the next Chapter, where we

[7] Contrary to $k = 0$, where the heat flow nullifies as well and thermal equilibrium is approached.

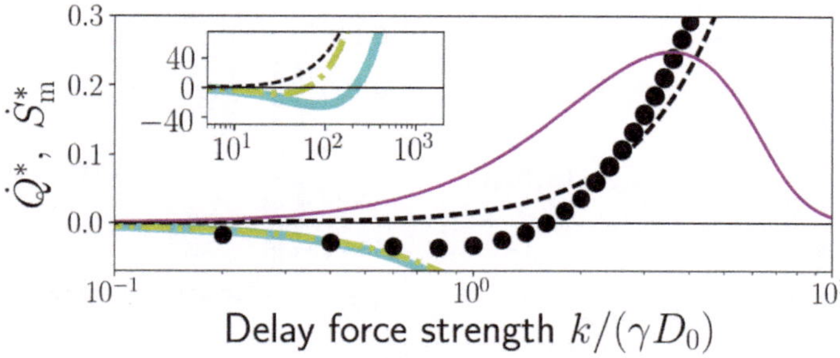

Fig. 9.5 Scaled NESS mean heat rate $\dot{Q}^* = \dot{Q}\,(\tau_{\rm B}/(k_{\rm B}\mathcal{T}))$ and medium entropy production rate $\dot{S}^*_{\rm m} = \dot{S}_{\rm m}\,\sigma^2/(k_{\rm B} D_0)$ versus delay force strength k over thermal energy γD_0; at $V_0 = 4\gamma D_0$, $\tau = \sigma^2/D_0$. Inset: zoom out. Color code as in Fig. 9.3. These results are also published in Ref. [3]

will reconsider the vanishing heat flow from an entropic perspective. An interesting objective of future work would be to see how the fluctuations of the heat and other thermodynamic quantities behave at this point.

9.3.2 *High Thermal Energy—Interwell Dynamics*

When the thermal energy is sufficiently high, jump processes between the potential wells dominate the dynamics. Then, the previously discussed approximations by construction all break down. The reason is that the "memory of a jump" expressed in the non-Markovian ρ_2 (and higher PDFs), is then what dominates the process and, e.g., triggers subsequent jumps; and precisely these higher PDFs are the (over)simplified quantities in all approaches. In particular, the FLC effectively assumes an harmonic potential well out of which no jumps can occur, while the PT and small delay expansion render Markovian systems which don't exhibit spontaneous oscillations, as the latter are a delay-induced phenomenon. This includes situations, where the interplay of noise and delay leads to the spontaneous oscillations of X. However, we can still treat this regime analytically via an alternative strategy, the *2-state reduction*, which bases on the discretized approach for multistable systems [15] (which we have mentioned earlier in Chap. 3).

To motivate this ansatz in the present context, let us first look at the waiting time distributions.

9.3.2.1 Waiting Time Distributions

The delay-induced oscillations manifest in the waiting time distributions, as we have already mentioned in Chap. 7, and exemplary shown in Figs. 7.6, 7.10 there. Here, we revisit this idea and consider in more detail waiting time distributions for

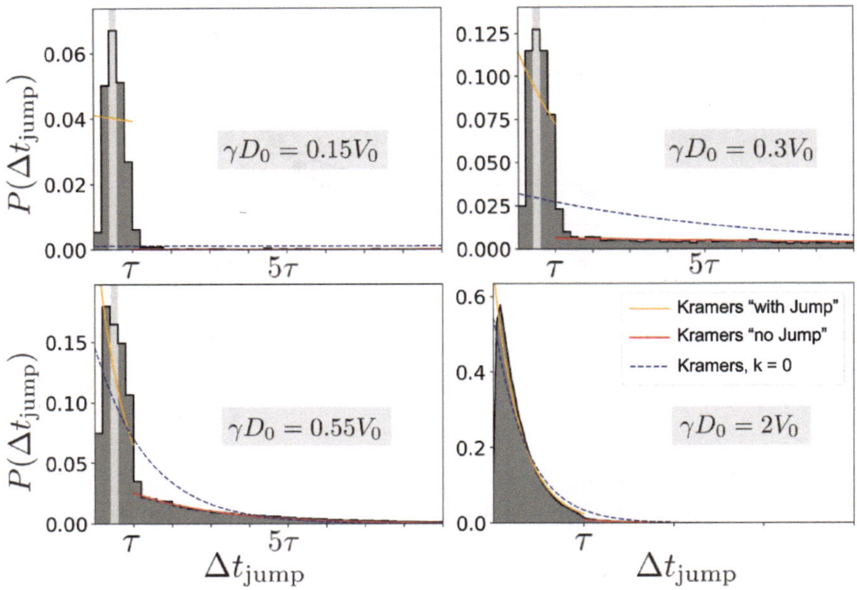

Fig. 9.6 Waiting time distribution for different ratios of thermal energy over barrier height $\gamma D_0/V_0$, below and above coherence resonance (CR). The value $\tau/2$ is marked by a vertical gray line as a guide to the eye. *Upper left:* Small thermal energy below CR. *Upper right:* Thermal energy in the range of CR, pronounced peak around $\tau/2$. *Lower left:* Thermal energy slightly above CR. *Lower Right:* Very high thermal energy. The distributions are numerically (BD) generated (normalized) histograms of the times between sequential jumps Δt_{jumps} in the doublewell potential with linear delay force at $k = 0.6V_0/\sigma$ and $\tau = 5\tau_B$. The orange, blue and red lines indicate estimates based on Kramers theory, see text for details

different thermal energies. Let us first remind the Reader that in the Markovian case, for example at $k = 0$, the distributions are exponential decays whose exponent is given by the Arrhenius formula (see Sect. 2.6.1, specifically the blue histograms in Fig. 7.6). Here, in Fig. 9.6, the corresponding waiting time distributions according to Kramers theory are plotted by blue dashed lines.

Figure 9.6 further shows numerically obtained distributions before, at and after coherence resonance. For very small thermal energy levels below CR (upper left panel), the jumps are very rare with extremely long waiting times before a jump occurs. Some jumps are followed by a quick subsequent jump back to the original valley which occurs within the time interval $[0, \tau]$, because in this time, the delay force drags the particle back. These subsequent jumps yield a small peak within $[0, \tau]$ (note the different y-axis-scales, which make this peak appear higher). We observe that these jumps typically occur two or three times in a row and then randomly stop again. When the thermal energy is increased, the number of escape events grows and also delay-induced *oscillations*, i.e., a sequence of regular jumps with period $\sim \tau$, are getting more likely. These oscillations manifest in a high jump probability around $\tau/2$. This value is marked by a gray vertical line in Fig. 9.6 as a guide to the eye. At thermal energy in the range of CR (upper right and lower left panel), a pronounced peak around $\tau/2$ is shown, caused by the delay-induced oscillations. For even higher thermal energies, there is still a pronounced peak around $\tau/2$, but the exponential background gets stronger. This is because, for high $\gamma D_0/V_0$, the effect of the delayed force on the dynamics becomes less important, as compared to the thermal interwell motion. In the lower left panel an exemplary value slightly above CR is depicted. Finally, for even higher thermal energies, the waiting time distribution is dominated by the exponential thermal decay, and the distributions is very similar to the case of a Markovian Kramers escape problem (lower right panel). The latter is depicted by blue lines in Fig. 9.6.

9.3.2.2 Quasistatic Approximation

As a crude estimation of the waiting time distributions in the non-Markovian case, one can use the following hand waving argument based on a quasistatic approximation. The quasistatic approximation is expected to be justified when the delay time is small compared to the intrawell relaxation time, $\tau \gg \tau_{\text{ir}}$, and $\gamma D_0 < V_0$.

The true system has a history-dependent energy landscape with dynamical barrier. However, when the particle is in the same valley as it was τ times ago, which we denote case 1, the effective energy landscape can roughly be estimated by fixing x_τ to the minima around which x can be found which yields the quasistatic potential $V_{\text{qs}}(x) = V_{\text{s}}(x) + V_{\text{d}}[x, x_\tau = \sigma\,\text{sign}(x)]$. This is approximately the situation for all detected jumps with $\Delta t_{\text{jump}} > \tau$. The red lines in Fig. 9.6 plot the corresponding Kramers estimate, which will be given below.

On the contrary, in case 2, that is, when the particle has jumped within the last τ times, the quasistatic potential reads $V_{\text{qs}}(x) = V_{\text{s}}(x) + V_{\text{d}}[x, x_\tau = -\sigma\,\text{sign}(x)]$. This is approximately the situation for all detected jumps with $\Delta t_{\text{jump}} < \tau$ (orange lines in Fig. 9.6).

The transition rates in cases 1 and 2 can be estimated by the two Kramers rates associated with V_{qs}, given by (2.60) with $r_{\text{K,j}} = r_{\text{K}}(x_{\text{min,j}})$ for escapes out of the two minima, $x_{\text{min},1} = \pm\sigma$ and $x_{\text{min},2} = \mp(\sigma/2)[1 + \sqrt{1-(k/V_0)}] \approx \mp\sigma$, over the barrier at $x_{\max} = \mp(\sigma/2)[1 - \sqrt{1-(k/V_0)}] \approx 0$. In the last step we have used twice $\sqrt{1-(k/V_0)} \approx 1$ for $k \ll V_0$. The corresponding second derivatives can be calculated exactly as $V''_{\text{qs}}(x_{\max}) = -(4V_0 - k)/\sigma^2$ and $V''_{\text{qs}}(x_{\text{min,j}\in\{1,2\}}) = (8V_0 + k)/\sigma^2$, while the potential barrier heights read $\Delta V_{\text{qs},1} = V_0 + k/2$, and $\Delta V_{\text{qs},2} = V_0 - 3k/2$. This finally yields the transition rates

$$r_{\text{K},j\in\{1,2\}} = \sqrt{(8V_0 + k)(4V_0 - k)}/\left(2\pi\gamma\sigma^2\right)\exp\left(-\Delta V_{\text{qs,j}}/\gamma D_0\right), \tag{9.36}$$

One can see in Fig. 9.6 that the Kramers distributions from the quasistatic approximations indeed roughly capture the characteristics of the true jump distributions, at least in the regime $\Delta t_{\text{jump}} > \tau$. Of course, the characteristic shape of the distributions in the interval $[0, \tau]$ is not described by this exponential ansatz, but the decay rates are in fact, surprisingly accurate.

9.3.2.3 2-State Reduction

We have seen that the waiting time distributions can roughly be approximated by two quasistatic situations, where we have fixed the delayed position to be either in the same, or in the opposite potential valley. Inspired by this finding we go one step further, and ask: *Is it possible to describe the interwell dynamics by only taking into account in which minima $X(t)$ and $X(t-\tau)$ are? And is this description sufficient to capture the characteristics from a thermodynamic perspective?* In the regime, where the dynamics is dominated by the jumps between the valleys, it seems somewhat natural to take the intrawell fluctuations out of the description.

This leads us to the state discretization proposed by Tsimring and Pikovsky [15], which is a type of coarse-graining of the system state space. In particular, the X-dynamics is reduced to switching processes between *two discrete* states $s = \pm\sigma$, corresponding to the two potential valleys. There are two different transition rates corresponding to the two possible scenarios that the delayed and instantaneous state have the same, or opposite sign, respectively, which are just given by $r_{\mathrm{K},\{1,2\}}$ from the quasistatic approximation (9.36).

This 2-state reduction brings two simplifications to the calculation of the heat rate from Eq. (9.5). First, various relevant temporal position-autocorrelation functions are significantly simplified, since the intrawell dynamics is omitted. In particular, (for $\tau > 0$)

$$\left\langle X(t)^2\right\rangle \rightarrow \langle s(t)s(t)\rangle = \left\langle \sigma^2\right\rangle = \sigma^2 \tag{9.37}$$

$$\left\langle X^i(t)X(t-\tau)\right\rangle \rightarrow \left\langle s^i(t)s(t-\tau)\right\rangle = \begin{cases} \sigma^i \, \langle s(t-\tau)\rangle = 0, & \forall \text{ even } i \\ \sigma^{i-1} \, \langle s(t)s(t-\tau)\rangle = C_1(\tau)\,\sigma^{i-1}, & \forall \text{ odd } i \end{cases}. \tag{9.38}$$

Thus, all relevant correlations can be expressed via $C_1(\tau)$. Consequentially, Eq. (9.5) simplifies to

$$\begin{aligned} \dot{W} &= (b/\gamma)\left(a_1 + \sigma^2 a_3\right) C_1(\tau) + \left(b^2/\gamma\right)\sigma^2 \\ &= k^2/\left(\gamma\sigma^2\right)\left[1 - C_1(\tau)/\sigma^2\right] = \dot{Q} = \dot{S}_{\mathrm{m}}\gamma D_0/k_{\mathrm{B}}, \end{aligned} \tag{9.39}$$

where we have used that the mean change of internal energy must vanish in the NESS.

Second, the full position-autocorrelation function $C_1(\Delta t)$ can now be calculated analytically, via

$$C_1(\Delta t) = \frac{\left(\sqrt{r_{\mathrm{K},1}} + \sqrt{r_{\mathrm{K},2}}\right) e^{-2\sqrt{r_{\mathrm{K},1} r_{\mathrm{K},2}}\,\Delta t} + \left(\sqrt{r_{\mathrm{K},1}} - \sqrt{r_{\mathrm{K},2}}\right) e^{-2\sqrt{r_{\mathrm{K},1} r_{\mathrm{K},2}}\,(\tau-\Delta t)}}{\sqrt{r_{\mathrm{K},1}} + \sqrt{r_{\mathrm{K},2}} + \left(\sqrt{r_{\mathrm{K},1}} - \sqrt{r_{\mathrm{K},2}}\right) e^{-2\sqrt{r_{\mathrm{K},1} r_{\mathrm{K},2}}\,\tau}}, \tag{9.40}$$

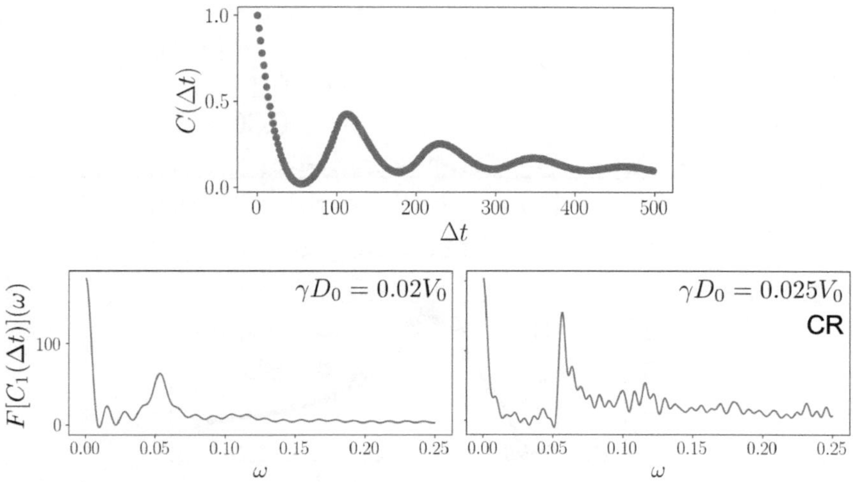

Fig. 9.7 Detecting CR in the temporal position-autocorrelation function $C_1(\Delta t) = \langle X(t)X(t - \Delta t)\rangle$ from numerical results. *Upper plot:* $C_1(\Delta t)$ itself (at CR), which shows peaks around multiples of τ. *Lower plot:* The power spectrum, i.e., the Fourier transformation of $F[C_1(\Delta t)](\omega)$, *left:* at a value of $\gamma D_0/V_0$ before, and *right*: at CR. The position of the main peak around $\omega = 0.057$ is close to $2\pi/\tau$. The height of this peak is used as order parameter to detect coherence resonance. Here, the parameters are $V_0 = 4$, $k = 0.1$, and $\tau = 100$. This corresponds to the parameters $k = 0.4V_0$, $\tau = 25\gamma\sigma^2/V_0$, which are also considered in the following Figs. 9.8 and 9.9

which was derived in [15]. We have plotted $C_1(\Delta t)$ for three exemplary values in Fig. 9.8. Thus, from Eqs. (9.40, 9.39), the mean heat rate and medium EP rate can readily be evaluated. Corresponding results are plotted in Fig. 9.9 by magenta lines.

9.3.2.4 Coherence Resonance

As we have already mentioned in Sect. 2.7, the delay-induced oscillations manifest in the position-autocorrelation function. The latter can be computed by numerical simulations, and from the 2-state reduction (9.39). The delay-induced oscillations have a period of about τ. Thus, the more regular and likely the oscillations are, the more pronounced are the peaks around τ, 2τ, 3τ, ...in $C_1(\Delta t)$. This shows up as a significant peak around a frequency of $\omega = 2\pi/\tau$ in the power spectrum, i.e., in the Fourier-transformation of $C_1(\Delta t)$. As suggested in [15], we use the peak's height as an order parameter to detect CR. Figure 9.7 shows a numerically obtained correlation function and the corresponding power spectrum, while Fig. 9.8 shows power spectra from the 2-state reduction. Figure 9.8 further displays the order parameter from both, simulation and approximation, which give very similar results, confirming the idea of employing a 2-state reduction. The order parameter has a maximum at $0.36\gamma D_0/V_0$ indicating CR.

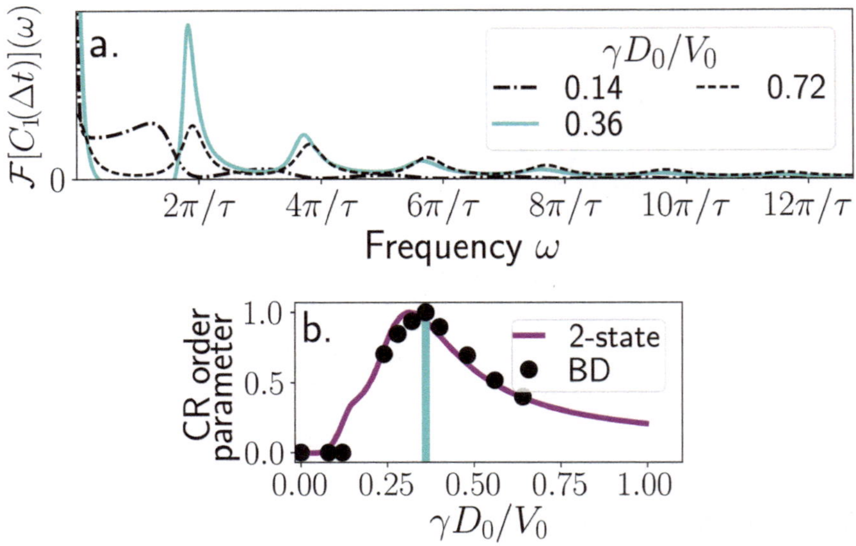

Fig. 9.8 Detecting coherence resonance (CR) in the temporal position-autocorrelation function in the bistable system with $k = 0.4V_0$, $\tau = 25\gamma\sigma^2/V_0$. **a** Analytical power spectrum (from 2-state model) given by the Fourier transform of $C_1(\Delta t) = \langle X(t)X(t-\Delta t)\rangle_{ss}$ [Eq. (9.40)], at three thermal energies: below (0.14), at (0.36) and above (0.72) the value of CR. Note that Fig. 9.7 shows corresponding results from numerical simulation, which are naturally a bit noisy. **b** Order parameter: Normalized main peak at $\omega \approx 2\pi/\tau$ of the power spectrum of C_1, from BD (symbols) and from the 2-state model (magenta line). The vertical line at $0.36\gamma D_0/V_0$ indicates the maximum around which the system is coherence-resonant. These results are also published in Ref. [3]

9.3.2.5 Heat and Medium Entropy Production

The numerical data in Fig. 9.9 reveals that, when the thermal energy is small compared to V_0, $\dot{Q}$ increases linearly with γD_0, as it also does for a linear delayed system [see Eq. (9.17)]. In this regime, approximating the doublewell by its linearized version (dashed line) even renders the correct slope. By contrast, the 2-state model yields $\dot{Q} = 0$. This is expected, since the intrawell dynamics is completely neglected. At larger $\gamma D_0/V_0$, the slope of $\dot{Q}$ abruptly changes and nonlinear behavior sets in. This occurs at the onset of the delay-induced oscillations, as reflected by a sudden increase of the CR order parameter from both, the 2-state reduction (9.40), and from numerical BD simulations (see Fig. 9.8). The actual value of $\gamma D_0/V_0$ where the nonlinear behavior sets in, is predicted more or less correctly by the 2-state reduction.

By further increasing $\gamma D_0/V_0$, one enters the regime where delay-induced oscillations dominate the dynamics. At a certain finite thermal energy of about $0.36\gamma D_0/V_0$ (indicated by cyan vertical lines in Figs. 9.8 and 9.9, and plotted with a cyan line in Fig. 9.8a.), the peak in the power spectrum is highest, resulting in a maximum of the order parameter. In this range of thermal energies, the delay-induced oscillations with mean period $\approx \tau$ are most pronounced.

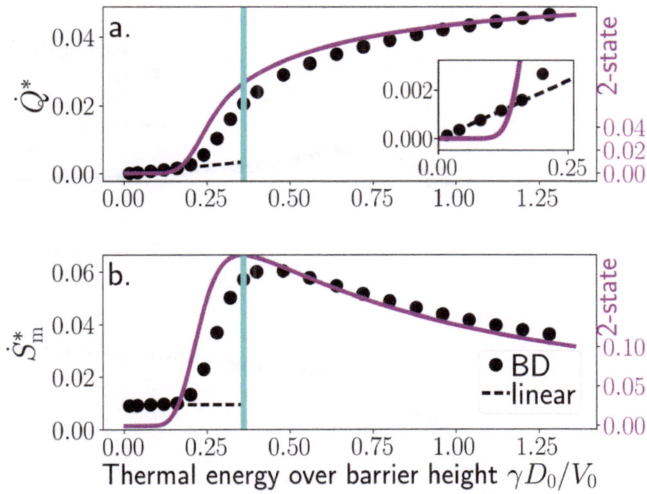

Fig. 9.9 Influence of the ratio of thermal energy γD_0 and barrier height V_0 on mean heat and medium entropy production, in the delayed bistable system at $k = 0.4\,V_0$, $\tau = 25\,\gamma\sigma^2/V_0$. a. Heat rate $\dot{Q}^* = \dot{Q}\,(\gamma\sigma^2/V_0^2)$, the inset gives a magnification, b. Medium EP rate $\dot{S}_m^* = \dot{S}_m\,(\gamma\sigma^2/V_0 k_B)$. Magenta lines and y-axes display the results from the 2-state reduction. Vertical cyan lines: Maximum of CR order parameter (see Fig. 9.8). The symbols are the (BD) simulations, the dashed lines correspond the the linearized system. These results are also published in Ref. [3]

We find that in this range of $\gamma D_0/V_0$, the 2-state reduction renders a qualitatively good approximation. In particular, it predicts accurately that a region of steep slope is followed by a lower slope of $\dot{Q}$ accompanied by a maximum of $\dot{S}_m$, and that the latter lies in the regime of CR. However, it yields an overestimation (note the different Y-axis scales).

9.3.2.6 Generality

To test the generality of the discussed behavior and of the approximations, we have tested several parameter settings. Figures 9.10 and 9.11 explicitly show the results for the case $k = 1\,V_0$, $\tau = 5\,\gamma\sigma^2/V_0$. We again observe the coincidence of CR order parameter maximum and maximum in the medium EP. The plots in Fig. 9.10 are complemented by the results for the corresponding linearized system. Like in Fig. 9.9, the linear system has a very comparable heat rate in the regime of small $\gamma D_0/V_0$, where the particle only "sees" its approximately quadratic surrounding (dashed cyan lines). This time, we have also considered the regime of *high* $\gamma D_0/V_0$, where the particle does not "see" the potential barrier (blue dashed lines). Here we also find that the heat rate is similar to a particle in the corresponding quadratic trap. The agreement with the numerical results for the doublewell potential is remarkable, especially given the fact that the fourth order potential is not well-describable by a

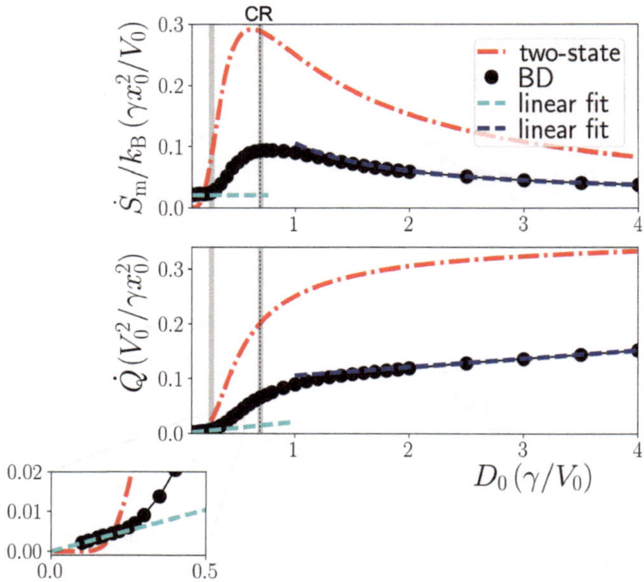

Fig. 9.10 Mean heat rate and medium EP versus thermal energy over barrier height $\gamma D_0/V_0$, for other parameters than in Fig. 9.9, here $k = V_0$, $\tau = 5\gamma\sigma^2/V_0$. Again, the maximum of the medium EP is around coherence resonance (CR), which is at $\gamma D_0 \approx 0.7V_0$ (dashed vertical lines). The 2-state model (red dashed-dotted line) correctly predicts the behavior of heat and medium EP, but overestimates them by a factor of three, as in the other parameter case. Moreover, the linear system yields again a good approximation for very small $\gamma D_0/V_0$ (cyan dashed line), where the intrawell dynamics dominate, as well as for the high $\gamma D_0/V_0$, where the particle does not see the barrier anymore (blue dashed line)

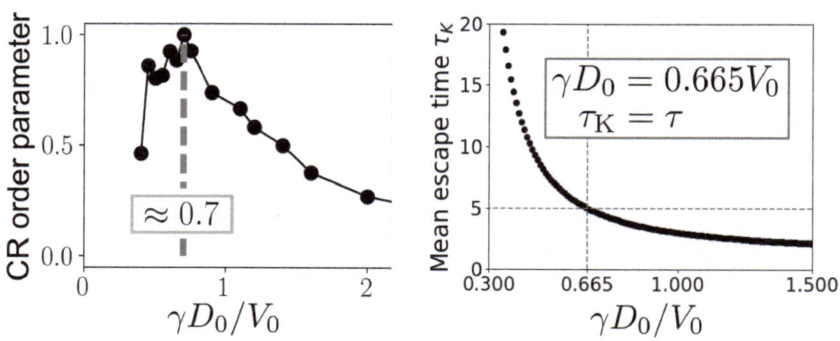

Fig. 9.11 *Left panel:* Coherence resonance (CR) order parameter, given by the (normalized) height of the first peak in the power spectrum of the temporal position-autocorrelation function, as explicitly shown in 9.7, where a different parameter setting is considered. The peak is highest around $\gamma D_0 \approx 0.7D_0$, indicating highly regular oscillations with mean period τ. *Right panel:* Kramers escape times, in the absence of delay force. In the $\gamma D_0/D_0$ regime of CR, the escape times are comparable to the delay time. Thus, one can say that the noisy system has a *resonant* response to the delay force. Parameters as in Fig. 9.10

quadratic trap. For this reason, we do not expect this agreement to generally hold. It is nevertheless an indication for the fact that a linearized system captures the dynamics quite well, except for the delay-induced oscillations, which dominate around CR.

For the second parameter case, Fig. 9.11 shows the escape times estimated by the Kramers rates in the corresponding system without delay ($\tau = 0$). Interestingly, these escape times are comparable to the delay time, $\tau_K \approx \tau$, at the very ratio of thermal energy and barrier height, where CR is observed (compare left and right panel in Fig. 9.11). This justifies the idea that the pronounced peak in the power spectrum signals a *resonant* response of the system to the delay.

We have tested several parameter settings confirming that the $\dot{S}_m$ is indeed maximal at CR conditions as predicted by the 2-state model. Quantitatively, we always observe an overestimation of the heat rate by a factor of about three, as can be seen in Figs. 9.9 and 9.10. This is somewhat surprising, since the phase-space reduction and coarse-graining inherent to the discretization is rather expected to yield an *underestimation* of the heat and medium EP. This was explicitly shown in Ref. [16], where, however, a different kind of coarse-graining was considered. A possible explanation lies in the involved quasistatic approximation, or it could be due to the details of this very coarse-graining, but the precise reason is unclear. This would be an interesting subject of future research.

9.3.2.7 Explanation for the Observed Maximum of Entropy Production at CR

The behavior of $\dot{Q}$ and $\dot{S}_m$ can be understood on the basis of the 2-state model, specifically Eq. (9.39), which neglects the intrawell fluctuations. If the particle rests in one potential valley, this yields $C_1(\tau) = 1$. Further, when the delay-induced oscillations occur, one would still expect $C_1(\tau) = 1$ for perfect oscillations, because they have mean period τ. However, due to the stochastic nature of their period (see, e.g., Fig. 9.6) and because they set in and pause randomly, sometimes after an odd number of jumps, $C_1(\tau)$ is on average lowered with the occurrence of oscillatory events. When the timescales of noise-induced escapes and oscillatory motion become comparable, the particle dynamics responds resonantly to the noise. A slight increase of $\gamma D_0/V_0$ then significantly increases the number of occurring oscillation periods, yielding a strong reduction of $C_1(\tau)$. Thus, $\dot{Q}$ steeply increases and $\dot{S}_m$ reaches high values [Eq. (9.39)]. At the CR maximum (vertical cyan lines in Figs. 9.8, 9.9), this effect saturates, resulting in a reduced slope of $\dot{Q}$ and a *maximum* of $\dot{S}_m$ at CR.

For even higher γD_0, the superimposing noise generates irregular, low correlated motion with $C_1(\tau) \to 0$, hence $\dot{S}_m \to 0$ in the 2-state model. In contrast, $\dot{S}_m$ of the full system approaches a (nonzero) constant (not shown here). The breakdown of the 2-state approximation in this limit is indeed expected since the state discretization then becomes meaningless.

In the next Chapter, we will again present analytical results for the mean heat flow in non-Markovian systems. There, we will focus on linear systems with distributed delay and colored noise. But let us first briefly consider the distributions of the heat, work and internal energy fluctuations in the presence of delay, which we have, so far, not discussed.

9.4 Preliminary Numerical Results for Fluctuation of Heat, Work and Internal Energy

So far, we have considered the ensemble averaged rates of the energy flows associated with the delay force. However, as these energy flows are fluctuation quantities, they are not only characterized by their mean, but by an entire probability distribution, quantifying the chances of measuring a particular value along a random trajectory. A systematic study of the distributions in the presence of time-delayed feedback force is an objective of future work and beyond the scope of this thesis. But let us here nevertheless show some preliminary numerical results to give an impression of how they look like. As a nonlinear example system, we again consider the particle in the doublewell potential with and without the delayed feedback force.

First, we consider the equilibrium approached in the absence of delayed feedback, shown in the left panel of Fig. 9.12. All mean values are zero in this case, as the system is in thermal equilibrium (ensemble averages are plotted with vertical lines). Please note that we here consider parameters, where escape events are seldom. Furthermore, as there are no external nonconservative forces acting on the particle, there is no work performed on it, thus, w is strictly zero along each fluctuating trajectory. In contrast, the internal energy is a fluctuating quantity, owing to the fact that the particle can have a different internal (potential) energy e depending on its position within the static

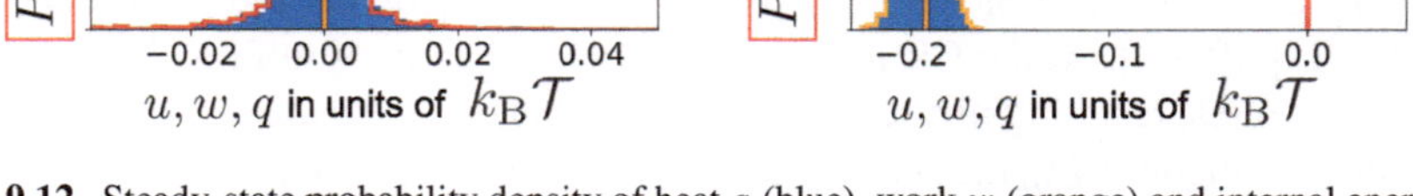

Fig. 9.12 Steady-state probability density of heat q (blue), work w (orange) and internal energy e (red) along fluctuating trajectories of length $\Delta t = 200\tau_B$ from numerical simulations with 5×10^4 realizations, in different situations. The averages are marked with vertical lines in the respective color. In steady states, one always has $\langle e \rangle = 0$. *Left:* Particle in doublewell potential (with barrier height $V_0 = 0.1k_B\mathcal{T}$) in the absence of delayed feedback force. The system is in an equilibrium, thus $w \equiv 0$, and $\langle q \rangle = \langle w \rangle = 0$. *Right:* In the presence of delayed feedback (with $k = 0.1k_B\mathcal{T}$, $\tau = 0.1\tau_B$), without external static potential, thus $e \equiv 0$. The system is in a NESS characterized by $\langle q \rangle = \langle w \rangle \neq 0$

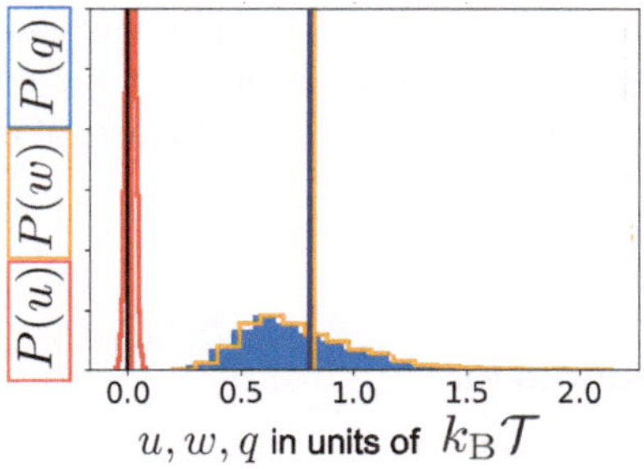

Fig. 9.13 Steady-state probability distributions of heat q (blue), work w (orange) and internal energy e (red), like in Fig. 9.12. Here, the doublewell potential and the delayed feedback are both present, thus, heat, work and internal energy are all fluctuating quantities. As always in a NESS: $\langle e \rangle = 0$ and $\langle q \rangle = \langle w \rangle \neq 0$. Parameters: $V_0 = k = 2k_B\mathcal{T}$, and $\tau = \tau_B$

doublewell potential. The particle can increase its potential energy due to a kick from the surrounding heat bath, associated with an energy flow from the bath to the particle, i.e., a negative heat flow. On the other contrary, when the particle slides down the static potential loosing its internal (potential) energy, it automatically transfers energy to the bath via friction, i.e., a positive heat flow. Thus, the distributions of internal energy and the heat are identical mirror images from each other, as can be seen in Fig. 9.12 (left panel). $P(q)$ resembles the equilibrium distribution in a linear system, which was analytically and numerically investigated in [17], see Fig. 4.1. However, the bistable potential considered here, renders a slightly more peaked distribution than the second order Bessel function observed in the linear case [17].

In the right panel of Fig. 9.12, the system in the absence of the bistable static potential is shown, but with feedback force. Here, the linear delayed feedback force induces nonzero mean heat and work, with distributions that are symmetric with respect to their mean value. The distributions appear to be Gaussian-like, yet they are non-Gaussian. We could not find literature results for the distributions of the stochastic energies induced by delay forces to compare with. In Markovian systems, a linear external force (for example, a constant drag of the particle by constantly moving the harmonic trap), yields a Gaussian heat distribution. This was also shown in [17].

Finally, when combining the nonlinear static forces from the bistable potential, with the linear delayed feedback force, the distributions of all fluctuating energies are non-symmetric with a pronounced skewness, as shown in Fig. 9.13. The question whether the mean heat and work are positive or negative, depends on the precise values of k and τ, as we have discussed earlier in this Chapter, see for example Fig. 9.5. At first sight, we could not detect apparent differences in the distributions between cases with $\dot{Q} > 0$ and $\dot{Q} < 0$.

In the next Chapter, we will come back to fluctuations of thermodynamic quantities, specifically, we will discuss the distribution of the total entropy for the fully linear system with feedback forces involving distributed delay. We will show that the entropy fulfills well-known fluctuation theorems, which impose universal symmetry conditions on the distributions. Finding corresponding statements for the distributions of stochastic work, heat and internal energy in the non-Markovian case considered here is an open problem, which has, to the best of our knowledge, not been tackled to date.

9.5 Concluding Remarks

In this Chapter, we have investigated the steady-state heat flow and medium entropy production induced by time delay, in linear and nonlinear systems. We have seen that the presence of a delay force is almost always accompanied by net heat flow, except for rare parameter settings. The associated entropy production implies that the system operates far from equilibrium. The heat flow can be negative corresponding to energy extraction of the heat bath. We have further shown that the overdamped limit can be problematic, if τ gets below the velocity-relaxation timescale, and linked the behavior at small τ to the phenomenon of entropy pumping. Moreover, we found that the medium entropy production due to the feedback is maximal, when the delay-induced oscillations are coherence-resonant.

Thus, by investigating the mean heat flow and medium EP, we could already answer several of the initially posed questions. Such an energetic investigation is an important first step towards an understanding of thermodynamic notions in delayed systems. In the next Chapter, we will discuss the complete entropy balance. While this turns out the be a much harder task, it will, at the same time, provide a more profound physical picture.

References

1. Sekimoto K (2010) Stochastic energetics, vol 799. Springer, Berlin
2. Xiao T (2016) Heat dissipation and information flow for delayed bistable Langevin systems near coherence resonance. Phys Rev E 94
3. Loos SAM, Klapp SHL (2019) Heat flow due to time-delayed feedback. Sci Rep 9:2491
4. Munakata T, Iwama S, Kimizuka M (2009) Linear stochastic system with delay: energy balance and entropy production. Phys Rev E 79
5. Frank TD, Beek PJ (2001) Stationary solutions of linear stochastic delay differential equations: applications to biological systems. Phys Rev E 64
6. Seifert U (2012) Stochastic thermodynamics, fluctuation theorems and molecular machines. Rep Prog Phys 75, 126001
7. Loos SAM, Klapp SHL (2017) Force-linearization closure for non-Markovian Langevin systems with time delay. Phys Rev E 96
8. Munakata T, Rosinberg ML (2014) Entropy production and fluctuation theorems for Langevin processes under continuous non-Markovian feedback control. Phys Rev Lett 112
9. Rosinberg ML, Munakata T, Tarjus G (2015) Stochastic thermodynamics of Langevin systems under time-delayed feedback control: second-law-like inequalities. Phys Rev E 91

10. Rosinberg ML, Tarjus G, Munakata T (2017) Stochastic thermodynamics of Langevin systems under time-delayed feedback control. II. Nonequilibrium steady-state fluctuations. Phys Rev E 95
11. Gardiner CW (2002) Handbook of stochastic methods, 2 edn. Springer, berlin
12. Kim KH, Qian H (2004) Entropy production of Brownian macromolecules with inertia. Phys Rev Lett 93
13. Kim KH, Qian H (2007) Fluctuation theorems for a molecular refrigerator. Phys Rev E 75
14. Blickle V, Bechinger C (2012) Realization of a micrometre-sized stochastic heat engine. Nat Phys 8:143
15. Tsimring LS, Pikovsky A (2001) Noise-induced dynamics in bistable systems with delay. Phys Rev Lett 87
16. Esposito M (2012) Stochastic thermodynamics under coarse graining. Phys Rev E 85
17. Imparato A, Peliti L, Pesce G, Rusciano G, Sasso A (2007) Work and heat probability distribution of an optically driven Brownian particle: Theory and experiments. Phys Rev E 76

Chapter 10
Entropy, Information and Energy Flows

> *[W]e must return to Maxwell's original concern about the second law and try to address the basic problems of statistical mechanics,*
> *such as the emergence of the macroscopic world and the subjectivity of entropy,*
> *in the light of a general physical theory of information.*
>
> Juan Parrondo, 2015 [1]

In the last chapter, we have studied one component of the steady-state entropy balance, the medium entropy production and associated heat flow. While this investigation has already provided interesting insights, open questions remain. First, we observed that the mean heat rate can vanish at some specific, rare parameter values, despite the presence of time-delayed feedback. On the basis of the energetic consideration alone, we cannot *clarify whether these points correspond to equilibrium.*

Second, we have seen that the medium EP can be negative. We recall that this would not be possible in a Markovian system with position-dependent driving, where the medium EP is the only entropic contribution and must be non-negative, as stated by the second law. The negative medium EP observed in the case of time-delayed position-dependent driving, thus, indicates *missing entropic terms*. While the amount of negative medium EP provides a lower bound to these missing entropy contributions, we could, so far, not get any further physical insights about what these terms look like and how they arise. For this purpose, we shall investigate the whole entropy balance.

Due to the presence of delay in our equations, we run into a major problem of stochastic thermodynamics, that is, the notion of entropy production of non-Markovian systems, which is connected to the role of hidden degrees of freedom and coarse-graining. We have introduced this problem in Chap. 4. We recall the

© The Author(s), under exclusive license to Springer Nature Switzerland AG 2021

S. A. M. Loos, *Stochastic Systems with Time Delay*, Springer Theses,
https://doi.org/10.1007/978-3-030-80771-9_10

technical problem associated with the acausality of the backward process, and we recall the different attempts to formulate an effective thermodynamic description, which yet comprises ambiguities. We will not focus on this debate, but rather go one step back, and reconsider the first step of the whole problem.

Main goal and general idea In this last Chap. 10, we will investigate a specific way to address the problem. The embedding technique introduced in Part II, yielded a *Markovian* network description of the delayed process. Let us recall the Eq. (6.1),

$$\gamma \dot{X}_0(t) = \quad -kX_n(t) + F_0[X_0] \quad +\sqrt{2k_B \mathcal{T}_0 \gamma}\, \xi_0(t) \tag{10.1a}$$

$$\gamma' \dot{X}_j(t) = (n/\tau)[X_{j-1}(t) - X_j(t)] + \sqrt{2k_B \mathcal{T}' \gamma'}\, \xi_j(t), \tag{10.1b}$$

with $j \in \{1, 2, \ldots, n\}$. This representation allows application of the standard formulae of stochastic thermodynamics. In this chapter, we will investigate whether the thermodynamic picture obtained by studying the total EP of the network (10.1) (i.e., of the "super-system"), instead of the delay LE, *is meaningful and interesting.*

We will obtain entropic contributions of the $X_{j>0}$, which are so far only abstract auxiliary variables. On this abstract level, these entropy terms can be considered *thermodynamic cost of the memory* [2–5]. However, the physical meaning of these terms is subject of an ongoing debate. To give an example, we will suggest an interpretation for the $X_{j>0}$ for a specific realization of a feedback controller. From the viewpoint of effective thermodynamics, our investigation represents a first step, which the effective thermodynamics could be build upon.

As mentioned above, one view is to consider the $X_{j>0}$ variables as degrees of freedom of the controller (i.e., the controller is now included in the model, contrary to the considerations in the rest of this thesis). At this point, one should stress that our approach is different from studies, where the memory of a feedback-controller is realized with a *tape* [1, 6–11]. In particular, in the interpretation presented here, all degrees of freedom are "physical", meaning that the forces they perform are associated with energy cost. To further differentiate from earlier studies, we note that the description employed here is a space-continuous, stochastic one, which contrasts earlier bipartite-models basing on Master equations, like [7, 12, 13]; and the controller acts time-continuously (different from a "Maxwell demon" or Szilárd engine), and is thus an autonomous control [7].

There is yet another way to tell the following story. As a second application of the investigation presented in this chapter serve systems of type (10.1), where all d.o.f. X_j are physical subsystems. In such a situation, the total EP calculated here has a clear meaning, it is the true thermodynamic dissipation and irreversibility measure.

The network (10.1) has unidirectional coupling, which is a specific type of *non-reciprocal* interactions. Such interactions are very uncommon in statistical physics and thermodynamics, as mechanical systems describable by Hamiltonians are always reciprocally coupled (usually the underlying equations of motion fulfill Newton's laws). Thus, a significant part of this chapter is dedicated to the discussion of NR interactions, which have not been studied in earlier literature from a thermodynamic perspective. We will show that non-reciprocity has various interesting implications.

Contrary to the rest of this thesis, we will here not only consider feedback with discrete delay, and LEs with white noise, but we will also consider more general cases with colored noise and memory kernels, which can be obtained, e.g., by keeping n finite in the description (see Sect. 6.1). The colored noise will be linked to *measurement errors* [6, 14, 15] of the controller, which we have also not considered so far. Stepping away from the concrete example of an optical tweezers feedback controller, distributed delay is indeed the more realistic scenario for many delayed processes, e.g., in biological contexts [16, 17]. Furthermore, we will specify all deterministic forces to be *linear*, i.e., $F \propto X_0$. This is because we are interested in the main idea, and further aim to discuss the concepts based on exact results. However, extensions to the nonlinear models considered earlier in this thesis would in principle be possible (at least numerically).

10.1 Emergence of Non-monotonic Memory

An important property of the time-delayed systems is that the memory is sharply peaked around a particular value of the past, i.e., $t - \tau$. This implies a non-monotonic memory kernel. Stepping away from the (idealized) case of a delta-peaked kernel, i.e, discrete delay, a non-monotonic kernel with a pronounced maximum seems a reasonable model.

Here we will investigate the connection between topological properties of the Markovian network representation and the non-monotonicity of the kernel in the non-Markovian projected equation, by using a simple example network with three d.o.f. In Sect. 6.1, we have already seen that the Markovian embedding of the time-delayed process yielded NR interactions. However, let us investigate here, whether we really need the non-reciprocity, or if we can obtain a similar kernel with a symmetrically coupled network.

Let us consider as an illustrative example an all-to-all coupled network of three nodes X_j with $j \in \{0, 1, 2\}$, described by

$$\begin{aligned} \dot{X}_0 &= -(p+\kappa)X_0 + pX_1 - \kappa X_2 + \xi_0, \\ \dot{X}_1 &= -\kappa X_0 - (p+\kappa)X_1 + pX_2 + \xi_1, \\ \dot{X}_2 &= +pX_0 - \kappa X_1 - (p+\kappa)X_2 + \xi_2, \end{aligned} \tag{10.2}$$

as schematically shown in Fig. 10.1. This network can also be written as

$$\dot{X}_j = -(p+\kappa)X_j + pX_{j+1} - \kappa X_{j-1} + \xi_j, \tag{10.3}$$

with the periodic index $j \in \{0, 1, 2\}$.

Projection Analogously to the projection in Chap. 6, we project this set of equations onto X_0. To this end, we perform a Laplace-transformation and iteratively solve the algebraic equations for X_0. In this way, we immediately obtain

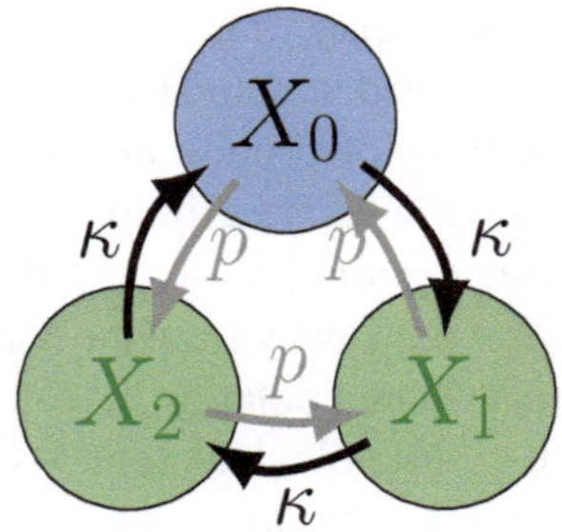

Fig. 10.1 All-to-all coupled network of three nodes. For $\kappa = p$, the network corresponds to a model for three colloids reciprocally coupled by springs of stiffness κ. For $p \to 0$, the network is unidirectionally coupled

$$\begin{aligned} s\hat{X}_0 =& p\left[\frac{p^2}{(s+p+\kappa)^2 - p\kappa} + \frac{\kappa(s+p+\kappa)}{(s+p+\kappa)^2 - p\kappa}\right]\hat{X}_0 \\ &+ \kappa\left[\frac{\kappa^2}{(s+p+\kappa)^2 - p\kappa} + \frac{p(s+p+\kappa)}{(s+p+\kappa)^2 - p\kappa}\right]\hat{X}_0 - \kappa\hat{X}_0 - p\hat{X}_0 + \mathcal{O}(\hat{\xi}_j). \end{aligned} \tag{10.4}$$

Transforming back to real space yields a non-Markovian description of the dynamics of X_0, in particular,

$$\dot{X}_0(t) = -pX_0(t) - \kappa X_0(t) + \int_0^t K(t-t')X_0(t')\mathrm{d}t' + \xi_0(t) + \nu(t), \tag{10.5}$$

with a colored noise ν, and the memory kernel

$$\begin{aligned} K(t-t') =& \frac{e^{\left(-\sqrt{p\kappa}+p+\kappa\right)t' - t\left(\sqrt{p\kappa}+p+q\right)}}{2\sqrt{p\kappa}} \times \\ & \left[\left(p^{3/2} + \kappa^{3/2}\right)^2 e^{2\sqrt{p\kappa}t} - \left(p^{3/2} - \kappa^{3/2}\right)^2 e^{2\sqrt{p\kappa}t'}\right]. \end{aligned} \tag{10.6}$$

In the same manner, we have derived the corresponding equation for a network, where all couplings are different, e.g., X_0 is coupled to X_1 via κ_{01}, and back via p_{01} and so on. However, because the results have the same overall characteristics, while yielding very cumbersome expressions; we omit to discuss this case here, and focus on this simpler case. Let us now consider the memory kernel (10.6) generated by the network (10.2) in more detail, for different cases.

Reciprocal coupling For $\kappa = p$, the network has *reciprocal* coupling. Then, it can be considered as a model for three overdamped particles connected by harmonic springs with spring constant κ, which has the corresponding Hamiltonian $H = (\kappa/2)\sum_{i\neq j}(X_i - X_j)^2$. Similar models have been considered from a thermodynamic perspective in, e.g., [2, 18].

In this case, the memory kernel (10.6) simplifies to an exponential decay,

$$K(T) = 2\kappa^2 e^{-\kappa T}. \tag{10.7}$$

We observe the same qualitative behavior for larger networks with $n > 2$, or more general coupling topologies.

Non-reciprocal coupling In any other case, $\kappa \neq p$, the nodes have NR interactions. In our model, the coupling is then stronger in one direction of the ring compared to the other direction. Considering (10.6), we observe that the memory automatically becomes *non-monotonic*. Specifically, a maximum at a finite $t - t'$ emerges. Now, let us consider an extreme case, where the equation simplifies.

In the limit $p \to 0$, the network is unidirectionally coupled. In this limit, it in fact becomes a specific realization of our embedded model (10.1) introduced in Sect. 6.1 (specifically, see the inset of Fig. 6.2). In this case, the memory kernel (10.6) converges to

$$K(T) = \kappa^3 T e^{-\kappa T}, \tag{10.8}$$

which is a Gamma-distribution with a maximum near $\kappa/3$. We have displayed such a kernel for $\kappa = 3$ in Fig. 6.2b.

Playing around with different coupling topologies, we generally find that reciprocal couplings yield monotonic memory kernels. We thus conclude that NR coupling is a crucial ingredient to generate non-monotonic memory.[1] However, we aim to note that we have not tried out nonlinear couplings, or nodes with nonlinear "local" dynamics.

In the following, we will consider our embedded model (10.1), which is essentially a generalization of the unidirectional ring network to arbitrary lengths. Before we proceed with exploring the thermodynamic properties, we discuss a possible interpretation of the d.o.f. $X_{j>0}$ in the case of the feedback controller.

10.1.1 Interpretation of $X_{j>0}$ in the Case of a Feedback Controller

Let us take a closer look at our prime example of non-Markovian systems considered throughout the thesis, i.e., feedback control of colloidal particles via optical traps.

In experimental setups, the delay is typically created within a memory device, e.g., the computer's memory, storing the past trajectory of length τ. To model the memory device, we step away from an actual computer (which would be hopelessly complicated to model and treat analytically). In a simplistic manner, the additional d.o.f. in the network (10.1) may be interpreted as the *memory cells* of the controller. Let us explain this a bit better.

[1] This observation seems reasonable. A non-monotonic memory indicates that the information is passing through the network as a "package". To realize this, one somehow has to break the symmetry of the dynamics.

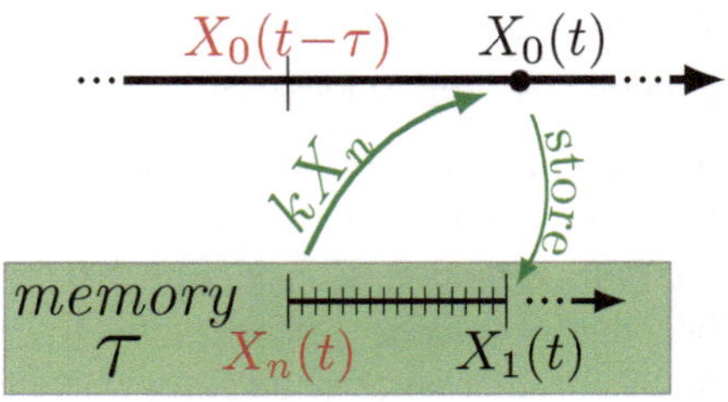

Fig. 10.2 Relation between the temporal evolutions of the system X_0 and the n "memory cells" $X_{j>0}$ of the controller

Imagine a memory device with a "shift register" type of architecture, as sketched in Fig. 10.2. At every instant in time, the information stored in each cell X_j is shifted one step further from j to $j+1$, while the new information (gathered by a measurement operation) is stored in the first cell X_1. Thus X_n stores the oldest information and is used to perform the feedback control force kX_n. Mathematically, this operation can be represented in discretized form as $\Delta X_j = \frac{\Delta t}{\tau/n}[X_{j-1} - X_j] + \epsilon_j$. The ϵ_j are random numbers accounting for errors, e.g., from measurement, rounding operations, or a finite temperature of the device. The updates of the memory cells are scaled with τ/n to enable storage of a trajectory of length τ in n cells.

Since we aim to model *continuously operating* feedback loops, we assume that the updates are performed (infinitely) fast, that is, $\Delta t \to 0$, yielding the dynamical equations of X_j (10.1b). For the sake of simplicity, we have set all $X_{j>0}$ to the same temperature, $\mathcal{T}'$. In the present context this implies the magnitude of error is identical in each memory cell.

We indeed found a similar consideration in [6]. As suggested there, such a memory device could be realized by n colloidal particles at positions X_j, each trapped in a harmonic trap of stiffness n/τ centered around the position of the proceeding colloid, X_{j-1}. This idea is, of course, rather an idealized thought experiment. Still, it is useful to illustrate how a memory kernel concentrated around a specific instant in time of the past (the delay time), can result from coupled exponential relaxation processes. It further demonstrates that an implementation of (non-monotonic) memory, with physical d.o.f. is inevitably associated with thermodynamic cost and dissipation, which we will discuss in the following.[2]

Complex bio-chemical systems as controller Turning away from artificial systems or computer models described above, and towards actual physical, biological, chemical systems, such a "memory device" could for example also be a cascade of chemical reactions.

In fact, in [19] a model for a *cellular sensor* (corresponding to X_1) with one "memory cell" (i.e., X_2) is proposed, which exactly corresponds to our unidirectionally coupled model (10.1) with $n = 2$ and $k = 0$ (i.e., no backcoupling from X_2 to X_0, since [19] doesn't consider feedback). The focus of Ref. [19] is to calculate information flows for $n = 1, 2$. We will as well consider our systems from that per-

[2] Interestingly, this thought experiment also illustrates that the friction terms $-\gamma' \dot{X}_i$ may be crucial to memorize, because they enable "forgetting" the old information.

spective below, but mostly focus on other thermodynamic quantities, particularly the heat and total EP, and study how they depend on the number of d.o.f., n.

The apparent connection between our feedback controller and this cellular sensor model, which is an active system, motivates us to elucidate the relation between memory, activity and feedback from a fundamental perspective. First, we will need to clarify the formal connection. This is the objective of the following Sect. 10.2.

10.2 The Role of Non-reciprocal Coupling—Connection to Active Matter

In Sect. 4.7.2 we have mentioned some general links between active matter and colloids subject to time-delayed feedback. Here, we resume this discussion, and bring it to a formal level.

10.2.1 A Generic Model with Non-reciprocal Coupling

As we have already mentioned in Sect. 4.7, Langevin-like equations are currently used by many researchers to describe the motion of active systems. In particular, we have introduced a very simple model for a microswimmer, the active Ornstein-Uhlenbeck particle, see Sect. 4.7.1. Like the cellular sensor, the AOUP model as well has a two-dimensional network representation, which also involves unidirectionally coupling. In fact, all three examples can be considered as special cases of the model

$$\gamma_j \dot{X}_j = a_{jj} X_j + a_{jj-1} X_{j-1} + \xi_j, \tag{10.9}$$

with periodic index $j \in \{0, 1, \ldots, n\}$, or with the corresponding projected equation

$$\dot{X}_0(t) = a_{00} X_0(t) + \int_0^t K(t-t') X_0(t') \mathrm{d}t' + \xi_0(t) + \nu(t). \tag{10.10}$$

ν is a colored noise, while ξ_j are Gaussian white noises with $\langle \xi_i(t) \xi_j(t') \rangle = 2k_B \mathcal{T}_j \gamma_j \, \delta_{ij} \delta(t-t')$. Note that the temperatures $\mathcal{T}_j$ are now scaled into the variance of the noise, for the sake of a shorter notation. Let us now make the connection with the aforementioned examples explicit.

Feedback controller with time delay First of all, our original feedback control model (10.1) is recovered as a special case at $a_{0n} = -k$, $a_{ii-1} = -a_{ii} = \gamma' n/\tau$, $\mathcal{T}_{j>0} = \mathcal{T}'$, $\gamma_{j>0} = \gamma'$. The time-delayed part of the force can be written in two ways as $-kX_n$ or, equivalently, as $-k \int_0^t K(t-t') X_0(t') \mathrm{d}t'$, corresponding to the two representations. n can be any natural number, giving the number of "memory cells", and yielding monotonically decaying colored noise and a Gamma-distributed memory kernel given in (6.3), (6.4), recall

$$C_\nu(T) = \langle \nu(t)\nu(t+T)\rangle = k^2 \frac{k_B\mathcal{T}'}{\gamma'} \sum_{p=0}^{n-1}\sum_{l=0}^{p} \frac{2^{l-2p}(2p-l)!}{p!(p-l)!\,l!} \frac{e^{-nT/\tau}T^l}{(\tau/n)^{l-1}},$$
$$K(T) = \frac{n^n\, T^{n-1}}{\tau^n(n-1)!} e^{-\frac{nT}{\tau}}. \tag{10.11}$$

Thus, n decides on the functional form of the distributed delay. The discrete delay, white noise LE is recovered in the limit $n \to \infty$, as we have explicitly shown in Sect. 6.1.2.

Cellular sensor, AOUP, feedback controller with exponential delay At $n = 1$, the cellular sensor model [19] is obtained for the unidirectional coupling with $a_{01} = 0$, $a_{10} > 0$, while the AOUP model[3] corresponds to reversed unidirectional coupling $a_{10} = 0$, $a_{01} > 0$. The corresponding projected equation has exponentially decaying colored noise and memory kernel

$$C_\nu(T) := \langle \nu(t)\nu(t+T)\rangle = k_B\mathcal{T}_1(a_{01}^2/a_{11})e^{a_{11}T/\gamma_1},$$
$$K(T) = (a_{10}/\gamma_1)e^{a_{11}T/\gamma_1}. \tag{10.12}$$

At *bi*directional, non-reciprocal coupling $a_{01} = k$, $a_{10} = -a_{11} = \gamma' n/\tau$, this model with $n = 1$, also reproduces the smallest version of our feedback controller model, i.e., where the entire memory device has one cell, or in other words, where the controller has one d.o.f. X_j (and the memory kernel is exponential). This case is readily included in (10.12).

Reciprocal coupling—Mechanical models Last, in the specific case of reciprocal coupling, i.e., $a_{ij} = a_{ji} \forall i, j \in \{0, 1, \ldots, n\}$, the network may model mechanical systems. For example, the $n = 1$ network

$$\gamma_0 \dot{X}_0 = -\kappa X_0 + \kappa X_1 + \xi_0,$$
$$\gamma_1 \dot{X}_1 = +\kappa X_0 - \kappa X_1 + \xi_j, \tag{10.13}$$

with $\mathcal{T}_0 \neq \mathcal{T}_1$ corresponds to a minimal model for *heat conducting* through mechanical motion, discussed in [18] (see p. 154). Here, the two degrees of freedom, $X_{0,1}$ could model the *angles of two vanes* that rotate in two different heat baths, and are coupled by a torsion spring with spring constant κ. More generally, (10.9) corresponds to a system of n colloidal particles trapped in quadratic potentials where the colloids are further coupled among each other by harmonic springs. We have already considered such a model for $n = 2$ in Sect. 10.1.

[3] In comparison with the notation in Eq. (4.38), here $\gamma_1 = 1/\tau_P$, $2k_B\mathcal{T}'\gamma_1 = 1$.

To summarize, depending on the symmetry properties of the coupling topology, the generic network (10.9) can be used to model various different systems. Specifically, the $n = 1$ case includes a model for a cellular sensor, a minimal version of a time-delayed feedback controller (with exponential delay), a mechanical system of two reciprocally coupled vanes, and the motion of an active swimmer. Figure 10.3 provides an overview over the different examples, and shows the corresponding coupling topologies. Importantly, the coupling is non-reciprocal in all these cases but the mechanical two-vane model. This motivates us to take a closer look at the *thermodynamic implications of NR coupling*. As the case $n = 1$ already includes very different, intriguing example systems, and further represents the smallest version of our model, with a single non-reciprocal link, we will focus on the generic model (10.9) at $n = 1$. At the end, we will also consider higher values of n to come back to the original example of the feedback controller with time delay.

Interpretation of non-reciprocal coupling To avoid confusion, we aim to emphasize the following, concerning the meaning of the NR interactions in (10.9). First, we would like to remind the Reader that the LE is a coarse-grained, i.e., effective

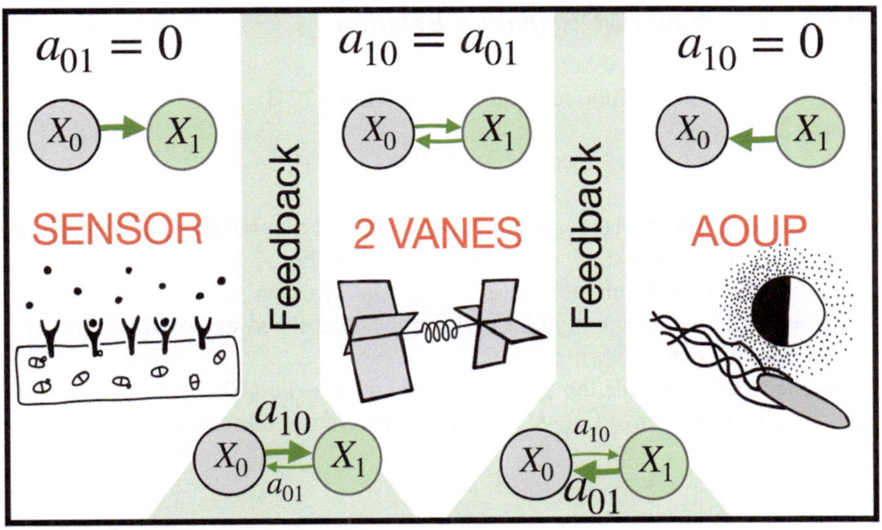

Fig. 10.3 Overview over different examples included in the generic network model (10.9) at $n = 1$ (with increasing ratio a_{01}/a_{10} from left to right). *Left:* If X_1 is unidirectionally coupled to the system d.o.f. X_0, i.e., $a_{01} = 0$, such that only X_1 "sees" X_0, but not *vice versa*, X_1 corresponds to a cellular sensor in the model [19]. *Center:* For reciprocal coupling $a_{01} = a_{10}$, X_1 represents a mechanically coupled d.o.f., and X_0, X_1 correspond to the angles of two vanes coupled by a torsion spring [18]. *Right:* If X_0 is unidirectionally coupled to X_1, i.e., $a_{10} = 0$, X_0 corresponds to the position of a microswimmer within the AOUP model, where X_1 models the effect of the flagella of a bacterium or the asymmetric flow field around a colloid. In the intermediate cases, where the coupling is bidirectional but non-reciprocal, X_1 corresponds to the controller d.o.f. and X_0 to the position of the feedback-controlled colloid in our model with exponentially distributed delay, and colored noise

equation, and the dynamical variable is a microscopic one. Fundamentally, physical interactions are typically reciprocal and symmetric, including mechanical coupling, as states Newton's third law. However, in complex systems, involving chemical reactions, multiscale processes, or coupling between a chemical component and a mechanical d.o.f., much more complicated relationships between subsystems can emerge [20–23], which might be representable as coupling between few mesoscopic, stochastic d.o.f. This is the situation considered here.

10.3 Non-reciprocal Coupling and Non-equilibrium

First we demonstrate that NR coupling alone "drives" the system out of equilibrium. To this end, we consider two typical measures, first the fluctuation-dissipation relation, then the total entropy production.

10.3.1 Fluctuation-Dissipation Relation

We recall the FDR (or Kubo relation of second kind) [24]

$$\langle \mu(t)\mu(s)\rangle = k_{\rm B}\mathcal{T}_0\,\gamma(|t-s|),$$

which we have introduced earlier in Part I. It describes a balance between friction kernel γ and thermal noise μ [note that contrary to Eq. (2.36), we have here used μ instead of ν to differentiate the different noises]. As it is well known for e.g., viscoelastic-fluids, the validity of a FDR implies that the system will equilibrate in the absence of external driving [24, 25].

To check the FDR for the present model, we rewrite (10.10) in the form of a generalized LE by converting K via partial integration into a friction-like kernel. This yields

$$\int_0^t \gamma(|t-s|)\dot{X}_0(s)\,\mathrm{d}s = a_{00}X_0 + \widetilde{K}(0)X_0(t) + \mu(t), \tag{10.14}$$

involving $\mu(t) = \xi_0(t) + \nu(t)$ and $\gamma(t-s) = 2\gamma_0\,\delta(t-s) + \widetilde{K}(t-s)$, with the integrated memory kernel $\widetilde{K}(t-s)$. In this way, an equation with a memory kernel in an integral over $X(t)$ can be transformed in a dynamical equation with a *time-nonlocal friction force*. The friction kernel then involves the integrated memory kernel $\widetilde{K}$. For the generic network (10.9) at $n = 1$, the latter reads

$$\widetilde{K}(T) = (a_{01}a_{10}/a_{11})e^{a_{11}T/\gamma_1}\gamma(t-s) = 2\gamma_0\,\delta(t-s) + \widetilde{K}(t-s). \tag{10.15}$$

By construction, the instantaneous friction $-\gamma_0\dot{X}$ and the white noise ξ_0 readily balance each other. Further, comparing the additional "friction" kernel (10.15) and the correlations of the colored noise ν given in (10.12), it can easily be seen that the FDR holds if

$$a_{01}\mathcal{T}_1 = a_{10}\mathcal{T}_0. \tag{10.16}$$

Remarkably, (10.16) suggests that the system with NR coupling can reach equilibrium despite $\mathcal{T}_0 \neq \mathcal{T}_1$. This is in sharp contrast to reciprocally coupled ("passive") systems, which generally never equilibrate in the presence of temperature gradients. For NR couplings, (10.16) can only be true when both coupling strengths are nonzero and have the same sign, i.e., $a_{01}a_{10} > 0$. We will later in Sect. 10.4 show that this case corresponds to a "mild" form of symmetry-breaking.

For the feedback controller model with $n > 1$ corresponding to (10.11), the integrated kernel reads $\widetilde{K}(T) = k\,\Gamma\,(n, nT/\tau)\,/(n-1)!$ with the upper incomplete Gamma function, and the FDR is violated. An explicit proof is given in the Appendix A.5.

10.3.2 Total Entropy Production

Now we turn to the total entropy production, which gives a more fundamental notion of non-equilibrium. While the broken FDR indicates non-equilibrium, it does not quantify the distance from equilibrium. In contrast, the total EP directly quantifies the irreversibility, i.e., time-reversal symmetry breaking.

However, as we have already argued in Chap. 4, the calculation of the EP for a non-Markovian process (10.10) is not straightforward [3–5, 26]. We recall that there is no closed Fokker-Planck equation to (10.10), and no standard path integral formalism , such that both standard routes [27] are not readily applicable. We will instead exploit from now on the Markovian version of our model (10.9), which allows application of the standard framework. We have introduced the formulae in Chap. 4, particularly, in Eq. (4.15).

We recall that if all X_j of the network are physical d.o.f., this is the usual total EP as traditionally defined in thermodynamics. For cases where the variables $X_{j>0}$ are auxiliary variables without clear physical meaning, the interpretation of the EP calculated in this way, is indeed not entirely clear. It comprises open questions and leaves room for discussion. At the end of this chapter, we will comment on this issue. But let us first explore, how much we can learn from this quantity about the considered processes.

Specifically, we apply the standard definition (4.15), to calculate the total EP along a fluctuating trajectory $\underline{\mathbf{X}} = \{X_0(t'), \ldots, X_n(t')\}$, $t' \in [t_s, t_f]$ belonging to the network (10.9). Recall

$$\frac{\Delta s_{\text{tot}}[\underline{\mathbf{X}}]}{k_{\text{B}}} = \ln \frac{\mathcal{P}[\underline{\mathbf{X}}]}{\hat{\mathcal{P}}[\hat{\underline{\mathbf{X}}}]} = \frac{\Delta s_{\text{sh}}}{k_{\text{B}}} + \ln \frac{\mathcal{P}[\underline{\mathbf{X}}|\underline{x}_s]}{\hat{\mathcal{P}}[\hat{\underline{\mathbf{X}}}|\underline{x}_f]}.$$

Due to the $(n+1)$–dimensionality of the network, the total EP involves the *multivariate (joint) Shannon entropy* $s_{\text{sh}} = -k_{\text{B}} \ln[\rho_{n+1}(\underline{x})]$ of the $(n+1)$-point joint PDF ρ_{n+1}.

Just as in the one-dimensional case, the total EP, by construction, fulfills an integral fluctuation theorem, $\langle e^{-\Delta s_{\text{tot}}/k_{\text{B}}} \rangle = 1$. In Sect. 4.4 we have explicitly discussed this theorem, and shown that it implies the second law of thermodynamics $\Delta S_{\text{tot}} \geq 0$, i.e., non-negativity of the mean total EP.

Furthermore, the path probabilities conditioned on the starting point $\underline{X}(t_s) = \underline{x}_s$, readily follow from the corresponding LEs, as we have given in Eqs. (2.28), (4.18). Specified to the present case, they read

$$\begin{aligned}
\mathcal{P}[\underline{\mathbf{X}}|\underline{x}_s] &\propto e^{-\sum_{j=0}^{n} \int_{t_s}^{t_f} \frac{\left(\gamma_j \dot{X}_j(t') - F_j[\underline{X}(t')]\right)^2}{4 k_{\text{B}} \mathcal{T}_j \gamma_j} \mathrm{d}t'}, \\
\hat{\mathcal{P}}[\hat{\underline{\mathbf{X}}}|\underline{x}_f] &\propto e^{-\sum_{j=0}^{n} \int_{t_s}^{t_f} \frac{\left(-\gamma_j \dot{X}_j(t') - F_j[\underline{X}(t')]\right)^2}{4 k_{\text{B}} \mathcal{T}_j \gamma_j} \mathrm{d}t'},
\end{aligned} \tag{10.17}$$

with $F_{j>0} = a_{jj} X_j + a_{j-1j} X_{j-1}$, and $F_0 = a_{00} X_0 + a_{0n} X_n$. Importantly, we here assume that $X_{j>0}$ are *even* under time-reversal, like positions, angles, or orientations. This is more suitable for the examples considered throughout the thesis (e.g., when $X_{j>0}$ model memory cells of a feedback-controller as we will argue below, odd parity would imply information storage in particle velocities, which seems less appropriate). However it should be noted that in some cases the parity of $X_{j>0}$ is indeed a nontrivial issue and subject of an ongoing debate, see e. g., [28] for active swimmers. This assumption implies

$$\ln \frac{\mathcal{P}[\underline{\mathbf{X}}|\underline{x}_s]}{\hat{\mathcal{P}}[\hat{\underline{\mathbf{X}}}|\underline{x}_f]} = \int_{t_s}^{t_f} \frac{\delta q_0}{\mathcal{T}_0} + \sum_{j=1}^{n} \int_{t_s}^{t_f} \frac{F_j \circ \mathrm{d}X_j}{\mathcal{T}_j}, \tag{10.18}$$

where δq_0 can be identified as the heat flow from X_0 to its bath [18], which reads $\delta q_0 = [a_{00} X_0 + a_{0n} X_n] \circ \mathrm{d}X_0$. In the steady state, $\Delta S_{\text{sh}} = 0$, and the ensemble average of the EP is thus

$$\dot{S}_{\text{tot}} = \frac{a_{0n}^2\langle X_n^2\rangle + a_{0n}\,a_{00}\langle X_0X_n\rangle}{\mathcal{T}_0} + \sum_{j=1}^{n}\frac{a_{jj-1}^2\langle X_{j-1}^2\rangle + a_{jj}a_{jj-1}\langle X_{j-1}X_j\rangle}{\mathcal{T}_j}, \tag{10.19}$$

where we have used that $X_j\dot{X}_j$-correlations vanish in the steady state, since $2\langle X_i\dot{X}_i\rangle = \mathrm{d}\langle X_i^2\rangle/\mathrm{d}t = 0$. The mean EP (10.19) has two contributions, that is, the dissipation of X_0

$$\dot{Q}_0 = a_{0n}^2\langle X_n^2\rangle + a_{0n}\,a_{00}\langle X_0X_n\rangle, \tag{10.20}$$

and a contribution from the additional d.o.f. $X_{j>0}$.

From the perspective of a non-Markovian process (10.10), the second contribution can be considered as the *entropic cost* of the memory. At this point it should be emphasized, however, that memory is *not* unequivocally connected to EP. For example, the reciprocal case $a_{01} = a_{10}$ of our network (10.12) also generates memory in X_0, but no net dissipation, thus no EP (we will explicitly show this below). We further remind the Reader that there are, of course, real-world examples of physical systems which involve complex memory but nevertheless reach thermal equilibrium (where the mean EP naturally vanishes), e.g., viscoelastic fluids. As we will later see, such *entropic cost*-free memory is, however, *not useful* in terms of extracting net information or for achieving goals of feedback control.

From (10.19), (10.20), analytical expressions for $\dot{Q}_0$ and the total EP can be derived for any n, by means of Green's functions. We will show this below in Sect. 10.3.3.

Let us here already give the results for the case $n = 1$ (which was also discussed in [26]),

$$\dot{S}_{\text{tot}} = -\frac{k_B}{\mathcal{T}_0\mathcal{T}_1}\frac{(a_{10}\mathcal{T}_0 - a_{01}\mathcal{T}_1)^2}{a_{00}/\gamma_0 + a_{11}/\gamma_1}, \tag{10.21}$$

$$\dot{Q}_0 = k_B\frac{a_{01}(a_{10}\mathcal{T}_0 - a_{01}\mathcal{T}_1)}{a_{00}\gamma_1 + a_{11}\gamma_0}. \tag{10.22}$$

Later, we will discuss the EP and the heat flow for cases $n \geq 1$.

From (10.21) one immediately sees that the EP vanishes if, and only if, the FDR (10.16) is fulfilled. In this case, also the mean heat flow (10.22) and associated medium EP vanish. Results for $\dot{Q}_0$ are illustrated in Fig. 10.4, which highlights the role of NR coupling. Since we are primarily interested in the role of NR coupling, we focus in Fig. 10.4 and in the following on *isothermal conditions*, i.e., $\mathcal{T}_j = \mathcal{T}$ for all $j > 0$. Furthermore, we see from Eq. (10.22) that the heat flow may nullify, although the total EP (10.21) is positive, but only if $a_{01} = 0$, $a_{10} \neq 0$.

We conclude that the notion of equilibrium based on the non-Markovian description (from FDR) and based on the total EP of the network representation with addi-

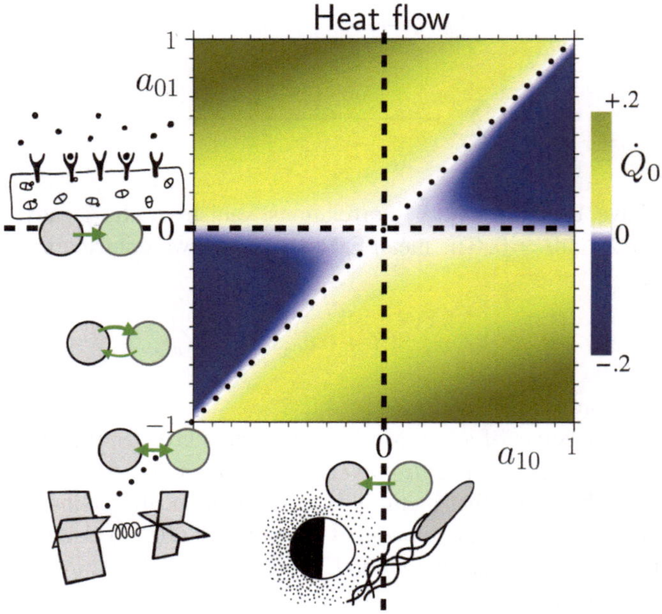

Fig. 10.4 Steady-state mean heat flow $\dot{Q}_0$ in units of $k_B\mathcal{T}/\tau_B$ from Eq. (10.22) in the minimal model (10.12) with $a_{11} = a_{00} = -\gamma_0/\tau_B$, at isothermal conditions $\mathcal{T}_0 = \mathcal{T}_1$, and $\gamma_0 = \gamma_1$. $\dot{Q}_0$ vanishes along the diagonal $a_{01} = a_{10}$, where the system is reciprocally coupled and, thus, "passive" (in this case, $X_{0,1}$ could, e.g., model the angles of vanes coupled by a torsion spring [18]). $\dot{Q}_0$ further vanishes on the unidirectional coupling line $a_{01} = 0$, where X_0 is "sensed" by X_1 (corresponding to a cellular sensor [19]). $\dot{Q}_0 > 0$ along the other unidirectional coupling line $a_{10} = 0$ (where X_0 corresponds to the position of an active swimmer in the AOUP model [28–32], X_1 modeling the effect of a flagellum of a bacterium, or the asymmetric flow field around a Janus particle). All other parameters are set to unity. These results are also published in [33]

tional d.o.f. are consistent. This finding is in accordance with [34], but contrasts the viewpoint in [26], where a different definition of the total EP was used. Besides being a consistency check, this also indicates that the total EP considered here gives some reasonable information in regard to the non-Markovian system.

At this point, we further aim to note an interesting observation concerning the so called "non-Markovian FDR" [35]

$$C_\nu(T) = \langle X_0^2 \rangle \, K(T). \tag{10.23}$$

In the derivation [35], it seems that this relation holds under fairly general conditions. However, being broken despite zero EP, it is found to be not suitable to correctly describe the equilibrium here. We suspect that this is due to the overdamped limit, which requires velocity relaxation as an additional assumption, absent in (10.23).

While the non-equilibrium nature of the non-Markovian Eq. (10.10) is indicated by the violation of FDR (10.16), the mechanism that prevents the system from equilibration, i.e., the driving, might not be very obvious. As we will show in Sect. 10.4, this point is clarified by studying the connection between network topology and energy flows in the corresponding Markovian description.

10.3.3 Analytical Solutions

As already indicated by (10.19), (10.20), various thermodynamic and dynamic quantities can be calculated on the basis of (cross-)correlations $\langle X_i X_j \rangle$, including the PDFs, and information flows [see (10.49) below]. For example, the steady-state PDF ρ_{n+1} is, due to the linearity of the model, a Gaussian-distribution with zero mean and the covariance matrix $(\underline{\underline{\Sigma}})_{ij} = \langle X_i X_j \rangle$. Thus, it is fully determined by all the (cross-)correlations.

Let us briefly sketch how analytical expressions for the correlations can be obtained for arbitrary network sizes n. To this end, we transform Eq. (10.9) via the Fourier-transformation $\tilde{X}_j(s) = \int_{-\infty}^{\infty} X_j(t) e^{-i\omega t} \mathrm{d}t$, which readily yields

$$i\omega \underline{\underline{\gamma}} \underline{\tilde{X}}(\omega) = \underline{\underline{a}} \underline{\tilde{X}}(\omega) + \underline{\tilde{\xi}}(\omega) \tag{10.24}$$

$$\Rightarrow \underline{\tilde{X}}(\omega) = \underbrace{\left(i\omega \underline{\underline{\gamma}} - \underline{\underline{a}} \right)^{-1}}_{=\underline{\underline{\tilde{\lambda}}}(\omega)} \underline{\tilde{\xi}}(\omega). \tag{10.25}$$

$\underline{\underline{\tilde{\lambda}}}(\omega)$ is the Green's function in frequency space, determined by the inverse of the topology matrix and the diagonal friction matrix $\underline{\underline{\gamma}}$ with elements γ_j. Using the well-known relationship (A.14) between spatial correlations and the Green's function from linear response theory [36] (see Appendix A.3 for a derivation) one readily finds

$$\langle X_j^2 \rangle = \frac{k_\mathrm{B}}{\pi} \sum_{p=0}^{n} \mathcal{T}_p \gamma_p \int_{-\infty}^{\infty} \tilde{\lambda}_{jp}(\omega) \tilde{\lambda}_{jp}(-\omega) \, \mathrm{d}\omega, \tag{10.26}$$

$$\langle X_j X_l \rangle = \frac{k_\mathrm{B}}{\pi} \sum_{p=0}^{n} \mathcal{T}_p \gamma_p \int_{-\infty}^{\infty} \tilde{\lambda}_{jp}(\omega) \tilde{\lambda}_{lp}(-\omega) \, \mathrm{d}\omega. \tag{10.27}$$

These are analytical expressions for all correlations for arbitrary network sizes. The matrix inversion required to find $\underline{\underline{\tilde{\lambda}}}(\omega)$ from (10.25) is indeed possible up to very large network sizes due to the sparse coupling. To find closed-form solutions for the integrals (10.26), the residue theorem can be used. However, this requires finding the roots of a polynomial of order $n+1$. Instead, we have evaluated them using computer algebra systems, specifically Mathematica10.1. Therewith, this can be done reasonably fast up to about $n = 10$. We also note that for the case $\mathcal{T}_{j>0} = 0$ solutions up to $n \sim 10^4$ can be found in this way.

10.4 Non-reciprocal Coupling and Activity

To further unravel the non-equilibrium nature of the systems with NR coupling, we consider the system from an energetic perspective. Since we regard all d.o.f. to be of physical nature, each force applied by, or to, a subsystem j is inevitably connected to an energy exchange.

Let us for a moment take the perspective that each one of the d.o.f. X_j was an entire system. Then, Sekimoto's framework [18] tells us that the fluctuating heat flows along a trajectory of length dt are defined as

$$\delta q_j = (\gamma_j \dot{X}_j - \xi_j) \circ \mathrm{d}X_j = F_j \circ \mathrm{d}X_j, \tag{10.28}$$

the internal energy changes read

$$\mathrm{d}u_j = a_{jj} X_j \circ \mathrm{d}X_j. \tag{10.29}$$

while the work contributions are

$$\delta w_j := a_{ij} X_{j-1} \circ \mathrm{d}X_j. \tag{10.30}$$

Recall the usage of Stratonovich calculus throughout this chapter (see Sect. 2.2.5). Now we return to the original perspective where the entire set of $X_{\{0,1,\ldots,n\}}$ is the system, and the X_j are subsystems. How does the energetic consideration change? The important difference is the notion of work.

From a mechanical point of view, the fluctuating work δw_j as given by Eq. (10.30) is the energy associated with the coupling term $a_{ij} X_{j-1}$. It corresponds to the work applied to X_j via the single interaction between X_j and X_i ("coupling force times displacement"). If the coupling is *reciprocal*, i.e., stems from an interaction Hamiltonian $H_{\text{interaction}}$, the sum of work applied to X_i and X_j corresponds to its total differential

$$\delta w_i + \delta w_j = -\frac{\partial H_{\text{interaction}}}{\partial X_i} \circ \mathrm{d}X_i - \frac{\partial H_{\text{interaction}}}{\partial X_j} \circ \mathrm{d}X_j = -\mathrm{d}H_{\text{interaction}}. \tag{10.31}$$

Thus, in the reciprocally coupled case the energy exchange (10.31) is, in this picture, actually part of the *internal energy change* of the network $X_{\{0,1,\ldots,n\}}$. For stability reasons, the latter must on average be conserved in steady states, hence, $\sum_{j=0}^{n} \dot{W}_j \equiv 0$.

In sharp contrast, a NR coupling turns out to be associated with a steady *energy input* into the system at rate

$$\dot{E}_{\text{input}} := \dot{W}_0 + \sum_{j=1}^{n} \dot{W}_j \geq 0. \tag{10.32}$$

Equality is reached for reciprocal coupling, and the positivity holds for isothermal conditions. The individual terms in (10.32) are given by Eq. (10.30). We recall that capital letters denote ensemble averages.

To prove the non-negativity of (10.32), we consider the energy balance. One shall expect that all picked up energy must ultimately dissipate. The validity of the first law of thermodynamics for each d.o.f: $\delta q_j = \delta w_j + \mathrm{d}u_j$ (with the energetic contributions given above), is indeed already implied in the model (10.9). As the average $\langle \mathrm{d}u_j \rangle = \langle a_{jj} X_j \circ \mathrm{d}X_j \rangle = a_{jj} \mathrm{d}\langle X_j^2 \rangle = 0$ is conserved in steady states, the first law implies $\dot{Q}_j = \dot{W}_j$. Thus, the energy input $\dot{E}_{\text{input}}$, which we define in (10.32), is indeed identical to the total dissipation rate $\sum_{j=0}^{n} \dot{Q}_j \geq 0$. The latter is non-negative, as directly follows from the connection to the EP [see (10.19)]. A (positive) energy input on the level of the individual constituents is considered a defining property of active systems [34, 37–40]. Hence, it is reasonable to use NR networks to model active matter, which is indeed commonly done.

We note that the energy flow through the system could be initiated by a (bio)chemical energy source, or some sort of external gradient, or force field. In any case, the source will be exploited, such that, looking at the entire system plus surrounding, this steady energy flow must result in free energy loss (or another respective thermodynamic potential). This can be seen by the fact that the energy flow is accompanied by total EP. For this reason, the energy input $\dot{E}_{\text{input}}$ (10.32) can be associated with "free energy consumption".

The NR networks discussed here also include, as a specific case, a temperature gradient as underlying driving mechanism. In particular, if $a_{01}a_{10} > 0$ (upper right and lower left quadrants in Figs. 10.4 and 10.5), the NR coupling is (potentially) ascribable to a (hidden[4]) temperature gradient, as can be shown by mapping the network onto a symmetric system. We will show this in the next subsection.

10.4.1 Mapping Non-reciprocity of the Coupling to a Temperature Gradient Between the Coupled Entities

We here show that for the specific case $a_{01}a_{10} > 0$, the NR coupling can potentially be traced back to a hidden temperature gradient. Hence, this case represents a "mild" form of intrinsic non-equilibrium. Consider the NR network of X_0 and X_1 with $a_{01}a_{10} \neq 0$

$$\begin{cases} \gamma_0 \dot{X}_0 &= a_{00} X_0 + a_{01} X_1 + \xi_0 \\ \gamma_1 \dot{X}_1 &= a_{10} X_0 + a_{11} X_1 + \xi_1 . \end{cases} \tag{10.33}$$

Via a scaling, (10.33) can be mapped onto another network, in particular, by introducing $\widetilde{X}_0 = X_0/\sqrt{|a_{10}|}$, $\widetilde{X}_1 = X_1/\sqrt{|a_{01}|}$, and $\widetilde{\mathcal{T}}_0 = \mathcal{T}_0/|a_{10}|$, $\widetilde{\mathcal{T}}_1 = \mathcal{T}_1/|a_{01}|$. (We

[4] If X_1 is hidden, i.e., when the non-Markovian description is employed.

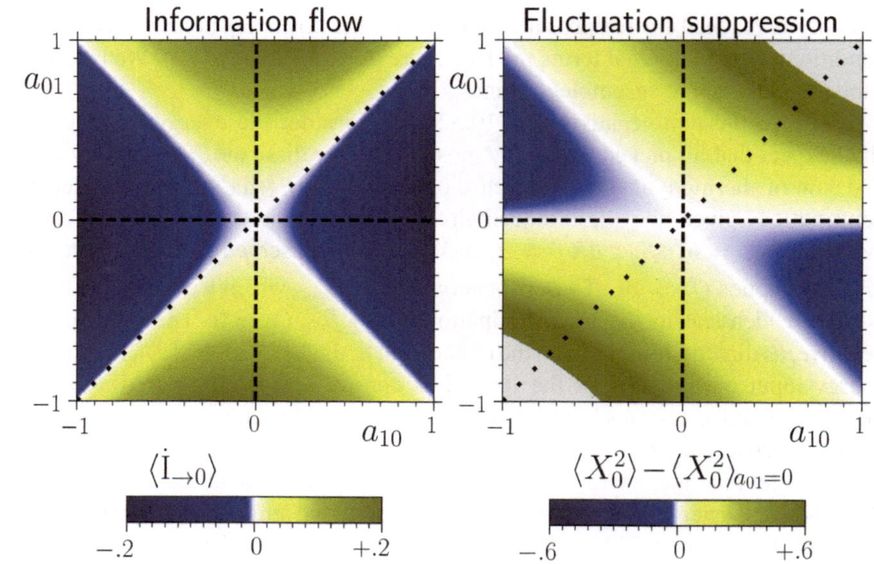

Fig. 10.5 Two d.o.f. X_0 and X_1, with NR coupling as in Fig. 10.4. *Left*: Information flow to X_0 in units of $1/\tau_B$. Blue areas indicate that X_1 "knows" more about X_0 than *vice versa* corresponding to feedback control regimes ($|a_{10}| > |a_{01}|$). *Right:* Thermal fluctuations of X_0 measured by the second moment compared to the uncoupled case ($a_{01} = 0$), $\langle X_0^2 \rangle - \langle X_0^2 \rangle_{a_{01}=0}$ in units of σ^2. Blue areas indicate thermal fluctuation suppression. The gray areas indicate unstable regions (where $\langle X_0^2 \rangle \to \infty$). All other parameters are set to unity and $a_{11} = a_{00} = -1$. These results are also published in [33]

note that scaling of positional d.o.f. shall indeed be accompanied with scaling of the temperatures due to the connection between temperatures and the time-derivative of the positions, e.g., think of the equipartition theorem). In this way, we find

$$\begin{cases} \dot{\widetilde{X}}_0 &= a_{00}\widetilde{X}_0 + \mathrm{sgn}(a_{01})\sqrt{a_{01}a_{10}}\widetilde{X}_1 + \widetilde{\xi}_0 \\ \dot{\widetilde{X}}_1 &= \mathrm{sgn}(a_{10})\sqrt{a_{10}a_{01}}\widetilde{X}_0 + a_{11}\widetilde{X}_1 + \widetilde{\xi}_1, \end{cases} \tag{10.34}$$

with $\langle \widetilde{\xi}_i(t)\widetilde{\xi}_j(t) \rangle = 2k_B\widetilde{\mathcal{T}}_j\gamma_j\delta_{ij}\delta(t-t')$. If $a_{01}a_{10} > 0$, this system has *reciprocal* coupling. Further, even if $\mathcal{T}_0 = \mathcal{T}_1$, it involves a temperature gradient ($\widetilde{\mathcal{T}}_0 \neq \widetilde{\mathcal{T}}_1$). This symmetric network (10.34) corresponds to the aforementioned model of two vanes at the angles $\widetilde{X}_0, \widetilde{X}_1$ in different heat baths, coupled by a torsion spring (see Sect. 10.2.1). As is well known, such a reciprocally coupled system equilibrates if, and only if, $\widetilde{\mathcal{T}}_1 = \widetilde{\mathcal{T}}_0$. This implies $|a_{01}|\mathcal{T}_1 = |a_{10}|\mathcal{T}_0$, which is indeed identical to the equilibrium-condition (10.16) found from FDR.

Now we turn to the impact of this scaling on the thermodynamic quantities. For the energy flows, we find the relations

$$\begin{aligned}\delta\tilde{w}_0 &= \mathrm{sgn}(a_{01})\sqrt{a_{01}a_{10}}\,\tilde{X}_1 \circ \mathrm{d}\tilde{X}_0\\ &= a_{01}|a_{10}|\,X_1 \circ \mathrm{d}X_0 = |a_{10}|\,\delta w_0,\\ \delta\tilde{q}_0 &= (\gamma_0\dot{\tilde{X}}_0 - \tilde{\xi}_0) \circ \mathrm{d}\tilde{X}_0 = |a_{10}|\,\delta q_0. \end{aligned} \tag{10.35}$$

Thus, the energy flows to and out of X_0 are both scaled with $|a_{10}|$. Likewise, $\delta\tilde{w}_1 = |a_{01}|\,\delta w_1$ and $\delta\tilde{q}_1 = |a_{01}|\,\delta q_1$. This further means

$$\Delta\tilde{s}_{\mathrm{tot}} = \frac{\delta\tilde{q}_0}{|a_{10}|\mathcal{T}_0} + \frac{\delta\tilde{q}_1}{|a_{01}|\mathcal{T}_1} = \frac{\delta q_0}{\mathcal{T}_0} + \frac{\delta q_1}{\mathcal{T}_1} = \Delta s_{\mathrm{tot}}, \tag{10.36}$$

i.e., the EP in the scaled model is *identical* to the EP in the original model, implying that equilibration of *both* models is expected, if (10.16) holds.

We conclude that the two "driving mechanisms", that is, NR coupling with $a_{01}a_{10} > 0$, or a temperature gradient, can formally not be distinguished on the level of EP. Hence, this represents a "mild" form of intrinsic non-equilibrium. Indeed, (hidden) temperature gradients have been discussed in the literature as possible mechanisms that fuel active motion, see, e.g., [41, 42].

It should be emphasized that a scaling as employed here cannot be found, if $a_{01}a_{10} \leq 0$. Thus, the NR coupling studied here is the more general case.

10.4.2 Reversed Heat Flow

Let us now consider the direction of the heat flow (10.22) induced by a NR coupled d.o.f. We recall that in a passive system a non-conservative force acting on X_0 can induce a steady-state heat flow. The latter is strictly *nonnegative*, as dictated by the second law of thermodynamics $\dot{Q}_0/\mathcal{T}_0 = \dot{S}_{\mathrm{tot}} \geq 0$. Thus, heat is never flowing (on average) from the bath to the system.

In contrast, we find that the coupled subsystem X_n, which also applies a non-conservative force onto X_0 (namely $a_{0n}X_n$), *can* induce a reversed heat flow $\dot{Q}_0 < 0$; implying a *negative dissipation* of X_0. This is possible due to the usage of extracted information, as we discuss below. This implies a steady extraction of energy from the bath, which is converted into work $\dot{W}_0$, i.e., a (potentially useful) form of energy. As we have already mentioned in Part I (Chap. 4), it is well known that such a reversed heat flow can be induced by "Maxwell-demon" type devices [43, 44]. In the following, we show that the $(n = 1)$-network considered here represents a minimal, time-continuous version of such a device.

Figure 10.4 reveals that there are parameter regimes, where the heat flow of the d.o.f. X_0 in the $(n = 1)$-network is reversed (blue regions in Fig. 10.4). Specifically, the energy extraction from the bath occurs if $a_{01}a_{10} > 0$. At this point, we recall the observation from the previous chapter that a time-delayed feedback control force, in a linear system can induce as well a negative heat flow. There, it occurred for negative feedback[5] only. To compare our present system with the one from Chap. 9, we consider the projected equation (10.10) for the case $n = 1$. Here, the feedback force is $a_{01}X_1 \sim a_{01}a_{10}/\gamma_1 \int e^{a_{11}(t-t')/\gamma_1} X_0(t')\mathrm{d}t'$, i.e., has exponentially distributed delay instead of discrete, delta-distributed delay. Using the feedback controller analogy, the reversed heat flow here occurs only if $a_{01}a_{10} = k/\tau > 0$, which corresponds to negative feedback, i.e., analogous to the discrete delay case. Thus, the functional form of the memory kernel and the presence of colored noise, do not seem to qualitatively change the "energy extraction" regime. Later in this chapter, we will return to this point and consider the behavior for larger n (see, e.g., Figs. 10.8 and 10.11).

Taking a closer look at Fig. 10.4 further reveals that a reversed heat current occurs, if the "sensing" is stronger than the "active force" applied to X_0, i.e., $|a_{10}| > |a_{01}|$. To elucidate this observation, we consider the information flows.

10.5 Non-reciprocal Coupling and Information

In the previous section, we have established that NR coupling implies an energy input on the level of individual constituents, which is a defining property of active matter. Now we change the focus onto information and feedback. We will address two questions. We have argued that a unidirectional coupling to another d.o.f. can model an active force onto X_i (of strength a_{ij}), i.e, "active propulsion"; or a sensing operation (of strength a_{ji}) [see example (10.12)]. *Do these interpretations conform with the information-thermodynamic properties?* Furthermore, a combination of both, i.e., sensing and applying an (active) force according to the measurement outcome, is in line with the intuitive notion of feedback. But, *is the existence of a single NR coupling sufficient to model a feedback loop?*

We clarify these questions by analyzing information flows and their connection to energy flows and entropy in detail. To this end, we use a framework similar to [45], where the reciprocally coupled case of (10.9) at $n = 1$ was studied in detail.

10.5.1 Information Flow and Generalized Second Law

In Sect. 10.3.2, we have considered the entropy balance of the entire network $X_{\{0,1,\ldots,n\}}$. Now we want to consider the *entropy balance* of a single d.o.f. X_j. The motivation is to go back to the original picture, where X_0 represents the colloidal position X and the other d.o.f. represent the feedback controller, which are usually left out in the description. We recall that we expect an information-theoretic contribution in the balance of X_0, as sketched in Fig. 4.3. Here we derive the unknown information-theoretic term from Fig. 4.3.

[5] I.e., a force directed *towards* the past trajectory.

We start by considering the total temporal derivative of the multivariate (joint) Shannon entropy (4.15) of the $(n+1)$-point PDF, i.e.,

$$\frac{\dot{s}_{\text{sh}}}{k_{\text{B}}} = \frac{-\partial_t \rho_{n+1}}{\rho_{n+1}} + \sum_{j=0}^{n} \frac{-\left(\partial_{x_j} \rho_{n+1}\right) \dot{X}_j}{\rho_{n+1}}. \tag{10.37}$$

In the steady state, the first term naturally vanishes. Therefore, we concentrate on the sum. To calculate the ensemble average of the latter, use $\langle \dot{X}_j A(x_j, t)\rangle = \int J_j A(x_j, t) \mathrm{d}x_j$ [27, 46] and apply the natural boundary conditions (see Chap. 3). This readily yields

$$\begin{aligned} \left\langle \frac{-\left(\partial_{x_j} \rho_{n+1}\right) \dot{X}_j}{\rho_{n+1}} \right\rangle &= \int \frac{-\left(\partial_{x_j} \rho_{n+1}\right) J_j}{\rho_{n+1}} \mathrm{d}\underline{x} \\ &= -\int \underbrace{\left[\ln(\rho_{n+1}) J_j\right]_{-\infty}^{\infty}}_{\to 0} \mathrm{d}\underline{x}_{i \neq j} + \int \ln \rho_{n+1}(\underline{x}) \partial_{x_j} J_j \, \mathrm{d}\underline{x} \\ &= \underbrace{-\int \ln \frac{\rho_1(x_j)}{\rho_{n+1}(\underline{x})} \partial_{x_j} J_j \, \mathrm{d}\underline{x}}_{=\dot{I}_{\to j}} + \int \ln \rho_1(x_j)\, \partial_{x_j} J_j \, \mathrm{d}\underline{x}. \end{aligned} \tag{10.38}$$

In the last step, we have identified the information flow $\dot{I}_{\to j}$, which we have introduced before in Sect. 4.5 Eq. (4.31). Importantly, the involved information flow is from *all* other d.o.f. $\{X_{l \neq j}\}$ to X_j. Even if not directly coupled with each other, two d.o.f. can exchange information through a third d.o.f.

To further proceed, we utilize the closed, multivariate Fokker-Planck equation for the $(n+1)$-point probability distribution $\rho_{n+1}(\underline{x}, t)$ of $\underline{x} = (x_0, \ldots, x_n)^T$, which we have given in Chap. 3 [in Eq. (3.14)]. We aim to stress that the existence of the closed FPE is a key benefit of employing a Markovian description.[6] More specifically, we use the FPE to replace the $\partial_{x_j} J_j$-term in the integral, and find

$$\begin{aligned} \left\langle \frac{-\left(\partial_{x_j} \rho_{n+1}\right) \dot{X}_j}{\rho_{n+1}} \right\rangle &= -\dot{I}_{\to j} + \int \ln \rho_1(x_j) \left\{ -\partial_t \rho_{n+1} + \sum_{i \neq j} \partial_{x_i} J_i \right\} \mathrm{d}\underline{x} \\ &= -\dot{I}_{\to j} + \frac{\dot{S}_{\text{sh}}^j}{k_{\text{B}}} - \sum_{i \neq j} \int \ln \rho_1(x_j) \underbrace{\left[J_i\right]_{-\infty}^{\infty}}_{\to 0} \mathrm{d}\underline{x}_{l \neq i}, \end{aligned} \tag{10.39}$$

where we have introduced the *Shannon entropy of the marginal PDF* $\rho_1(x_j)$: $\dot{S}_{\text{sh}}^j = -k_{\text{B}} \int \ln \rho_1(x_j) \partial_t \rho_1(x_j) \, \mathrm{d}x_j$. To summarize, by considering the ensemble average of

[6] There is in general no closed FPE for non-Markovian systems of type (10.10), as we have discussed in detail in Part II.

each summand of (10.37), we have found that the total Shannon entropy change is given by $\dot{S}_{\text{Sh}} = -k_{\text{B}} \sum_{j=0}^{n} \dot{I}_{\to j} + \dot{S}^{j}_{\text{sh}}$.

As we have shown in Sect. 4.5, the involved information flows are connected to the multivariate generalization of the mutual information, which is a conserved quantity in steady states. Specifically, $\dot{\mathcal{I}} = \sum_{j=0}^{n} I_{\to j} = 0$. This immediately tells us that the sum over all information flows nullifies in the steady state, which means that we do not have to take them into account. However, when an *individual* subsystem is considered, they turn out to be an important contribution to the *entropy balance*.

To see this, we reconsider the summands of (10.37), and now rewrite them using the FPE (3.14) as

$$\frac{-\left(\partial_{x_j} \rho_{n+1}\right) \dot{X}_j}{k_{\text{B}}^{-1} \rho_{n+1}} = \underbrace{\frac{\gamma_j\, J_j(\underline{x}, t) \dot{X}_j}{\mathcal{T}_j\, \rho_{n+1}}}_{:= k_{\text{B}} \dot{s}^{j}_{\text{tot}}} - \frac{\dot{q}_j}{\mathcal{T}_j}. \tag{10.40}$$

Combining (10.37), (10.39), (10.40) readily yields

$$\dot{S}^{j}_{\text{tot}} = \dot{S}^{j}_{\text{sh}} - k_{\text{B}} \dot{I}_{\to j} + \frac{\dot{Q}_j}{\mathcal{T}_j} = \int \frac{\gamma J_j(\underline{x}, t)^2}{k_{\text{B}} \mathcal{T}_j\, \rho_{n+1}} \, \mathrm{d}\underline{x} \geq 0, \tag{10.41}$$

$$\dot{S}_{\text{tot}} = \sum_{j=0}^{n} \frac{\dot{Q}_j}{\mathcal{T}_j} + \dot{S}_{\text{sh}} = \sum_{j=0}^{n} \dot{S}^{j}_{\text{tot}} \geq 0. \tag{10.42}$$

With Eq. (10.42) we have recovered the mean *total EP*, which we have equivalently deduced from the path probabilities in Eq. (4.15). From Eq. (10.41) we see that the definition (10.40) of a fluctuating "total EP per subsystem", $\dot{s}^{j}_{\text{tot}}$, is meaningful, as it has the non-negative ensemble average. Equation (10.41) is a *generalized second law* for each d.o.f., which involves the information flow.

Returning to the picture of the colloid controlled by an external feedback controller, this equation gives the appropriate entropy balance (with $j = 0$)

$$\dot{S}^{0}_{\text{tot}} = \dot{S}^{0}_{\text{sh}}[\rho_1(X_0)] - k_{\text{B}} \dot{I}_{\to 0} + \frac{\dot{Q}_0}{\mathcal{T}_0} \geq 0, \tag{10.43}$$

as depicted in Fig. 10.6. Besides the medium EP and Shannon entropy change (of the marginalized PDF), it contains the information flow. The latter represents the aforementioned "missing entropic term".

In steady states $\dot{S}^{j}_{\text{sh}} = 0$. Thus, (10.41) simplifies to

$$\dot{Q}_j \geq k_{\text{B}} \mathcal{T}_j\, \dot{I}_{\to j}, \tag{10.44}$$

in accordance with [6, 45]. Equation (10.44) implies that a reversed steady heat flow, $\dot{Q}_0 < 0$, is only possible, if $\dot{I}_{\to 0} < 0$, i.e., *information is flowing from the X_0 to the*

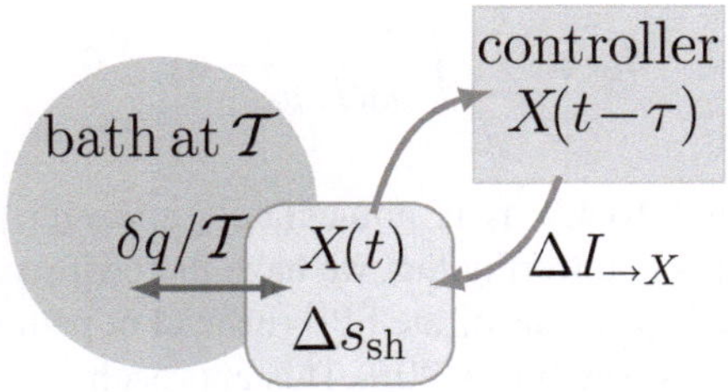

Fig. 10.6 Fluctuating entropy balance of the system $X(t)\hat{=}X_0(t)$: the Shannon entropy may vary upon a change of the PDF or $X(t)$. It corresponds to $s_{\mathrm{sh}}\hat{=}s^0_{\mathrm{sh}}$. In the steady state, this quantity is on average preserved. Further, the medium entropy might change due to a heat flow $\delta q\hat{=}\delta q_0$. Last, due to measurement and control action, an additional entropic term arises in the entropy balance (10.43) given by the information flow (4.31). The corresponding energetic picture is given in Fig. 4.2. In Fig. 4.3, we have formulated the expectations for the entropy balance

rest of the system. The more information about X_0 is gathered by the other $X_{j>0}$ (the controller d.o.f., see Sect. 10.1.1), the more heat can be extracted from the bath.

Figure 10.7 plots the information and heat flows, as well as the total EP of our network at $n = 1$, which are all connected via (10.41), (10.42). It also illustrates that in the reciprocal case, the memory is free of "entropic cost" (zero EP), but, at the same time, no net information extraction is achieved, nor is a heat flow induced. This further implies that the "missing entropic term" is specific to the NR case.

In this section, we have derived a generalized second law for individual subsystems X_j. At the same time, we have shown that the ensemble average of the total EP for the *entire* system $\{X_0, \ldots, X_n\}$

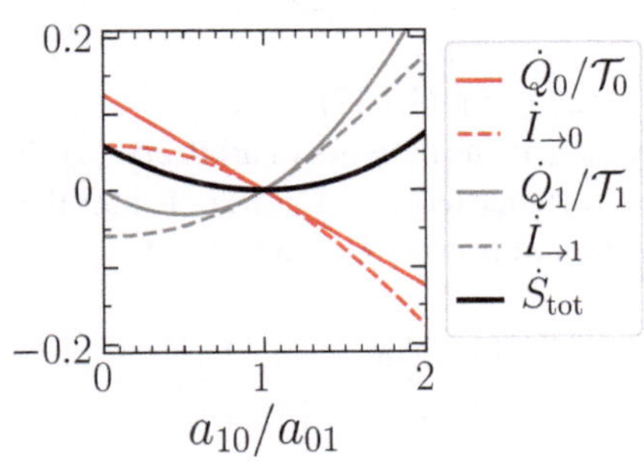

Fig. 10.7 Information $\dot{I}_{\to j}$ (10.49) in units of $1/\tau_{\mathrm{B}}$ and medium EP $\dot{Q}_j/\mathcal{T}_j$ (10.22) in units of $k_{\mathrm{B}}/\tau_{\mathrm{B}}$, for two NR coupled d.o.f., vs. the ratio of coupling strength, a_{01}/a_{10}. At $a_{01} = a_{10}$, the system is reciprocally coupled and displays no net heat and information flow, and has zero total EP (black line). The medium EP (solid red and gray lines) of each d.o.f. is bounded from below by the information flow (dashed lines), as predicted by the generalized second law (10.44). The total EP is given by the sum over all other quantities. $a_{11} = a_{00} = -\gamma_0/\tau_{\mathrm{B}}$, all other parameters are unity. These results are also published in [33]

$$\dot{S}_{\text{tot}} = \sum_{j=0}^{n} \int \frac{\gamma J_j(\underline{x}, t)^2}{k_B \mathcal{T}_j \, \rho_{n+1}} \, d\underline{x} \geq 0, \tag{10.45}$$

fulfills the second law (10.42) as is immediately apparent from the nonnegative expression (10.45). It is noteworthy that we have therewith also seen that the mean total EP can be derived by means of the FPE (instead of path integrals) [as we have already noted in Chap. 4, see Eq. (4.20)]. This approach is possible for Markovian systems, where a closed FPE is known [27].

Explicit expressions for information flows Let us now calculate the information flow. As already mentioned in Sect. 10.3.3, due to the linearity of our model the steady-state PDFs are *multivariate Gaussians* with zero mean and with the covariance matrix $(\underline{\underline{\Sigma}})_{ij} = \langle X_i X_j \rangle$. To derive explicit expressions for the steady-state averaged information flows, it turns out to be most convenient to use

$$\dot{I}_{\to j} = \left\langle \frac{\left(\partial_{x_j} \rho_{n+1}\right) \dot{X}_j}{\rho_{n+1}} \right\rangle = -\left\langle (\underline{\underline{\Sigma}}^{-1} \underline{X})_j \dot{X}_j \right\rangle, \tag{10.46}$$

which stems from Eq. (4.31). In the last step, we have utilized a general property of normal distributions $\partial_{x_j} \rho_{n+1}(\underline{x}) = -(\underline{\underline{\Sigma}}^{-1} \underline{X})_j \rho_{n+1}$. Substituting (10.9), we find the general formula

$$\begin{aligned} \dot{I}_{\to j} &= -\sum_{l \neq j}^{n} (\underline{\underline{\Sigma}}^{-1})_{jl} \langle X_l \dot{X}_j \rangle = -\sum_{i=0}^{n} \sum_{l \neq j}^{n} \frac{a_{ji}}{\gamma_j} \left(\underline{\underline{\Sigma}}^{-1}\right)_{jl} \left(\underline{\underline{\Sigma}}\right)_{li} \\ &= -\frac{a_{jj}}{\gamma_j} + \frac{1}{\gamma_j} \left(\underline{\underline{\Sigma}}^{-1}\right)_{jj} \left(\underline{\underline{a}\Sigma}\right)_{jj}. \end{aligned} \tag{10.47}$$

In combination with (10.26), Eq. (10.47) represents an analytic expression for the information flow to any node in the network of arbitrary size. The explicit closed-form expressions are, however, rather tedious. Let us consider the results for $n = 1, 2$.

For $n = 1$, the information flow to X_0 simplifies to

$$\begin{aligned} \dot{I}_{\to 0} &= -\frac{a_{00}}{\gamma_0} - \frac{\langle X_1^2 \rangle (a_{00} \langle X_0^2 \rangle + a_{01} \langle X_0 X_1 \rangle)}{\gamma_0 (\langle X_0 X_1 \rangle^2 - \langle X_0^2 \rangle \langle X_1^2 \rangle)} \\ &= \frac{[a_{01} \mathcal{T}_1 - a_{10} \mathcal{T}_0][a_{00} a_{01} / \mathcal{T}_0 + a_{11} a_{10} \gamma_0 / (\mathcal{T}_1 \gamma_1)]}{\mathcal{T}_0 (a_{00} \gamma_1 + a_{11} \gamma_0)^2 + \frac{a_{01}^2 \mathcal{T}_1}{\gamma_0 \gamma_1} - 2 \mathcal{T}_0 a_{01} a_{10} + a_{10}^2 \frac{\mathcal{T}_0^2}{\mathcal{T}_1}}. \end{aligned} \tag{10.48}$$

From the last expression we can see that the information flow trivially vanishes if X_0 and X_1 are uncorrelated, as one would expect. It also vanishes for reciprocally coupled systems under isothermal conditions. More generally, at non-isothermal conditions, it nullifies at $a_{10} \mathcal{T}_0 = a_{01} \mathcal{T}_1$. These two cases correspond to thermal equilibrium situations, as we have shown above.

If $a_{01} \neq 0$, the information flow can be expressed as

$$\dot{I}_{\rightarrow 0} = \left(\underline{\underline{\Sigma}}^{-1}\right)_{01} \frac{\dot{Q}_0}{\gamma_0\, a_{01} \mathcal{T}_0} = \frac{-\langle X_1 X_0 \rangle}{\langle X_1 X_0 \rangle^2 - \langle X_1^2 \rangle \langle X_0^2 \rangle} \frac{\dot{Q}_0}{\gamma_0\, a_{01} \mathcal{T}_0}, \tag{10.49}$$

revealing that $\dot{I}_{\rightarrow 0}$ is automatically zero, if $\dot{Q}_0 = 0$. This implication $\dot{Q}_0 = 0 \Rightarrow \dot{I}_{\rightarrow 0} = 0$ was already noticed in Ref. [45], where the reciprocal case of our model was considered. However, we see here that this implication is only true if $a_{01} \neq 0$.

For $n = 2$, Eq. (10.47) yields

$$\dot{I}_{\rightarrow 0} = \frac{-a_{00}}{\gamma_0} + \frac{(a_{00}\langle X_0^2 \rangle + a_{01}\langle X_0 X_1 \rangle + a_{02}\langle X_0 X_2 \rangle)(\langle X_1 X_2 \rangle^2 - \langle X_1^2 \rangle \langle X_2^2 \rangle)}{\gamma_0\, \psi} \tag{10.50}$$

$$\psi = \langle X_0^2 \rangle \langle X_1 X_2 \rangle^2 + \langle X_1^2 \rangle \langle X_0 X_2 \rangle^2 + \langle X_2^2 \rangle \langle X_0 X_1 \rangle^2 - \langle X_0^2 \rangle \langle X_1^2 \rangle \langle X_2^2 \rangle - 2\langle X_0 X_1 \rangle \langle X_0 X_2 \rangle \langle X_1 X_2 \rangle \tag{10.51}$$

This expression is quite cumbersome and, admittedly, not very enlightening. However, (10.50) suggests that in this case the information flow can be nonzero, despite $\dot{Q}_0 = 0$ and $a_{02} \neq 0$. For even higher n, the results get even more tedious.

Information-theoretic perspective on NR coupling We are now in the position to clarify the meaning of unidirectional coupling from an information-thermodynamic perspective. Inspecting Figs. 10.4 and 10.5 we see that along the unidirectional coupling axis $a_{01} = 0$, there is net information flow from X_0 to X_1, but no net work applied onto X_0 ($\dot{Q}_0 = \dot{W}_0 = 0$). Thus, it is indeed very reasonable to consider X_1 a "sensor" and the unidirectional coupling a *sensing interaction*. On the other hand, if the unidirectional coupling is reversed ($a_{10} = 0$), the heat flow is always positive, $\dot{Q}_0 > 0$. This suits the idea that X_0 models an active swimmer within the AOUP model. In particular, the swimmer heats up the surrounding fluid but never has a net cooling effect, as shall be expected. In this case, there is as well a nonzero information flow, which is directed from the source of propulsion (i.e., the flagella or flow field) to the particle. This is indeed reasonable as the propulsion force "carries" information (one could, on average, reconstruct the position of the flagella by only monitoring X_0).

10.5.2 Information-Theoretic Perspective on Feedback Control

According to the results from the last section, a (unidirectional) interaction from one d.o.f., say X_0, to another, X_1, can model a measurement operation, while the backwards coupling can model an active (propulsion) force. Thus, a NR interaction can be considered as a combination of a measurement and a force depending on the measurement outcome, or, in other words, feedback. In this sense, we can state that our networks with NR bidirectional coupling model feedback loops.

However, it seems intuitive to consider X_0 a feedback-*controlled* system, if and only if the net information flow out of X_0 is positive, i.e., the controller "knows" more about X_0 than *vice versa*.

According to this definition, the *control regime* for the model (10.12), is given if $|a_{10}| > |a_{01}|$, see the blue regions in Fig. 10.5. The generalized second law (10.44) states that only in this regime, a reversed heat flow is possible, as indeed confirmed by Fig. 10.4. Interestingly, we find that another intricate phenomenon is limited to the control regime, namely, the *suppression of thermal fluctuations* of X_0, measured by a reduced second moment $\langle X_0^2 \rangle < \langle X_0^2 \rangle_{a_{01}=0}$ (see Fig. 10.5). Such a suppression, i.e., noise-reduction (which can be interpreted as feedback cooling) is desired in various experimental setups, and indeed one of the main applications of feedback control [47–49].

We thus conclude that characteristic features of feedback-controlled systems are described by a single NR coupling, e.g., by (10.9) with $n = 1$ and $|a_{10}| > |a_{01}|$. X_1 represents a controller that utilizes information about X_0 extracted via a sensing operation. The resulting system is characterized by positive information flow from X_0 to X_1, reversed heat flow, and suppression of thermal fluctuations (depending on the strength of the "feedback force" $a_{01}X_1$).

For further discussions about information flow, energetics and control, we refer the interested Reader to [45].

10.6 Total Entropy Production and Heat Flow in the Presence of Non-monotonic Memory

Let us return to the picture where X_0 is the position of a colloidal particle, while $X_{\{1,2,\dots,n\}}$ represent feedback controller degrees of freedom, and now focus on the impact of the number of n. To this end, we revisit the formulae derived in the previous sections. First, we consider the mean heat flow from X_0 to its bath from Eq. (10.20), which is given by $\dot{Q}_0 = k^2\langle X_n^2 \rangle + ka_{00}\langle X_0 X_n \rangle$, with k being the strength of the feedback kX_n, and $a_{00}X_0$ being the restoring force. A comparison of the cases $n = 1, 2$ is instructive, since the networks are of comparable size while the memory kernels $K(T)$ have distinct characteristics, i.e., exponential decay with maximum at $T = 0$ ($n = 1$) vs. a non-monotonic kernel with a minimum at $T = 0$ ($n = 2$), see Fig. 6.2a, b. Figure 10.8 reveals that the heat flow $\dot{Q}_0$ is qualitatively and quantitatively similar for $n = 1$ and 2. The (blue) area of reversed heat flow is slightly smaller for $n = 2$, but lies in the same region of the (τ, k)-plane. For $k = 0$, there is trivially no heat flow (in this case, X_0 is a passive d.o.f. which does not "see" the other d.o.f.). For both $n = 1, 2$, there is a second line along which the heat flow vanishes. For $n = 1$, this line corresponds to parameters where the FDR (10.16) is fulfilled (dashed line). As shown in Fig. 10.9, the total EP is then zero as well. For $n = 2$, the FDR is broken for all (τ, k) and the system never reaches thermal equilibrium. Correspondingly, the EP is strictly positive, quantifying the "thermodynamic cost" of the non-monotonic

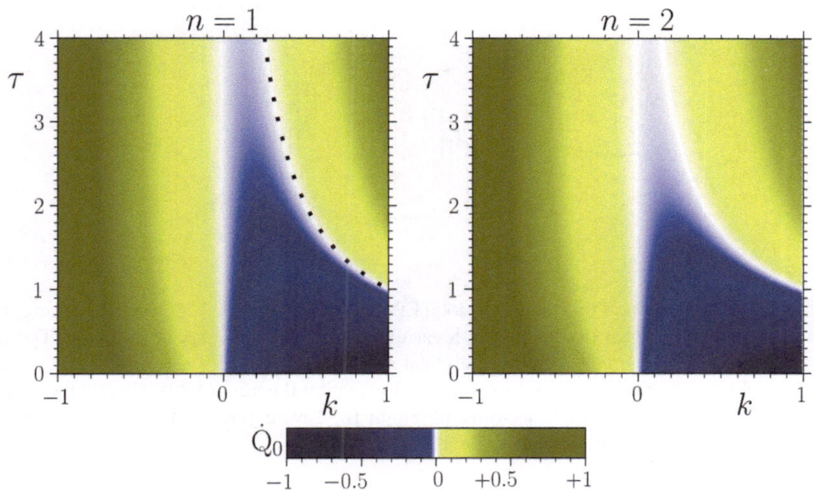

Fig. 10.8 Heat flow from X_0 to its bath [from Eqs. (10.20), (10.22)], in networks of type (10.11) of different sizes; *left*: $n = 1$ (one "memory cell", exponential memory) and *right*: $n = 2$ (two "memory cells", non-monotonic memory). For $n = 1$, the FDR (10.16) is fulfilled at $k = 1/\tau$ (dotted line). Since this system equilibrates, $\dot{Q}_0 = 0$. In the uncoupled case ($k = 0$), the subsystem X_0 equilibrates as well (for arbitrary n). All other parameters and k_B are set to unity. These results are also published in [33]

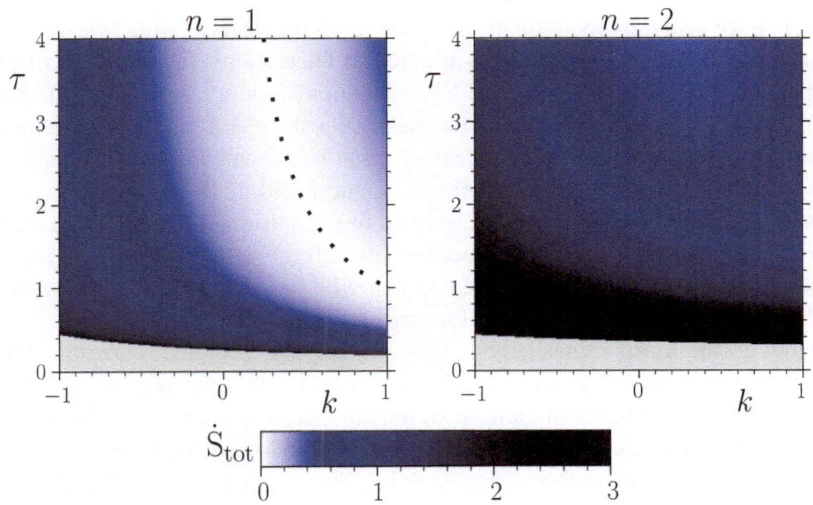

Fig. 10.9 Total EP $\dot{S}_{tot}$ [from Eqs. (10.19), (10.21)] in units of k_B/τ_B, in networks of sizes $n = 1, 2$, as in Fig. 10.8. Along the dashed line, FDR (10.16) is fulfilled, and the total EP is zero, showing consistency between the non-Markovian description (10.10) and the Markovian network description (10.9). The gray areas indicate unstable regions (it is slightly larger in the case $n = 2$)

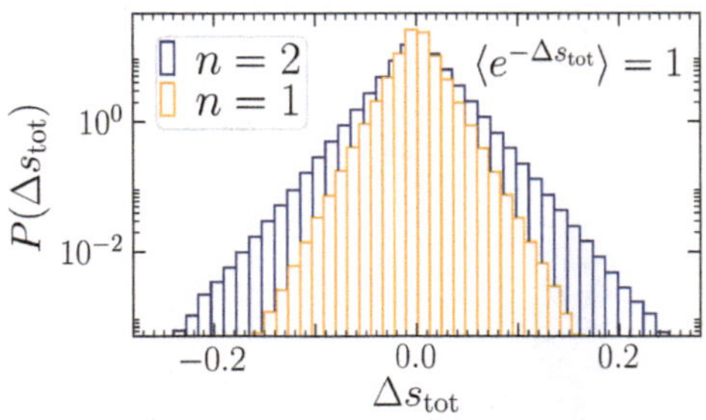

Fig. 10.10 Distributions of the fluctuating total EP Δs_{tot} (4.15) for $n = 1, 2$ from Brownian dynamics simulations. The linear decays in this logarithmic plots indicate exponential tails. The mean values are in good agreement with the analytical results ($\Delta S_{tot} = 2$ for $n = 1$, and $\Delta S_{tot} = 5$ for $n = 2$), and both distributions fulfill the integral fluctuation theorem (4.22). To calculate the distributions, 5×10^6 steady-state trajectories of length 10^{-4} were generated. $k = -1$, and all other parameters and k_B are set to unity

memory, or, the dissipation of $X_{j>0}$. For $n > 1$, there are (τ, k) values where the system is out of equilibrium, but nevertheless $\dot{Q}_0 = 0$. Thus, the vanishing medium EP is not sufficient to signal equilibrium. We recall that in the last chapter, we also observed points of vanishing heat flux in the presence of feedback with discrete delay (see Chap. 9). Note that the case $n = 1$ is significantly different in this regard, and zero dissipation always coincides with equilibrium.

While we generally focus on the mean values of the total EP, which we can access analytically, let us briefly take a look onto the fluctuations. Figure 10.10 displays probability distributions of the total EP fluctuations (obtained by BD simulations). The distributions have very similar characteristics for $n = 1$ and 2, despite the distinct memory kernels (exponential decay vs. non-monotonic memory). In particular, they both have exponentially decaying tails, and fulfill the IFT $\langle e^{-\Delta s_{tot}/k_B} \rangle = 1$, as expected (see Sect. 4.4). The characteristics of the entropy fluctuations are, in fact, identical for all cases, whether the network models an active microswimmer, a cellular sensor, or a feedback-controlled colloid with time delay. In the following, we will again return to the discussion of the mean values.

How do the mean values behave upon further increase of n, corresponding to more sharply peaked memory kernels appearing in the dynamics of X_0? In our interpretation, this means an increase of the number of memory cells. Figure 10.11 shows analytical results for the mean total EP and heat flow as functions of n, for three different values of the feedback strength k, i.e., the coupling between X_n and X_0. We observe that the total EP generally increases quadratically with n. This observation is robust against the variation of system parameters, e.g., the temperatures, coupling strengths, or delay time. In fact, it appears to hold even in nonlinear systems. To check this, we have added a nonlinear force F_0 in the dynamical equation of X_0, Eq. (10.9), and performed BD simulations to calculate the EP (not shown here). For all cases, we find quadratic increase with n.

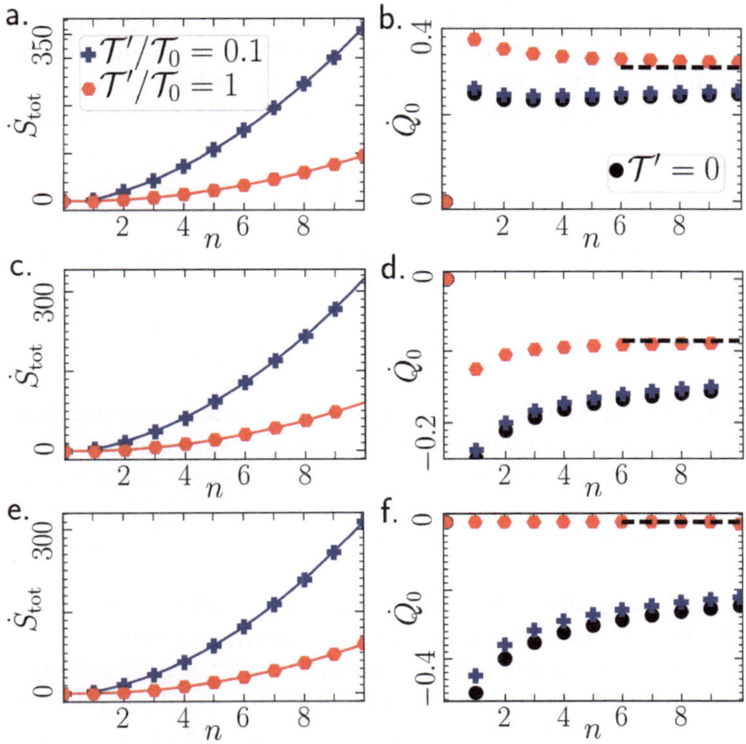

Fig. 10.11 (a.+c.+e.) Total EP [from Eq. (10.19)], and (b.+d.+f.) heat flow $\dot{Q}_0$ [from Eq. (10.20)], vs. the number n of $X_{j>0}$. The EP plots are complemented by quadratic fits $\dot{S}_{\rm tot} \sim n^2$ (solid lines). For all temperature ratios $\mathcal{T}'/\mathcal{T}_0$ (different colors), the heat flow converges in the limit $n \to \infty$ to the discrete delay solution (dashed lines). At $\mathcal{T}'=0$ (black disks) $\dot{S}_{\rm tot}$ diverges. (a.+b.) $k = -1/2$ (positive feedback), (c.+d.) $k = 1/2$ (negative feedback). (e.+f.) $k = 1$, where the system with $n = 1$ is reciprocally coupled. All other parameters and $k_{\rm B}$ are set to unity

In sharp contrast, the heat flow saturates with n, as shown in the right panel of Fig. 10.11. Moreover, in agreement with our previous observation for the case $n = 1$ (e.g., see Fig. 10.4), we again find a reversed heat flow ($\dot{Q}_0 < 0$) for negative feedback, i.e., when the feedback force is directed towards the past trajectory.

10.6.1 Limit of Discrete Delay

In realistic setups, a controller has a finite memory capacity, i.e., the amount of stored information is limited. This implies, on the one hand, that $\mathcal{T}'$ is finite (finite precision), and on the other hand, that the number of memory cells n should be finite. The resulting memory kernel has a peak around τ with non-zero width (distributed delay).

However, in theoretical studies, the delay is often assumed to be discrete, i.e., *infinitely sharp*. In our approach, this corresponds to the limit $n \rightarrow \infty$, where $K(T) \rightarrow \delta(T-\tau)$. In this limit, we find a surprising result for the colored noise. As we have shown in Chap. 6 (see Sect. 6.1.2), the noise correlations completely *vanish*

$$\lim_{n \rightarrow \infty} C_{\xi} = 0 \tag{10.52}$$

(irrespective of the value of $\mathcal{T}'$). Hence, a delay LE with white noise is recovered from (10.10), that is,

$$\dot{X}_0(t) = a_{00}\, X_0(t) + k\, X_0(t-\tau) + \xi_0(t). \tag{10.53}$$

This implies that the heat flow (10.20) approaches the value of the heat flow for discrete delay and white noise, which we have intensely studied in the previous chapter, for both, linear and nonlinear systems. The corresponding values [the formula is given in Eq. (9.17)] are plotted as dashed lines in Fig. 10.11. Indeed, the analytical results for finite n seem to converge to this value. However, the convergence is relatively slow.

Remarkably, we generally find that $\dot{Q}_0$ approaches the limit fastest for isothermal conditions $\mathcal{T}_0 = \mathcal{T}'$. We will come back to this observation below.

As we have discussed above, the total entropy production increases quadratically with n, which implies a *divergent* EP in the limit $n \rightarrow \infty$. Taking the perspective that the network models a memory device, where each memory cell can contribute to dissipation, a divergent EP is indeed somewhat expected due to the infinite system size. Also from an information-theoretical viewpoint, this result is in fact not too surprising. To realize (10.53), a random trajectory of length τ needs to be memorized. However, the information content of a Brownian trajectory is infinite (even if it is arbitrarily short), since it is governed by *white noise*, which yields infinitely many jumps within arbitrarily short time intervals. This infinite amount of information results in a divergent EP.

To avoid confusion, we aim to differentiate our findings from [3–5], where a finite measure for irreversibility was reported. We have briefly reviewed this approach earlier in Sect. 4.6.4. Specifically, in this line of work, an irreversibility measures based on X_0 alone is employed, for the case of underdamped motion. In contrast, we here embed the delay problem into a higher dimensional structure and then use the standard framework of thermodynamics (and do not use a specifically adapted formalism). Therefore, our result is very different. We find a diverging EP owing to the infinite-dimensional character of this case, which is apparent from the infinite network size, as well as the infinite-dimensional nature of the delta-distribution kernel.

10.6.2 Impact of Measurement Errors

Finally, we briefly consider the impact of $\mathcal{T}'$, that is, the strength of the noise terms ξ_j in Eq. (10.9) (accounting for measurement or rounding errors). As visible in Fig. 10.11, the heat flow converges to the previously obtained result as $n \to \infty$ (dashed lines), fastest for isothermal conditions $\mathcal{T}_0 = \mathcal{T}'$. This implies that the details of the memory kernel (which depends on n) have a larger impact, if the temperatures are inhomogeneously distributed over the network. This is somehow puzzling, especially when one considers the colored noise. We recall that the colored noise is present at finite n, but vanishes in the limit $n \to \infty$ for all cases $\mathcal{T}'$. At $\mathcal{T}' = 0$, it is not present even for finite n, and its impact increases with $\mathcal{T}'$. Given these qualitative arguments, it is very unexpected that the case $\mathcal{T}' = \mathcal{T}_0$ converges fastest to the limit. So far, we have not resolved this seeming contradiction, but it would be very interesting to study this in more detail in the future (especially, because the finite n case may be used to approximate the discrete delay case).

Figure 10.12 shows that the heat flow $\dot{Q}_0$ increases linearly with $\mathcal{T}'/\mathcal{T}_0$. Due to the specific parameter choice in Fig. 10.12, where $n = 1$ corresponds to a reciprocally coupled network, the heat flow vanishes at $\mathcal{T}_0 = \mathcal{T}'$ for all n. Thus, at $\mathcal{T}_0 = \mathcal{T}'$, all lines cross, and the number n matters least. For different parameter settings, we also observe that the lines cross near this point (not shown here).

On the contrary, the total EP has for each n a pronounced minimum at a finite $\mathcal{T}'/\mathcal{T}_0$. Interestingly, at some values of $\mathcal{T}'/\mathcal{T}_0$ above the minima, the total EP for

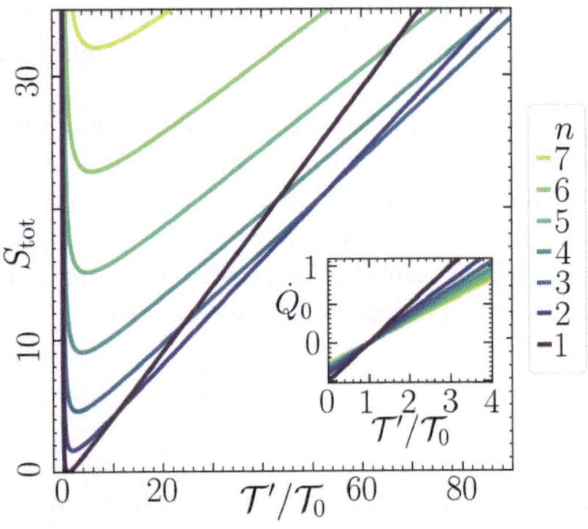

Fig. 10.12 Total EP $\dot{S}_{tot}$ [from Eq. (10.19)] as a function of the ratio of the temperatures of $X_{j>0}$ and X_0, for networks with different numbers n of d.o.f. $X_{j>0}$. The parameters are identical to Fig. 10.11 (e.+f.). The network with $n = 1$ is reciprocally coupled, and fulfills FDR (10.16) if $\mathcal{T}'/\mathcal{T}_0 = 1$. In accordance, the total EP vanishes. For each n, the EP has a minimum at a finite $\mathcal{T}'/\mathcal{T}_0$ of order 1, and diverges in the limits $\mathcal{T}'/\mathcal{T}_0 \to 0$ or $\mathcal{T}'/\mathcal{T}_0 \to \infty$. The inset shows the heat flow $\dot{Q}_0$, which for all n increases linearly and is zero at $\mathcal{T}'/\mathcal{T}_0 = 1$

different n are *identical* (i.e., the lines in Fig. 10.12 cross). This implies that two networks at the same temperature but with a different number of nodes can produce the same amount of entropy. This is somewhat surprising, as one would expect that a larger system automatically has a higher EP.

The EP diverges if one of the temperatures goes to zero, corresponding to the cases of vanishing or infinite measurement errors. This is in accordance with the result from [50], whose limit of "precise and infinitely fast" control corresponds in our time-continuous model to the limit $\mathcal{T}' \to 0$. We also note that when we interpret X_0 as the position of an active swimmer, $\mathcal{T}_0 \to 0$ corresponds to the limit of zero translational ("passive") noise (and pure "active" fluctuations). The divergence of EP in this limit was noted in [51] for a similar active swimmer model.

10.7 Irreversibility and Coarse-Graining

In this last section, we elucidate the relationship between our network model and the representation via X_0 alone from a more fundamental perspective in the context of *irreversibility* and *effective thermodynamics*. We have already mentioned several times that the investigation of the network model in this chapter can apply to two situations. First, if all X_j are d.o.f. with a direct physical meaning, we have considered the total EP Δs_{tot} of that actual model, and the interpretation of all entropic terms is straightforward. Here, we now consider the second, more subtle situation, i.e., if the $X_{j>0}$ are only considerable as auxiliary variables, but have no direct physical interpretation. As we have seen, we can indeed get an interesting and reasonable view on the process by considering the total EP Δs_{tot}, although the additional entropic terms have a somewhat debatable meaning.

As mentioned in Sect. 4.6.4, finding thermodynamic notions for systems where X_0 is the only observable (non-Markovian) process (10.10), is the objective of several recent publications aiming for an effective thermodynamics, see e.g., [3–5, 28–30]. Indeed, in many situations of particular interest, X_0 is the only experimentally accessible quantity. Then, a natural irreversibility measure, which can directly be calculated from the observed trajectories, would be [29]

$$\frac{\Delta s^0}{k_{\text{B}}} := \ln \frac{\mathcal{P}[\mathbf{X}_0]}{\hat{\mathcal{P}}[\hat{\mathbf{X}}_0]}, \tag{10.54}$$

[as we have already given in Eq. (4.37)]. With this picture in mind, one may question the meaning of the formula for the total EP Δs_{tot}, which explicitly involves $X_{j>0}$. However, our framework also yields a statement about Δs^0.

First, we notice that Δs^0 corresponds to one part of the total EP (4.15). To see this, we use the definition of conditional path probabilities, i.e., $\mathcal{P}[\mathbf{X}_0; \underline{\mathbf{X}}_{\text{c}}] = \mathcal{P}[\underline{\mathbf{X}}_{\text{c}}|\mathbf{X}_0]\mathcal{P}[\mathbf{X}_0]$, with $\underline{\mathbf{X}}_{\text{c}} = \{\mathbf{X}_1, \ldots, \mathbf{X}_n\}$, to split up the total EP

$$\frac{\Delta s_{\text{tot}}[\underline{\mathbf{X}}]}{k_{\text{B}}} = \ln \frac{\mathcal{P}[\underline{\mathbf{X}}]}{\hat{\mathcal{P}}[\hat{\underline{\mathbf{X}}}]} = \frac{s^0[\mathbf{X}_0]}{k_{\text{B}}} + \underbrace{\ln \frac{\mathcal{P}[\mathbf{X}_{\text{c}}|\mathbf{X}_0]}{\hat{\mathcal{P}}[\hat{\underline{\mathbf{X}}}_{\text{c}}|\hat{\mathbf{X}}_0]}}_{=s^{\text{c}|0}/k_{\text{B}}}. \tag{10.55}$$

Further, by exploiting our Markovian representation of the process, we can show that the fluctuations of Δs^0 fulfill an integral fluctuation theorem, as do those of Δs_{tot} (see Sect. 4.4, Fig. 10.10),

$$\begin{aligned}\left\langle e^{-\frac{\Delta s^0}{k_{\text{B}}}} \right\rangle &= \int .. \int \frac{\hat{\mathcal{P}}[\hat{\mathbf{X}}_0]}{\mathcal{P}[\mathbf{X}_0]} \mathcal{P}[\underline{\mathbf{X}}]\, \mathfrak{D}\underline{\mathbf{X}} \\ &= \int \hat{\mathcal{P}}[\hat{\mathbf{X}}_0] \underbrace{\int \mathcal{P}[\mathbf{X}_1, ..., \mathbf{X}_n|\mathbf{X}_0]\, \mathfrak{D}\mathbf{X}_1..\mathfrak{D}\mathbf{X}_n}_{=1}\, \mathfrak{D}\mathbf{X}_0 = \int \hat{\mathcal{P}}[\hat{\mathbf{X}}_0]\, \mathfrak{D}\mathbf{X}_0 = 1.\end{aligned} \tag{10.56}$$

This implies $\Delta S^0 \geq 0$. The same can be shown for the other contribution in (10.55), i.e., the entropy-like quantity $\Delta s^{\text{c}|0}$, because

$$\begin{aligned}\left\langle e^{-\Delta s^{\text{c}|0}/k_{\text{B}}} \right\rangle &= \iint \frac{\hat{\mathcal{P}}[\hat{\underline{\mathbf{X}}}_{\text{c}}|\hat{\mathbf{X}}_0]}{\mathcal{P}[\underline{\mathbf{X}}_{\text{c}}|\mathbf{X}_0]} \mathcal{P}[\underline{\mathbf{X}}]\, \mathfrak{D}\underline{\mathbf{X}} \\ &= \int \mathcal{P}[\mathbf{X}_0] \left[\int \hat{\mathcal{P}}[\hat{\underline{\mathbf{X}}}_{\text{c}}|\hat{\mathbf{X}}_0]\, \mathfrak{D}\underline{\mathbf{X}}_{\text{c}} \right] \mathfrak{D}\mathbf{X}_0 = \int \mathcal{P}[\mathbf{X}_0]\, \mathfrak{D}\mathbf{X}_0 = 1.\end{aligned} \tag{10.57}$$

These considerations yield the inequality

$$\Delta S_{\text{tot}} \geq \Delta S^0 \geq 0. \tag{10.58}$$

Thus, the irreversibility of X_0 alone underestimates the irreversibility of the entire system, consistent with what one would expect [52]. Furthermore, (10.58) states that the mean total EP considered in this thesis yields an upper bound to the average of the irreversibility measure Δs^0.

By considering Δs^0, we have addressed entropy-like quantities based on marginalized path probabilities. An obvious alternative to get rid of the additional d.o.f. $X_{j>0}$ would be to study instead the marginalized EP: $\iint \Delta s_{\text{tot}}[\underline{\mathbf{X}}]\, \mathfrak{D}\mathbf{X}_1 \ldots \mathfrak{D}\mathbf{X}_n$. We stress that this quantity is different from Δs^0 [as can be seen from Eq. (10.55)] and there is no direct relation between both. The relevance of such a marginalized EP is, in fact, questionable. It does *not* solely rely on measurable quantities (as opposed to Δs^0), neither does it have a direct link to the well-understood framework of statistical mechanics.

In contrast, the total EP Δs_{tot} is the central quantity in statistical mechanics with a direct connection to the energy flows, the first and second law of thermodynamics, and the underlying microscopic laws of motion [53].

In this chapter, we have investigated thermodynamic and information-theoretic notions of systems with time-delay. We have focused on a specific question relevant in this context, that is, the consequences of non-reciprocal coupling. By application of standard definitions, we were able to obtain a consistent picture, and gained some interesting insights. However, concerning the underlying bigger problem, various important open questions remain.

In the following, last Part of this thesis, we will give an outlook on some further objectives of future research.

References

1. Parrondo JMR, Horowitz JM, Sagawa T (2015) Thermodynamics of information. Nat Phys 11:131–139
2. Puglisi A, Villamaina D (2009) Irreversible effects of memory. EPL 88:30004
3. Munakata T, Rosinberg ML (2014) Entropy production and fluctuation theorems for Langevin processes under continuous non-Markovian feedback control. Phys Rev Lett 112:180601
4. Rosinberg ML, Munakata T, Tarjus G (2015) Stochastic thermodynamics of Langevin systems under time-delayed feedback control: second-law-like inequalities. Phys Rev E 91:042114
5. Rosinberg ML, Tarjus G, Munakata T (2017) Stochastic thermodynamics of Langevin systems under time-delayed feedback control. II. Nonequilibrium steady-state fluctuations. Phys Rev E 95:022123
6. Horowitz JM, Sandberg H (2014) Second-law-like inequalities with information and their interpretations. New J Phys 16:125007
7. Horowitz JM, Esposito M (2014) Thermodynamics with continuous information flow. Phys Rev X 4:031015
8. Sartori P, Granger L, Lee CF, Horowitz JM (2014) Thermodynamic costs of information processing in sensory adaptation. PLoS Comput Biol 10:e1003974
9. Mandal D, Quan HT, Jarzynski C (2013) Maxwell's refrigerator: an exactly solvable model. Phys Rev Lett 111:030602
10. Mandal D, Jarzynski C (2012) Work and information processing in a solvable model of Maxwell's demon. PNAS 109:11641–11645
11. Barato AC, Seifert U (2013) An autonomous and reversible Maxwell's demon. EPL 101:60001
12. Hartich D, Barato AC, Seifert U (2014) Stochastic thermodynamics of bipartite systems: transfer entropy inequalities and a Maxwell's demon interpretation. J Stat Mech Theor Exp 2014:P02016
13. Boukobza E, Tannor DJ (2006) Thermodynamics of bipartite systems: application to light-matter interactions. Phys Rev A 74:063823
14. Munakata T, Rosinberg ML (2013) Feedback cooling, measurement errors, and entropy production. J Stat Mech 13:06014
15. Argun A, Volpe G (2016) Better stability with measurement errors. J Stat Phys 163:1477–1485
16. Rateitschak K, Wolkenhauer O (2007) Intracellular delay limits cyclic changes in gene expression. Math Biosci 205:163–179
17. Longtin A (2010). In: Atay FM (ed) Complex time-delay systems: theory and applications. Springer, Berlin Heidelberg
18. Sekimoto K (2010) Stochastic energetics, vol 799. Springer
19. Hartich D, Barato AC, Seifert U (2016) Sensory capacity: an information theoretical measure of the performance of a sensor. Phys Rev E 93:022116
20. Agudo-Canalejo J, Golestanian R (2019) Active phase separation in mixtures of chemically interacting particles. Phys Rev Lett 123:018101

21. Durve M, Saha A, Sayeed A (2018) Active particle condensation by non-reciprocal and time-delayed interactions. Eur Phys J E 41:49
22. Kompaneets R, Vladimirov S, Ivlev A, Morfill G (2008) Reciprocal interparticle attraction in complex plasmas with cold ion flows. New J Phys 10:063018
23. Ivlev AV et al (2015) Statistical mechanics where Newton's third law is broken. Phys Rev X 5:011035
24. Kubo R (1966) The fluctuation-dissipation theorem. Rep Prog Phys 29:255
25. Maes C, Safaverdi S, Visco P, Van Wijland F (2013) Fluctuation-response relations for nonequilibrium diffusions with memory. Phys Rev E 87:022125
26. Crisanti A, Puglisi A, Villamaina D (2012) Nonequilibrium and information: the role of cross correlations. Phys Rev E 85:061127
27. Seifert U (2012) Stochastic thermodynamics, fluctuation theorems and molecular machines. Rep Prog Phys 75:126001
28. Shankar S, Marchetti MC (2018) Hidden entropy production and work fluctuations in an ideal active gas. Phys Rev E 98:020604(R)
29. Dabelow L, Bo S, Eichhorn R (2019) Irreversibility in active matter systems: fluctuation theorem and mutual information. Phys Rev X 9:021009
30. Caprini L, Marconi UMB, Puglisi A, Vulpiani A (2019) The entropy production of Ornstein-Uhlenbeck active particles: a path integral method for correlations. J Stat Mech Theor Exp 2019:053203
31. Mandal D, Klymko K, DeWeese MR (2017) Entropy production and fluctuation theorems for active matter. Phys Rev Lett 119:258001
32. Bonilla L (2019) Active Ornstein-Uhlenbeck particles. Phys Rev E 100:022601
33. Loos SAM, Klapp SHL (2020) Irreversibility, heat and information flows induced by non-reciprocal interactions. New J Phys 22:123051
34. Nardini C et al (2017) Entropy production in field theories without time-reversal symmetry: quantifying the non-equilibrium character of active matter. Phys Rev X 7:021007
35. Zwanzig R (1973) Nonlinear generalized Langevin equations. J Stat Phys 9:215–220
36. Hänggi P, Thomas H (1982) Stochastic processes: time evolution, symmetries and linear response. Phys Rep 88:207–319
37. Ramaswamy S (2010) The mechanics and statistics of active matter. Annu Rev Condens Matter Phys 1:323–345
38. Ramaswamy S (2017) Active matter. J Stat Mech Theor Exp 2017:054002
39. Fodor É et al (2016) How far from equilibrium is active matter? Phys Rev Lett 117:038103
40. Dauchot O, Löwen H (2019) Chemical physics of active matter. J Chem Phys 151:114901
41. Roldán É, Barral J, Martin P, Parrondo JM, Jülicher F (2018) Arrow of time in active fluctuations. ArXiv:1803.04743
42. Netz RR (2018) Fluctuation-dissipation relation and stationary distribution of an exactly solvable many-particle model for active biomatter far from equilibrium. J Chem Phys 148:185101
43. Koski JV, Maisi VF, Sagawa T, Pekola JP (2014) Experimental Observation of the Role of Mutual Information in the Nonequilibrium Dynamics of a Maxwell Demon. Phys Rev Lett 113:030601
44. Maxwell JC (1871) Theory of heat. Dover Publications inc
45. Allahverdyan AE, Janzing D, Mahler G (2009) Thermodynamic efficiency of information and heat flow. JSTAT 2009:P09011
46. Reimann P (2002) Brownian motors noisy transport far from equilibrium. Phys Rep 361
47. Steck DA, Jacobs K, Mabuchi H, Habib S, Bhattacharya T (2006) Feedback cooling of atomic motion in cavity QED. Phys Rev A 74:012322
48. Cohadon P-F, Heidmann A, Pinard M (1999) Cooling of a mirror by radiation pressure. Phys Rev Lett 83:3174
49. Vinante A et al (2008) Feedback cooling of the normal modes of a massive electromechanical system to submillikelvin temperature. Phys Rev Lett 101:033601
50. Strasberg P, Schaller G, Brandes T, Esposito M (2013) Thermodynamics of a physical model implementing a Maxwell demon. Phys Rev Lett 110:040601

51. Pietzonka P, Seifert U (2017) Entropy production of active particles and for particles in active baths. J Phys A Math Theor 51:01LT01
52. Esposito M (2012) Stochastic thermodynamics under coarse graining. Phys Rev E 85:041125
53. Wittkowski R, Löwen H, Brand HR (2013) Microscopic approach to entropy production. J Phys A 46:355003

Part IV
Concluding Remarks

Thermodynamik ist ein komisches Fach! Das erste Mal, wenn man sich damit befasst, versteht man nichts davon. Beim zweiten Durcharbeiten denkt man, man hätte nun alles verstanden, mit Ausnahme von ein oder zwei kleinen Details. Das dritte Mal, wenn man den Stoff durcharbeitet, bemerkt man, dass man fast garnichts davon versteht, aber man hat sich inzwischen so daran gewöhnt, dass es einen nicht mehr stört.

—Arnold Sommerfeld, as quoted in [361]

In this last Part IV, we summarize the main results presented in this thesis. We further put our ideas into a wider context, and critically challenge our findings. At the end, we draw the attention of the Reader to open questions building upon our results, and give some perspectives for future research on stochastic systems with time delay.

Chapter 11
Summary

This thesis is concerned with the non-Markovian dynamics of stochastic, fluctuation-dominated systems with time delay. The prime example studied here is a colloidal (Brownian) particle subject to a position-dependent, time-delayed feedback force and nonlinear static forces stemming from a potential. We describe its noisy motion by a Langevin equation, which is, due to the delay, non-local in time and infinite-dimensional. Being experimentally realizable and analytically treatable (by means of comparably simple equations), this system can be considered as a paradigm for a non-Markovian, nonequilibrium system.

While nonequilibrium systems are in general a major focus of modern statistical physics, the incorporation of time delays, or generally memory, in the theoretical description is still in its infancy. We have focused on two concepts that play a central role in statistical physics: the probabilistic treatment and thermodynamic notions. Both concepts are, so far, poorly understood for systems with time delay, especially in the presence of nonlinearities.

Probabilistic Treatment

First, we have discussed the probabilistic treatment of delayed stochastic systems. It was already known that, in the presence of delay, the Fokker-Planck equation for the one-time probability density is not closed, but explicitly involves the two-time (joint) probability density. In fact, an infinite hierarchy of coupled Fokker-Planck equations emerges [1–3], reflecting the infinite-dimensionality of the underlying Langevin equation.

While this was, in principle, already known, we could not find explicit expressions for the higher members of this hierarchy in the literature. One goal of this thesis was to investigate these higher members. As a starting point, we have reviewed an earlier derivation based on Novikov's theorem, and extended this approach to derive the second member of the Fokker-Planck hierarchy. Secondly, we have suggested a novel strategy, by means of a *Markovian embedding*, from which we derive an alternative representation of the Fokker-Planck hierarchy.

© The Author(s), under exclusive license to Springer Nature Switzerland AG 2021

S. A. M. Loos, *Stochastic Systems with Time Delay*, Springer Theses,
https://doi.org/10.1007/978-3-030-80771-9_11

Let us summarize the idea and technical details of this embedding procedure here. Mainly, one changes the system state from $X(t)$, which has a non-Markovian dynamics, to the entire interval $\{X(s)\}$, $s \in [t-\tau, t]$, which has a Markovian dynamics (τ being the delay time). More precisely, by introducing n additional (auxiliary) variables, which represent n timesteps between the times t and $t-\tau$, we embed the one-variable delayed process $X(t)$ in an $(n+1)$-variable space $\{X_0(t), X_1(t), ..., X_n(t)\} \hat{=} \{X(t), X(t-\mathrm{d}t), ..., X(t-\tau)\}$ ($\mathrm{d}t$ corresponds to τ/n). The original delay process is represented as one variable $X_0 \hat{=} X$ in this $(n+1)$-dimensional space. Importantly, this new representation is only exact in the limit $n \to \infty$. This is because the original dynamics is time-continuous, thus, there are infinitely many timesteps between t and $t-\tau$. Considering the equations of motion, this "blown up" Markovian representation is given by a ring network with $n+1$ nodes which are unidirectionally coupled among each other, and follow linear equations, even if the original process itself is nonlinear.

Due to the Markovianity, this $(n+1)$-variable process has a corresponding closed, $(n+1)$-variate Fokker-Planck equation. By projecting this Fokker-Planck equation onto a lower dimensional subspace (e.g., onto X_0 to derive the first member), and taking the limit $n \to \infty$ in the projected equations, we derived a hierarchy of Fokker-Planck equations. While the first member is identical, the representation of the higher members derived in this way differs from the one obtained by Novikov's theorem. This is because the underlying mathematical representation is different. Specifically, the Novikov-approach is based on an interpretation of the delay process where $\{X(t), X(t-\tau), X(t-2\tau), ...\}$ are the relevant variables, rather than $\{X(t), X(t-\mathrm{d}t), ..., X(t-\tau)\}$ like here.

The main theoretical advancements due to this thorough investigation are the following two aspects: On the one hand, we have derived explicit expressions for the higher Fokker-Planck hierarchy members, on the other hand, we have gained a more profound understanding of the mathematical structure of the delay process.

Next, we have turned to a more practical issue, that is, the development of suitable *approximation* schemes. At this point we should recall that, by not being closed, the Fokker-Planck equation for the one-time probability density alone is insufficient to find solutions for the probability density (an exception being the case of an underlying linear delay Langevin equations). Specifically, we have asked ourselves in the beginning whether it is possible to find a scheme, which yields an approximate, simplified, yet non-Markovian, description; contrasting previous approaches from the literature, which effectively render Markovian descriptions. The two most established earlier strategies are the "small delay expansion", which is a Taylor expansion of the Langevin equation in τ [4, 5], and a perturbation theory on the Fokker-Planck level [1, 6], where the entire delay force is treated as a small deviation from the Markovian dynamics. To this end, we have here proposed a new scheme, called the "force-linearization closure". This is a closure of the infinite Fokker-Planck hierarchy, achieved by linearizing the total deterministic force in all members of the hierarchy starting from the second. In this way one obtains an approximate steady-state one-time probability density of nonlinear delay systems.

Applying the force-linearization closure to two exemplary model systems, a bistable and a periodic system with linear delay force, we present numerical evidence that our approach generates indeed a very good approximation of the one-time probability density. In fact, the force-linearization closure performs better than the small delay expansion and the perturbation theory. We also note that the perturbation theory has a further technical disadvantage as it actually relies on an analytical expression for the conditional probability for the corresponding system without delay force, which is usually not available and requires additional approximations.

A further aspect touched upon is the calculation of transport-related quantities, such as escape times or waiting time distributions. For non-Markovian processes, an exact calculation would require analytical expressions of the two-time probability or even higher n-time densities. The force-linearization closure does not provide access to these probabilistic quantities. Still, we have demonstrated that a reasonable estimate of the escape times is possible by combining the force-linearization closure and the Kramers theory. This estimate, however, breaks down when the delay times are in the range of (or larger than) the jump duration times. Under such conditions, the interplay between the dynamics of the system state variable and the time-dependent energy landscape causes oscillatory motion, i.e., stochastic particle oscillations between two adjacent minima, whose description is clearly beyond this approximation. This exemplifies that an extension of the theory towards higher members of the Fokker-Planck hierarchy would be worthwhile.

Thus, while we have achieved our goal of finding a "true non-Markovian approximation" of the one-time probability density, we have actually learned along the way that the insights provided by this probabilistic quantity alone are somewhat limited. This is due to the fact that the dynamics is crucially determined by the memory, and, subsequently, the temporal correlations, particularly between the times t and $t - \tau$. The latter are given directly by the two-time probability density, which, in turn, is the solution of the second Fokker-Planck hierarchy member. In this spirit, we have reconsidered the second member of the Fokker-Planck hierarchy, which we have derived by means of the Markovian embedding technique. Based on this equation, we have introduced an approach that yields an approximate two-time probability density. This approximation finally enables us to describe truly non-Markovian, dynamical effects. We have demonstrated this by again considering the delay- and noise-induced interwell oscillations in our bistable example system. In particular, the proposed scheme renders very good results, for both cases, feedback forces that are directed towards, or away from the delayed position, capturing the main features of the two-time probability density related to the oscillations in the regime of large delay times. Moreover, the corresponding approximate one-time probability density (which can be obtained by marginalization) outperforms the approaches discussed before (small delay, perturbation theory, force-linearization closure). To the best of our knowledge, this is the first approach to calculate the two-time probability density, which is indeed a central quantity in the probabilistic description of stochastic systems with time delay.

Thermodynamics

The second main topic of this thesis is the search for physically consistent thermodynamic descriptions of systems with time delay. The generalization of the concepts of (stochastic) thermodynamics towards non-Markovian systems turns out to be a major problem that is not addressed in its full extent within the scope of this work. However, by pointing out fundamental theoretical issues, deriving some concrete solutions for the heat flow, discussing information flows and the total entropy production, and further adding some new ideas and arguments, we hope that our investigations have advanced the theoretical understanding of this thriving new field in statistical physics. Let us summarize our research questions and main findings.

To begin with, we have considered the net steady-state heat flow induced by a delay force focusing on nonlinear systems (with polynomial potentials). To gain analytical access to this quantity, we have first derived a formula for the heat rate that only involves positional moments, and can be combined with the approximations for the one-time probability density discussed before. Thus, despite the inherent memory, no expressions for the temporal correlations are needed to evaluate the mean heat flow.

Initially, we have asked ourselves whether the feedback may induce a negative heat flow, corresponding to a steady energy extraction from the heat bath. We have seen that such a reversed heat flow is indeed encountered in linear systems as well as nonlinear systems. As a concrete example, we have again considered the bistable model with linear delay force. Furthermore, we have presented analytical results for the Markovian limits and thereby observed a *growing* net heat flow when the delay time is reduced towards zero. We have argued that this pronounced heat flow at small τ above and around the (ballistic) velocity-relaxation timescale, is a consequence of the interplay of noise and delay linked to the phenomenon of "entropy pumping". To be more specific, in this regime of small delay times, the feedback becomes effectively dependent on the *velocity*. Further on, velocity-dependent forces are known to automatically induce heat flows, as they "pump entropy into the system".

Another point that we wanted to clarify concerns the consequences of the overdamped limit in the thermodynamic description. Indeed, we found that, when the inertia term is neglected in the Langevin equation, the regime of τ values below the (ballistic) velocity-relaxation timescale becomes problematic. In particular, we have uncovered discontinuous behavior in the limit $\tau \to 0$, which is absent in the corresponding underdamped description. The discontinuity in the limit of vanishing delay times is a consequence of the white noise assumption in combination with the overdamped limit. From a technical perspective, this is a notable result of our investigation, as the limit of small delay times, is, in fact, a very common regime to look at in the context of dynamical and probabilistic quantities of overdamped systems. As an example, recall the small delay expansion, which is a quite conventional approach to approximate (overdamped) delay systems. On the level of dynamical and probabilistic quantities, this limit is usually well-behaved for overdamped and underdamped descriptions, and rather unproblematic. Here we see that the thermodynamic quantities are more sensitive and one indeed needs to be careful with combining different approximations (white noise, the overdamped limit, the limit of small delay).

Physically, we suspect that the reason for the apparent offset at $\tau \to 0$ is related to the fact that the overdamped limit implies that the kinetic energy is not described. The latter could yield an additional contribution in the energy balance and thus affect the calculated heat flow.

For experimental realizations of feedback traps these theoretical results imply that the (unavoidable) delay causes a finite steady-state heat flow unless the delay is significantly smaller than the (ballistic) relaxation timescale. This might, for example, be relevant in the following context. Feedback loops offer a way to precisely control the effective temperature of a particle by suppressing its thermal fluctuations [7]. When this technique is used to change the temperature, e.g., during the cyclic process of a heat engine [8], the heat flow discussed here, which is caused even by tiny delays, might play a non-negligible role and should therefore be taken into account in the theoretical calculations of efficiency and entropy production.[1]

We have further explicitly shown that the direction of induced heat flow is tunable by adjusting the delay time and the (positive) strength of the feedback force (which would correspond to the laser intensity in an experimental setup with an optical trap). This might be a useful feature in view of possible applications.

Lastly, another interesting result of this investigation concerns the heat flow and the closely connected medium entropy production in the parameter regimes where the delay-induced interwell oscillations are prominent. As was already known, these oscillations are especially pronounced and regular around a certain finite temperature, a phenomenon which is called *coherence resonance* [9–16]. Here we have shown that the medium entropy production has a maximum right at the point where the oscillations are coherent-resonant; linking a nonlinear-dynamical effect to a thermodynamic phenomenon. By combining a discretized (Master) equation approach, called 2-state reduction [13], with the framework of stochastic energetics, we have found a way to approximate the thermodynamic quantities in this parameter regime and further provided an analytical explanation for this maximum, which was previously only detected numerically [16].

At the end of this thesis, we have turned to the question of how to quantify the "irreversibility" of delay processes, i.e., their distance from thermal equilibrium. For Markovian systems, the total entropy production provides the entire entropy balance of the process. Thereby, the total entropy production of a single fluctuating trajectory can be calculated based on the path probabilities of the observed trajectory and the path probability of its time-reversed picture in the backward process. This is a very interesting concept directly linking the notion of irreversibility and the thermodynamic arrow of time to the laws of thermodynamics. However, a generalization of this concept to *non-Markovian* systems is not straightforward, and in fact subject of many recent debates in stochastic thermodynamics.

Leaving aside the technical and conceptual challenges implied by the non-Markovianity, we have here pursued the following question: What do we learn about

[1]Although the investigations presented here are restricted to static energy landscapes, we indeed expect the results to also hold for time-dependent external potentials, at least, if their changes are slow as compared to the other dynamics.

the thermodynamics of the time-delayed process, if we study the thermodynamic properties of the corresponding Markovian network representation with degrees of freedom $\{X_0(t), X_1(t), ..., X_n(t)\}$ obtained by the Markovian embedding?

There is also another way to tell the story. One could also say that we have now included the feedback-controller explicitly into our model. In particular, we have added n additional variables $\{X_1(t), ..., X_n(t)\}$ representing the "memory cells" of the controller, and in this way obtained a Markovian description for a specific representation of the feedback controller. An important feature of the memory device is the *non-reciprocal* coupling between the cells, which corresponds to a "shift register architecture" and reflects that the updates of the device are a one-way operation in time.

We have noticed that such non-reciprocal interactions between mesoscale degrees of freedom are, in fact, also common in the context of *active* systems, examples being the active Ornstein-Uhlenbeck particle model [17–21], or the cellular sensor model [22, 23]. Indeed, this immediately matches our intuition. The state of one component (the "sensor"), say X_1, depends on the state of the other component ("the measured system"), say X_0, but not *vice versa*. In reverse, in the case of active self-propelled motion, the temporal evolution of the particle's position is assumed to be affected by the noisy orientation, but there is no backcoupling. Thus, non-reciprocal coupling is an important and quite universal concept, motivating us to discuss the connections between coupling topology and thermodynamic properties of the model in a general sense. We have shown that a single non-reciprocal interaction already implies that the system is intrinsically out of equilibrium as indicated by a positive total entropy production and a broken fluctuation-dissipation relation. A single unidirectional interaction can be viewed as an active "propulsion force", which will ultimately always have a heating effect on the environment, and never cool it down. A reversed unidirectional interaction can be considered as a "sensing" operation, inducing a net information flow from the measured system to the sensor. Furthermore, feedback can be viewed as a combination of measurement and active force. When the measurement is the dominating operation (i.e., the controller "knows" more about the sensed system than *vice versa*, indicated by a positive information flow out of the system) the following goals of feedback control can be achieved: thermal fluctuation suppression (i.e., feedback cooling), and energy extraction from the heat bath (a reversed heat flow). Taken together, from our analysis emerges a unifying perspective on activity, feedback and memory.

For the special case of discrete delay with white noise, which emerges as the limit of an infinitely large ring network, we confirm that the heat exchange of the system of interest, X_0, approaches the expected result for linear white noise delay Langevin equations. On the contrary, the total entropy production diverges in this limit, suggesting that the cost of storing a Brownian trajectory (of arbitrary length) is unbounded. We can understand this from an information-theoretic perspective. Recall that the white noise is a process with zero temporal correlation (yielding infinitely many jumps on each trajectory, no matter how short it is), thus, resulting in an unbounded information content. An interesting way to think about the approach discussed in this thesis is that it points out a connection between irreversibility and symmetry-broken coupling topology. The entropy production, which gives the *thermodynamic arrow of time*, directly measures the breaking of time-reversal symmetry, i.e., how sure can we be that a system's evolution is running from past to future, and not *vice versa*. In a way, memory breaks that symmetry in a fundamentally different way than, e.g., an external field, because memorizing is *per se* only one-way in time.

References

1. Frank TD (2005) Delay Fokker-Planck equations, Novikov's theorem, and Boltzmann distributions as small delay approximations. Phys Rev E 72:011112
2. Loos SAM, Klapp SHL (2017) Force-linearization closure for non-Markovian Langevin systems with time delay. Phys Rev E 96:012106
3. Rosinberg ML, Munakata T, Tarjus G (2015) Stochastic thermodynamics of Langevin systems under time-delayed feedback control: Second-law-like inequalities. Phys Rev E 91:042114
4. Guillouzic S, L'Heureux I, Longtin A (1999) Small delay approximation of stochastic delay differential equations. Phys Rev E 59:3970
5. Guillouzic S, Heureux L, I Longtin A (2000) Rate processes in a delayed, stochastically driven, and overdamped system. Phys Rev E 61:4906
6. Frank TD (2005) Delay Fokker-Planck equations, perturbation theory, and data analysis for nonlinear stochastic systems with time delays. Phys Rev E 71:031106
7. Lee DY, Um J, Paneru G, Pak HK (2018) An experimentally-achieved information-driven Brownian motor shows maximum power at the relaxation time. Sci Rep 8:12121
8. Martínez IA, Roldán É, Dinis L, Rica RA (2017) Colloidal heat engines: a review. Soft Matter 13:22–36
9. Schöll E, Schuster HG (eds) (2008) Handbook of chaos control. Wiley, Hoboken
10. Zakharova A, Feoktistov A, Vadivasova T, Schöll E (2013) Coherence resonance and stochastic synchronization in a nonlinear circuit near a subcritical hopf bifurcation. EPJ ST 222:2481–2495
11. Gang H, Ditzinger T, Ning C, Haken H (1993) Stochastic resonance without external periodic force. Phys Rev Lett 71:807
12. Geffert PM, Zakharova A, Vüllings A, Just W, Schöll E (2014) Modulating coherence resonance in non-excitable systems by time-delayed feedback. Eur Phys J B 87:291
13. Tsimring LS, Pikovsky A (2001) Noise-Induced Dynamics in Bistable Systems with Delay. Phys Rev Lett 87:250602
14. Masoller C (2002) Noise-Induced Resonance in Delayed Feedback Systems. Phys Rev Lett 88:034102
15. Masoller C (2003) distribution of residence times of time-delayed bistable systems driven by noise. Phys Rev Lett 90:020601
16. Xiao T (2016) Heat dissipation and information flow for delayed bistable Langevin systems near coherence resonance. Phys Rev E 94:052109
17. Shankar S, Marchetti MC (2018) Hidden entropy production and work fluctuations in an ideal active gas. Phys Rev E 98:020604(R)
18. Dabelow L, Bo S, Eichhorn R (2019) Irreversibility in active matter systems: fluctuation theorem and mutual information. Phys Rev X 9:021009
19. Caprini L, Marconi UMB, Puglisi A, Vulpiani A (2019) The entropy production of Ornstein-Uhlenbeck active particles: a path integral method for correlations. J Stat Mech Theor Exp 2019:053203
20. Mandal D, Klymko K, DeWeese MR (2017) Entropy production and fluctuation theorems for active matter. Phys Rev Lett 119:258001

21. Bonilla L (2019) Active Ornstein-Uhlenbeck particles. Phys Rev E 100:022601
22. Bo S, Del Giudice M, Celani A (2015) Thermodynamic limits to information harvesting by sensory systems. J Stat Mech Theor Exp 2015:P01014
23. Hartich D, Barato AC, Seifert U (2016) Sensory capacity: An information theoretical measure of the performance of a sensor. Phys Rev E 93:022116

Chapter 12
Outlook—Open Questions and Further Perspectives

We close with some comments putting the findings presented in this thesis into a broader perspective and giving an outlook on future research. We begin with formulating a few obvious follow-up questions and take a wider view towards the end.

Probability densities In this thesis, we have derived explicit equations for the second member of the Fokker-Planck hierarchy, and based thereon formulated a new approximation scheme, by relatively crude simplifications. This approximation is much simpler as compared to the multi-step procedure of the force-linearization closure, but still yields better results in the parameter regime where the particle motion is dominated by the interplay of nonlinearities, noise and memory. Thus, we have seen that an approximation scheme that operates on this second member of the Fokker-Planck hierarchy is, in some sense, superior as opposed to approaches based on the first one. This is due to the fact that the starting equation is already more sophisticated and, in particular, explicitly addresses the temporal correlations essential to capture non-Markovian behavior.

For this reason, we think it would be worthwhile to further explore the approximation proposed here or alternative approaches aiming for the two-time probability density, or other quantities that directly describe the temporal correlations. To this end, it might also be beneficial to inspect even higher members of the Fokker-Planck hierarchy.

Fluctuations An obvious follow-up question concerns the *fluctuations* of the thermodynamic quantities. In this thesis, we have mainly discussed the mean rates. However, another central topic in stochastic thermodynamics is the description of the fluctuations around those mean values. For Markovian systems, the probability distributions in non-equilibrium steady states are already quite well-understood, an example being the knowledge of several universal fluctuation theorems [1, 2]. Corresponding statements for general systems with memory have not been developed so far. Only for some specific non-Markovian processes, fluctuation theorems have been found in the past, e.g., [3–8], including linear Langevin equations with delay [9, 10]. However,

© The Author(s), under exclusive license to Springer Nature Switzerland AG 2021
S. A. M. Loos, *Stochastic Systems with Time Delay*, Springer Theses,
https://doi.org/10.1007/978-3-030-80771-9_12

we could not find general results applicable to the delay processes considered here, especially not for nonlinear systems. Our preliminary, numerical investigations have shown that the heat distributions are non-Gaussian, and violate fluctuation relations. Deriving analytical expressions for the probability distributions, and searching for universal symmetry properties of the heat fluctuations are open problems for future research.

Quantum effects Stepping away from the (classical) colloidal example, and reconsidering the stochastic delay processes and the meaning of non-reciprocal coupling, one notices some connections to quantum systems. Indeed, also in quantum systems, researchers have recently started to analyze, on the one hand, feedback with memory [11, 12], and, on the other hand, non-reciprocal interactions [13, 14].

In the present work, we have developed a unifying viewpoint on these two concepts, delayed feedback and non-reciprocity, for classical (stochastic) models. It would be very interesting to generalize this perspective to the quantum world, where entropy, information and energy play an analogous role, while measurement and stochasticity have different notions.

Uniqueness of the Markovian embedding In this thesis, we have employed a Markovian embedding to represent the time delayed process by means of a Markovian network, specifically, a unidirectionally coupled ring network, where all variables but X_0 follow linear equations of motion.

Focusing on such linear networks, we have shown that non-reciprocal interactions are an essential ingredient to generate non-monotonic memory kernels, and are thus essential to represent the memory kernels characteristic for delayed feedback. However, one important question that we have not yet addressed in this context, is the *uniqueness* of the Markovian (network) representation: Is it possible to find the exact same memory in another network of different size, coupling topology, or with another type of dynamics? This question is particularly relevant in situations, where only some of the degrees of freedom are *observable*, e.g., only X_0, while all others, $\{X_1, \ldots, X_n\}$, are "auxiliary" variables without physical interpretation.

We indeed think that the representation is *not* unique. In fact, exploring more complex network architectures (with nonlinear couplings, other coupling topologies, or nonlinear local dynamics) is an interesting question and shall be subject of future work. However, by considering linearly coupled, overdamped nodes with linear dynamics, we have, in fact, studied the most simple possibility.

In view of the approach to thermodynamics of delayed systems taken in this thesis, one should clarify whether and how the calculated entropic and information-theoretic quantities depend on the (probably ambiguous) representation. We aim to stress that we already know that the heat flow is independent (which follows from the fact that it can equivalently be derived from the non-Markovian Langevin equation alone). However, the entropic picture and information flows will probably not be.

We have also noted the possibility to build something like an effective thermodynamics upon our findings. For example, one may consider the total entropy production $\dot{S}_{\text{tot}}(X_0, X_1, \ldots, X_n)$ of the "super-system" consisting of all degrees of freedom, $\{X_0, X_1, \ldots, X_n\}$, and then in a second step trace out all "auxiliary" degrees of freedom $\{X_1, \ldots, X_n\}$. It will be particularly interesting to see whether the resulting effective "irreversibiltiy measure" will be independent on the chosen Markovian representation, or not.

A conceptionally interesting aspect regarding the way of modeling memory employed in this work, is that it can be considered as a "discretization of the past" [15]. This is a delicate point, as the dynamics itself is time-*continuous*. As opposed to that, a discretization of the entire dynamics was proposed in [16]. Surprisingly, the approach taken here yields a simpler thermodynamic framework, despite the apparent contradiction of its discrete–continuous nature.

Further physical systems with memory and activity Finally, we aim to discuss the Markovian embedding in view of possible further applications. So far, we have seen that a unidirectionally coupled ring network, where all auxiliary variables follow linear equations of motion, automatically generates a Gamma-distributed kernel and monotonically decaying colored noise (recall Fig. 6.2 from the Part II). Indeed, besides yielding discrete (delta-distributed) delay in the limit of an infinitely large network, such Gamma-distributed delays themselves have many applications. For example, in biological contexts they appear in gene regulatory networks [17–20], or in the dynamics of virus spreading between cells [21]. Furthermore, by combining multiple Gamma-distributed kernels, one can obtain models for a very general type of memory. From the perspective of the Markovian network, this amounts to generalizing the coupling topology. As a concrete example, we have depicted in Fig. 12.1 the (bimodal) kernel and noise correlations arising from a linear, unidirectional ring network and an additional, separately coupled degree of freedom. As a matter of fact, any kernel that eventually decays exponentially can be approximated in this way. We find that a memory kernel with N extrema can be represented by N non-reciprocally

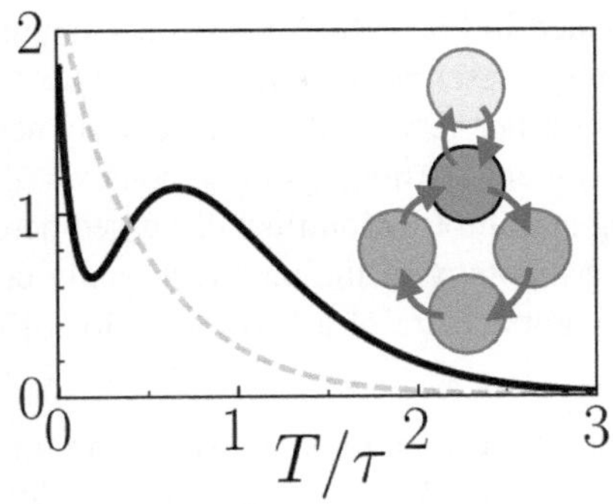

Fig. 12.1 Sketch of a Markovian network of a colloidal particle (blue disk) coupled to four (auxiliary) degrees of freedom (green disks). Due to the coupling, the corresponding non-Markovian, one-variable Langevin equation for the colloid features a (bimodal) memory kernel and a colored noise with monotonically decaying correlations (gray line). Also compare with Fig. 6.2, to see the corresponding kernels and noise correlations for simpler coupling topologies (specifically, for ring networks with 2, 3, 4 nodes)

coupled subsystems (while reciprocally coupled degrees of freedom always yield monotonically decaying memory kernels).

This further implies that more general types of stochastic processes with memory can, in principle, be treated with this approach. For example, one may think of other sources of memory like viscoelasticity of the surrounding fluid, and hydrodynamic backflow [22]. While there exist already sophisticated theories to treat these two kinds of memory (e.g., the Basset force for hydrodynamic backflow [23, 24]), the network representation employed here yields a simplistic, effective description with relatively few variables. Furthermore, the established theories usually describe systems that can (in the absence of external time-dependent drivings) reach thermal equilibrium. However, in various real-world scenarios, the heat bath does not only have memory, but at the same time is an active environment. Think for example of biocellular processes, where motor proteins drive actin filaments [25], creating a highly active and complex environment with memory effects. Having all this in mind, it appears to be an interesting and fruitful direction of future research to investigate whether a low-dimensional Markovian network with non-reciprocal couplings can be used as a simplistic model for the motion of a colloid in such a complex, active surrounding that is far from thermal equilibrium. We have undertaken first steps in this direction in [26], where we have shown that a whole zoo of different types of correlation functions can be generated by a linear (intrinsically nonequilibrium) model with only two auxiliary variables.

Computer science The Markovian network representation and the information-thermodynamic analysis from Part III might further be interesting from the viewpoint of computer science. On an abstract level, our thermodynamic investigation might be useful to incorporate the thermodynamic cost of memory in the theoretical description of computation in living [27] and artificial systems [28]. A more concrete link is the similarity between the ring network architecture studied here and the reservoir computers investigated in [29, 30]. A reservoir computer of this type may be experimentally realized by a laser network [31, 32], or by coupled RC circuits [33, 34]. The ring network is further very similar to a recurrent neural network [35], used for example for reinforcement learning. In these contexts, the connection between non-reciprocal coupling and information flow discussed here might be particularly interesting. Notably, the topology of the unidirectionally coupled ring considered here also resembles the architecture of a Brownian clock [36], which, in contrast, has discrete dynamics.

Collective behavior Lastly, we want to come back to an initial thought formulated in the introduction of this thesis. We have considered (colloidal) systems subject to time-delayed feedback. We have seen that these particles display complex dynamical behavior already on the level of individual particles. Therefore, one shall expect intriguing collective effects in a collection of multiple systems with delay (or delayed interactions). In this thesis, we have pointed out fundamental connections between the thermodynamic properties of individual active and feedback-controlled systems with memory. Given these connections, one should also expect common aspects of their collective behaviors. For active systems, particularly microswimmer suspensions [37–40], clustering and pattern formation are well known phenomena. Indeed, recent studies report similar phenomena in collections of feedback-controlled systems [41–44].

Reconsidering this from a broader perspective, we have argued that within the framework of modern statistical physics, single colloids under delayed feedback can be considered as a prototype of a "novel type of particle". Representing important phenomena of the complex world, which are, so far, not incorporated in the framework of statistical physics, like feedback, intrinsic energy and information flows, they are suitable building blocks for the construction of new, generalized theories. Understanding the (thermo-)dynamic properties of individual constituents is a first step in this direction.

References

1. Sekimoto K (2010) Stochastic energetics 799
2. Seifert U (2012) Stochastic thermodynamics, fluctuation theorems and molecular machines. Rep Prog Phys 75:126001
3. Roche B et al (2015) Harvesting dissipated energy with a mesoscopic ratchet. Nat Commun 6:6738
4. Whitney RS (2018) Non-Markovian quantum thermodynamics: laws and fluctuation theorems. Phys Rev B 98:085415
5. Mai T, Dhar A (2007) Nonequilibrium work fluctuations for oscillators in non-Markovian baths. Phys Rev E 75:061101
6. Schmidt R, Carusela MF, Pekola JP, Suomela S, Ankerhold J (2015) Work and heat for two-level systems in dissipative environments: strong driving and non-Markovian dynamics. Phys Rev B 91:224303
7. Vaikuntanathan S, Jarzynski C (2009) Dissipation and lag in irreversible processes. EPL 87:60005
8. Speck T, Seifert U (2007) The Jarzynski relation, fluctuation theorems, and stochastic thermodynamics for non-Markovian processes. J Stat Mech Theory Exp 2007:L09002
9. Rosinberg ML, Munakata T, Tarjus G (2015) Stochastic thermodynamics of Langevin systems under time-delayed feedback control: second-law-like inequalities. Phys Rev E 91:042114
10. Hasegawa H (2011) Classical open systems with nonlinear nonlocal dissipation and state-dependent diffusion: Dynamical responses and the Jarzynski equality. Phys. Rev. E 84:051124
11. Strasberg P, Schaller G, Brandes T, Esposito M (2013) Thermodynamics of quantum-jump-conditioned feedback control. Phys Rev E 88:062107
12. Carmele A, Kabuss J, Schulze F, Reitzenstein S, Knorr A (2013) Single photon delayed feedback: a way to stabilize intrinsic quantum cavity electrodynamics. Phys Rev Lett 110:013601
13. Fang K et al (2017) Generalized non-reciprocity in an optomechanical circuit via synthetic magnetism and reservoir engineering. Nat Phys 13:465
14. Metelmann A, Clerk AA (2015) Nonreciprocal photon transmission and amplification via reservoir engineering. Phys Rev X 5:021025
15. Loos SAM, Klapp SHL (2019) Fokker-Planck equations for time-delayed systems via Markovian embedding. J Stat Phys 177:95–118
16. Kwon C, Um J, Park H (2017) Information thermodynamics for a multi-feedback process with time delay. EPL 117:10011
17. Lai X, Wolkenhauer O, Vera J (2016) Understanding microrna-mediated gene regulatory networks through mathematical modelling. Nucl Acids Res 44:6019–6035

18. Josić K, López JM, Ott W, Shiau L, Bennett MR (2011) Stochastic delay accelerates signaling in gene networks. PLoS Comput Biol 7:e1002264
19. Rateitschak K, Wolkenhauer O (2007) Intracellular delay limits cyclic changes in gene expression. Math Biosci 205:163–179
20. Friedman N, Cai L, Xie XS (2006) Linking stochastic dynamics to population distribution: an analytical framework of gene expression. Phys Rev Lett 97:168302
21. Mittler JE, Sulzer B, Neumann AU, Perelson AS (1998) Influence of delayed viral production on viral dynamics in HIV-1 infected patients. Math Biosci 152:143–163
22. Franosch T et al (2011) Resonances arising from hydrodynamic memory in Brownian motion. Nature 478:85
23. Johnson RW (2016) Handbook of fluid dynamics. Crc Press
24. Basset AB (1888) A treatise on hydrodynamics: with numerous examples, vol 2. Deighton, Bell and Company
25. Guo M et al (2014) Probing the stochastic, motor-driven properties of the cytoplasm using force spectrum microscopy. Cell 158:822–832
26. Doerries TJ, Loos SA, Klapp SH (2021) Correlation functions of non-Markovian systems out of equilibrium: analytical expressions beyond single-exponential memory. J Stat Mech (in press)
27. Kempes CP, Wolpert D, Cohen Z, Pérez-Mercader J (2017) The thermodynamic efficiency of computations made in cells across the range of life. Philos Trans Royal Soc A 375:20160343
28. Wolpert DH (1992) Memory systems, computation, and the second law of thermodynamics. Int J Theor Phys 31:743–785
29. Larger L et al (2017) High-speed photonic reservoir computing using a time-delay-based architecture: million words per second classification. Phys Rev X 7:011015
30. Li J, Bai K, Liu L, Yi Y (2018) A deep learning based approach for analog hardware implementation of delayed feedback reservoir computing system. In: 2018 19th international symposium on quality electronic design (ISQED). IEEE, pp 308–313
31. Röhm A, Jaurigue L, Lüdge K (2019) Reservoir computing using laser networks. IEEE J Sel Top Quant 26:1–8
32. Röhm A, Lüdge K (2018) Multiplexed networks: reservoir computing with virtual and real nodes. J Phys Commun 2:085007
33. Kish LB, Granqvist C-G (2012) Electrical Maxwell demon and Szilard engine utilizing Johnson noise, measurement, logic and control. PloS One 7:e46800
34. Snider GL et al (2011) Minimum energy for computation, theory versus experiment. In: 2011 11th IEEE international conference on nanotechnology. IEEE, pp 478–481
35. Xu X, Lu Y, Liang Y (2004) Time-delay recurrent neural networks for dynamic systems control. Springer, Berlin
36. Barato AC, Seifert U (2016) Cost and precision of Brownian clocks. Phys Rev X 6:041053
37. Ramaswamy S (2010) The mechanics and statistics of active matter. Annu Rev Condens Matter Phys 1:323–345
38. Ramaswamy S (2017) Active matter. J Stat Mech Theor Exp 2017:054002
39. Nagai KH, Sumino Y, Montagne R, Aranson IS, Chaté H (2015) Collective motion of self-propelled particles with memory. Phys Rev Lett 114:168001
40. Kaiser A, Snezhko A, Aranson IS (2017) Flocking ferromagnetic colloids. Sci Adv 3:e1601469
41. Tarama S, Egelhaaf SU, Löwen H (2019) Traveling band formation in feedback-driven colloids. Phys Rev E 100:022609
42. Mijalkov M, McDaniel A, Wehr J, Volpe G (2016) Engineering sensorial delay to control phototaxis and emergent collective behaviors. Phys Rev X 6:011008
43. Khadka U, Holubec V, Yang H, Cichos F (2018) Active particles bound by information flows. Nat Commun 9:3864
44. Geiss D, Kroy K, Holubec V (2019) Brownian molecules formed by delayed harmonic interactions. New J Phys 21:093014

Appendix

A.1 Numerical Methods

In this appendix, we lay out all relevant information for the specific simulations used to generate the results presented in this thesis. We refer the Reader who is interested in detailed descriptions and background information about simulation technique for stochastic systems to [1], where extensive discussion can be found.

In order to test our theoretical results against quasi-exact data, we perform Brownian dynamics (BD) simulations of the delayed LE within C++. We use the Euler–Maruyama integration scheme [1, 2], with a varying temporal discretization (see below). We perform each simulation multiple times (with different random number seed), and take the average over at least $N = 10^4$ realizations (usually more, as occasionally stated in the main text). The (pseudo) random numbers are generated with the algorithm "Mersenne Twister" [3] and the Box–Müller method [1].

Density profiles from the BD simulations are obtained from histograms with a spatial resolution of $\Delta x = 0.005\,\sigma$. In general, steady-state ensemble averages are obtained after appropriate transient times have been cut off. That means, before we measure, we let at least $100\,\tau_{\rm B}$ times and at least 20τ (whichever is higher) pass, in the presence of the delay force, such that the system has reached a steady state. The simulation time thereafter is more than $200\,\tau_{\rm B}$. Furthermore, steady-state ensemble averages are obtained by using a sufficiently high number of realizations N. Both is checked by insensitivity against further increase of simulated time or sample size.

We have tested different initial conditions, including equidistantly distributed, as well as delta-distributed ones, both with fixed position within the interval $[-\tau, 0]$. We found that an equilibrium configuration in the respective static potential without delay force often yields the fastest relaxation towards the steady state.

The temporal discretization is varied between $\Delta t = 10^{-6}$ and 10^{-4}, such that the typical time scales are all properly resolved. In particular, the delay time $\tau > 1000\Delta t$ and the intrawell relaxation time $\tau_{\rm ir} > 1000\Delta t$. Indeed, calculating properties of a system with very short delay times is therefore very expensive, as the step size

© The Editor(s) (if applicable) and The Author(s), under exclusive license to Springer Nature Switzerland AG 2021

S. A. M. Loos, *Stochastic Systems with Time Delay*, Springer Theses,
https://doi.org/10.1007/978-3-030-80771-9

must be extremely small. For example, in Fig. 9.3, where $\tau < 10^{-2}\tau_B$, we have used $\Delta t = 10^{-6}$.

For the confining potentials (quadratic trap, doublewell potential), we did not have to pay special attention to the boundary conditions, as the probability to find the particle quickly vanishes for $|x| \gg 0$. We have just made sure that no particle may escape by stopping the simulations in case a position far away from zero occurs. On the contrary, in the case of the washboard potential, the particle is not confined. We have implemented a comoving simulation box. Note that whenever moving the box, the delayed particle position must be transformed accordingly.

A.2 Derivation of Novikov's Theorem

In this appendix, we sketch a derivation of the Novikov's theorem (5.3), which reads

$$\big\langle \Lambda[\xi]\xi(t)\big\rangle = \left\langle \frac{\delta\Lambda[\xi]}{\delta\xi(t)}\right\rangle, \tag{A.1}$$

for a functional Λ of a *Gaussian white noise* ξ [4]. It links the functional derivative w.r.t. the noise to the cross-correlation between functional and noise. First, we consider the case where the ensemble is w.r.t. a *fixed* initial condition $\phi = X(-\tau \leq t \leq t)$, yielding $\Lambda(0) = \Lambda_0$. In a second step we will generalize towards initial conditions drawn from an arbitrary distribution.

We express the ensemble average $\langle .. \rangle_{\Lambda_0}$ of the l.h.s. of (5.3) via the path integral over all possible paths ξ between the times 0 and t, accounting for all possible realizations of the random process ξ at each instant in time

$$\langle\Lambda[\xi]\xi(t)\rangle_{\Lambda_0} = \int_{\xi_0}^{\xi_t} \xi\ \Lambda\left[\xi|\Lambda_0\right]\ \mathcal{P}[\xi]\,\mathcal{D}[\xi], \tag{A.2}$$

with arbitrary but fixed $\xi_0 := \xi(t_0)$ and $\xi_t := \xi(t)$. Please note that specifying the noise process at the boundaries does *not* impose a restriction on the generality, as we deal with *white* noise. We recall that the weight $\mathcal{P}[\xi]$ of each white noise realization is given by the Gaussian path probability (for arbitrary $t > 0$)

$$\mathcal{P}[\xi] = \mathcal{J}e^{-\frac{1}{2}\int_0^t \xi(t')^2 \mathrm{d}t'}, \tag{A.3}$$

with Jacobian $\mathcal{J}$ [as already given in Eq. (2.4) in the main text, note that we there use bold symbols for paths, $\boldsymbol{\xi}$, different than here]. Now, we rewrite the integrand in (A.2) using that the functional derivative of $\mathcal{P}[\xi]$ w.r.t. ξ is simply $(-\xi)\mathcal{P}[\xi]$, yielding

$$\begin{aligned}\langle \Lambda[\xi]\xi(t)\rangle_{\Lambda_0} &= -\int_{\xi_0}^{\xi_t} \Lambda\left[\xi|\Lambda_0\right] \frac{\delta}{\delta\xi}\left\{e^{-\frac{1}{2}\int_0^t \xi(t')^2 \mathrm{d}t'}\right\} \mathcal{D}[\xi] \\ &= -\int_{\xi_0}^{\xi_t} \frac{\delta\left\{\Lambda\left[\xi|\Lambda_0\right]\mathcal{P}[\xi]\right\}}{\delta\xi}\mathcal{D}[\xi] + \int_{\xi_0}^{\xi_t} \frac{\delta\Lambda[\xi]}{\delta\xi}\mathcal{P}[\xi]\,\mathcal{D}[\xi]. \quad \text{(A.4)}\end{aligned}$$

In the last step we have simply used the product rule. The functional derivative in the first path integral yields

$$\int_{\xi_0}^{\xi_t} \frac{\delta\left\{\Lambda\left[\xi|\Lambda_0\right]\mathcal{P}[\xi]\right\}}{\delta\xi}\mathcal{D}[\xi] = \int_{\xi_0}^{\xi_t} \Lambda\left[\xi+\delta\xi|\Lambda_0\right]\mathcal{P}[\xi+\delta\xi]\mathcal{D}[\xi] - \int_{\xi_0}^{\xi_t} \Lambda\left[\xi|\Lambda_0\right]\mathcal{P}[\xi]\mathcal{D}[\xi]. \quad \text{(A.5)}$$

Now we use that the variations of the paths $\xi + \delta\xi$ are already contained in the integral over *all* paths, and find

$$\int_{\xi_0}^{\xi_t} \frac{\delta\left\{\Lambda\left[\xi|\Lambda_0\right]\mathcal{P}[\xi]\right\}}{\delta\xi}\mathcal{D}[\xi] = \int_{\xi_0}^{\xi_t} \Lambda\left[\xi|\Lambda_0\right]\mathcal{P}[\xi]\mathcal{D}[\xi] - \int_{\xi_0}^{\xi_t} \Lambda\left[\xi|\Lambda_0\right]\mathcal{P}[\xi]\mathcal{D}[\xi] = 0, \quad \text{(A.6)}$$

(as long as $\int \Lambda[\xi]\mathcal{P}[\xi]\mathcal{D}[\xi] < \infty$), see also [5] (on p. 273f). Hence, we obtain from (A.4),

$$\langle \Lambda[\xi]\xi(t)\rangle_{\Lambda_0} = \int_{\xi_0}^{\xi_t} \frac{\delta\Lambda\left[\xi|\Lambda_0\right]}{\delta\xi}\mathcal{P}[\xi]\,\mathcal{D}[\xi] = \left\langle \frac{\delta\Lambda[\xi]}{\delta\xi}\right\rangle_{\Lambda_0}, \quad \text{(A.7)}$$

for a specific initial condition Λ_0.

This result can readily be generalized to ensembles where the initial conditions are instead drawn from an arbitrary, normalized distribution $P(\Lambda_0)$. In particular, Eq. (A.7) implies

$$\langle \Lambda[\xi]\xi(t)\rangle = \int \langle \Lambda[\xi]\xi(t)\rangle_{\Lambda_0} P(\Lambda_0)\mathrm{d}\Lambda_0 = \int \left\langle \frac{\delta\Lambda[\xi]}{\delta\xi}\right\rangle_{\Lambda_0} P(\Lambda_0)\mathrm{d}\Lambda_0 = \left\langle \frac{\delta\Lambda[\xi]}{\delta\xi}\right\rangle. \quad \text{(A.8)}$$

This is the relation (5.3), which is often referred to as *Novikov's theorem.*

A.3 Green's Function Method

In Sect. 5.2.2, we have derived a closed-form expression of the autocorrelation function using the method of steps. An alternative strategy to obtain analytical expressions for correlation functions is via the the Green's function method. To this end, one transforms the LE by Laplace-Transformation $\tilde{X}_j(\omega) = \int_0^\infty X_j(t)e^{-i\omega t}\mathrm{d}t$, which readily yields

$$\omega\gamma\tilde{X}(\omega) = -\alpha\tilde{X}(\omega) - \beta e^{-i\tau\omega}\tilde{X}(\omega) + \sqrt{2D_0}\tilde{\xi}(\omega) \quad \text{(A.9)}$$

$$\Rightarrow \tilde{X}(\omega) = \underbrace{\left(i\omega\gamma + \alpha + \beta e^{-i\omega\tau}\right)^{-1}}_{=\tilde{\lambda}(\omega)} \sqrt{2D_0}\tilde{\xi}(\omega). \tag{A.10}$$

with the Green's function in Fourier-space $\tilde{\lambda}(\omega)$. As we are mainly interested in steady states, we have set the entire history function to zero, to facilitate the calculation. In Ref. [6] it is shown how this calculation in principle can be generalized to arbitrary initial conditions. If initial conditions matter, it is more handy to use the Laplace-Transformation instead. Transforming back to real space this yields

$$X(t) = \int_0^t \lambda(t-t')\sqrt{2D_0}\xi(t')\mathrm{d}t', \tag{A.11}$$

describing $X(t)$ as a functional of the noise process $\boldsymbol{\xi} = \{\xi(0 \leq t' \leq t)\}$. In this form, the Green's function λ tells, how the system state X responds to the external perturbation, and is hence also called a (linear) response function.

From Eq. (A.11) one can deduce a well-known relationship between position-autocorrelation and Green's function. Due to its importance, and because we will later apply it (in Chap. 10), let us briefly show how one can derive it [7]. Consider $\Delta t \geq 0$,

$$\begin{aligned} C(\Delta t) = \langle X(t)X(t+\Delta t)\rangle &= (2D_0)\int_0^t\int_0^{t+\Delta t} \lambda(t-t')\lambda(t+\delta t-t'')\langle\xi(t')\xi(t'')\rangle\,\mathrm{d}t'\mathrm{d}t'' \\ &= \int_0^t \lambda(t-t')\lambda(t+\Delta t-t')\,\mathrm{d}t' = 2D_0\int_0^t \lambda(T)\lambda(T+\Delta t)\,\mathrm{d}T \\ &\overset{\text{ss}}{=} 2D_0\int_0^\infty \lambda(T)\lambda(T+\Delta t)\,\mathrm{d}T. \end{aligned} \tag{A.12}$$

In the last step we have specified to steady states. Now we use $\lambda(t<0)=0$, and again transform to Fourier-space

$$\begin{aligned} \tilde{C}(\omega) &= 2D_0\int_{-\infty}^{\infty}\int_{-\infty}^{\infty} \lambda(T)\lambda(T+\Delta t)e^{-i\omega\Delta t}\,\mathrm{d}T\,\mathrm{d}\Delta t \\ &= 2D_0\int_{-\infty}^{\infty}\int_{-\infty}^{\infty} \lambda(T)e^{i\omega T}\lambda(T+\Delta t)e^{-i\omega(\Delta t+T)}\,\mathrm{d}T\,\mathrm{d}\Delta t \\ &= 2D_0\int_{-\infty}^{\infty}\int_{-\infty}^{\infty} \lambda(T)e^{i\omega T}\lambda(T')e^{-i\omega T'}\,\mathrm{d}T\,\mathrm{d}T' = 2D_0\tilde{\lambda}(\omega)\tilde{\lambda}(-\omega) \end{aligned} \tag{A.13}$$

Transforming back to real space, this implies

$$C(\Delta t) = \frac{D_0}{\pi}\int_{-\infty}^{\infty} \tilde{\lambda}(\omega)\tilde{\lambda}(-\omega)e^{-i\omega\Delta t}\,\mathrm{d}\omega. \tag{A.14}$$

While this strategy in principle yields analytical expressions for various (linear) systems (which can, e.g., be numerically integrated), explicit closed-form solutions are only available for specific cases, where the inverse Fourier-Transformation of the Green's function (A.10) is known, as well as the integral in Eq. (A.14) [see [6, 8] for some explicit results]. We could not find general closed-from solutions for the problem considered here.

A.4 Connection to Fokker–Planck Hierarchy from Novikov's Theorem

In this appendix, we consider the two representations of the second members of the Fokker-Planck hierarchy derived and discussed in Chap. 6. We recall that, while the FPE for ρ_1 obtained by our approach is equivalent to the first member of the FP hierarchy from Novikov's theorem (5.3), the second members differ. However, the apparent disagreement can be resolved. Indeed, as we show here, our result (6.44) can be transformed to the representation obtained from Novikov's theorem [given in (5.7)]. We focus on the term which differs, i.e., the drift term $\partial_{x_\tau}[\rho_2(x,t;x_\tau,t-\tau)\langle\dot{X}(t-\tau)|x=X(t),x_\tau=X(t-\tau)\rangle]$. To this end, we first use the definition of ρ_2 via delta-distributions, plug in the LE (2.43), which yields

$$\begin{aligned}
&\partial_{x_\tau}\Big[\langle\dot{X}(t-\tau)|x=X(t),x_\tau=X(t-\tau)\rangle\rho_2(x,t;x_\tau,t-\tau)\Big]\\
&=\partial_{x_\tau}\big\langle\dot{X}(t-\tau)\,\delta[X(t)-x]\delta[X(t-\tau)-x_\tau]\big\rangle\\
&\overset{\text{LE}}{=}\partial_{x_\tau}\langle F[X(t-\tau),X(t-2\tau)]\,\delta[X(t)-x]\delta[X(t-\tau)-x_\tau]\rangle\\
&\quad+\partial_{x_\tau}\Big\langle\sqrt{2D_0}\xi(t-\tau)\,\delta[X(t)-x]\delta[X(t-\tau)-x_\tau]\Big\rangle. \qquad\text{(A.15)}
\end{aligned}$$

At this point, we have two correlations to deal with. The first one (involving F) can easily be simplified by expressing the ensemble average with the help of the PDF ρ_3, i. e.,

$$\begin{aligned}
&\partial_{x_\tau}\Big\langle F\left[X(t-\tau),X(t-2\tau)\right]\delta[X(t)-x]\delta[X(t-\tau)-x_\tau]\Big\rangle\\
&\qquad=\partial_{x_\tau}\iiint_\Omega F(x_\tau,x_{2\tau})\,\delta[X(t)-x]\delta[X(t-\tau)-x_\tau]\times\\
&\qquad\quad\rho_3(x,t;x_\tau,t-\tau;x_{2\tau},t-2\tau)\mathrm{d}x\mathrm{d}x_\tau\mathrm{d}x_{2\tau}\\
&\qquad=\partial_{x_\tau}\int_\Omega F(x_\tau,x_{2\tau})\rho_3(x,t;x_\tau,t-\tau;x_{2\tau},t-2\tau)\mathrm{d}x_{2\tau}.
\end{aligned} \qquad\text{(A.16)}$$

The remaining term on the right side of (A.15) (involving ξ) is simplified with the help of Novikov's theorem (5.3) (see Appendix A.2). Specifically, we define $\Lambda[\xi]:=\delta[x-X(t)]\,\delta[x_\tau-X(t-\tau)]$, which yields

$$
\begin{aligned}
&\partial_{x_\tau}\left\langle\sqrt{2D_0}\xi(t-\tau)\,\delta[X(t)-x]\delta[X(t-\tau)-x_\tau]\right\rangle\\
&\qquad= \sqrt{2D_0}\partial_{x_\tau}\left\langle\xi(t-\tau)\Lambda[\xi]\right\rangle,\\
&\qquad= \sqrt{2D_0}\partial_{x_\tau}\left\langle\frac{\delta\Lambda[\xi]}{\delta\xi(t-\tau)}\right\rangle\\
&\qquad= \sqrt{2D_0}\partial_{x_\tau}\left[\left\langle\frac{\delta\Lambda[\xi]}{\delta X(t)}\frac{\delta X(t)}{\delta\xi(t-\tau)}\right\rangle+\left\langle\frac{\delta\Lambda[\xi]}{\delta X(t-\tau)}\frac{\delta X(t-\tau)}{\delta\xi(t-\tau)}\right\rangle\right]\\
&\qquad= -\sqrt{2D_0}\partial_{x_\tau}\left\langle\partial_x\delta[x-X(t)]\,\delta[x_\tau-X(t-\tau)]\frac{\delta X(t)}{\delta\xi(t-\tau)}\right\rangle\\
&\qquad\quad-\sqrt{2D_0}\partial_{x_\tau}\left\langle\partial_{x_\tau}\delta[x-X(t)]\,\delta[x_\tau-X(t-\tau)]\underbrace{\frac{\delta X(t-\tau)}{\delta\xi(t-\tau)}}_{=\sqrt{\frac{D_0}{2}}}\right\rangle.
\end{aligned}
\tag{A.17}
$$

In the last step we have plugged in the definition of Λ, employed the chain rule, and used Eq. (5.4) (derived in the main text, Chap. 5). We note that the functional derivative of the trajectory X w.r.t. the *earlier* noise ξ which also appears in (A.17) has no such general solution, as discussed in Chap. 5. Based on these considerations, we can rewrite (A.17) as

$$
\begin{aligned}
\sqrt{2D_0}\,\partial_{x_\tau}\left\langle\xi(t-\tau)\Lambda[\xi]\right\rangle = -\sqrt{2D_0}\,\partial_{x_\tau}\partial_x\left\langle\frac{\delta X(t)}{\delta\xi(t-\tau)}\bigg|_{\substack{X(t)=x\\X(t-\tau)=x_\tau}}\right\rangle\rho_2(x,t;x_\tau,t-\tau)\\
-D_0\,\partial_{x_\tau}^2\rho_2(x,t;x_\tau,t-\tau).
\end{aligned}
\tag{A.18}
$$

In combination with (A.15), (A.16), we obtain the identity

$$
\begin{aligned}
&\partial_{x_\tau}\left\langle\dot{X}(t-\tau)\,\delta[X(t)-x]\delta[X(t-\tau)-x_\tau]\right\rangle\\
&\quad= -\sqrt{2D_0}\partial_{x_\tau}\partial_x\left\langle\frac{\delta X(t)}{\delta\xi(t-\tau)}\bigg|_{\substack{X(t)=x\\X(t-\tau)=x_\tau}}\right\rangle\rho_2(x,t;x_\tau,t-\tau)\\
&\quad\quad-D_0\partial_{x_\tau}^2\rho_2(x,t;x_\tau,t-\tau)+\partial_{x_\tau}\int_\Omega F(x_\tau,x_{2\tau})\rho_3(x,t;x_\tau,t-\tau;x_{2\tau},t-2\tau)\mathrm{d}x_{2\tau}\\
&\quad= \partial_{x_\tau}\left[\langle\dot{X}(t-\tau)|x=X(t),x_\tau=X(t-\tau)\rangle\rho_2(x,t;x_\tau,t-\tau)\right],
\end{aligned}
\tag{A.19}
$$

proving that the second member of the here presented FP hierarchy, Eq. (6.44), is identical to the corresponding one (5.7) obtained from Novikov's theorem.

A.5 Fluctuation-Dissipation Relation for Unidirectional Ring of Arbitrary Length *n*

In this Appendix, we generally consider the existence of the FDR for a network of type (10.11) with arbitrary length n. The FDR is also discussed in the main text in Sects. 2.3.1 and 10.3.1. We need to show the equivalence of the following two terms

$$k_B \mathcal{T}_0 \gamma(\Delta t) = k_B \mathcal{T}_0 \left[2\gamma_0\, \delta(\Delta t) + \frac{k\Gamma\left(n, \frac{n\Delta t}{\tau}\right)}{(n-1)!} \right] \tag{A.20}$$

and

$$C_\mu(\Delta t) = k_B \mathcal{T}_0\, 2\gamma_0 \delta(\Delta t) + \frac{k_B k^2}{\gamma'} \sum_{p=0}^{n-1} \sum_{l=0}^{p} \mathcal{T}' \frac{2^{l-2p}(2p-l)!}{p!l!(p-l)!} \frac{e^{-n\Delta t/\tau}\, \Delta t^l}{(\tau/n)^{l-1}}. \tag{A.21}$$

The first summands on the right hand side of (A.20) and (A.21) are already equivalent. Further, the other term in (A.20) an be rewritten by partially integrating the incomplete Gamma function as

$$\Gamma(n, n\Delta t/\tau) = (n-1)! \sum_{p=0}^{n-1} \left(\frac{n\Delta t}{\tau} \right)^p \frac{e^{-n\Delta t/\tau}}{p!}. \tag{A.22}$$

Thus, the FDR holds, if at all $\Delta t \geq 0$

$$\sum_{p=0}^{n-1} \sum_{l=0}^{p} \frac{\mathcal{T}'}{\mathcal{T}_0} \left[\frac{k\, 2^l (2p-l)!}{n\, 2^{2p} l!(p-l)!} \right] \frac{\tau (n\Delta t)^l e^{-n\Delta t/\tau}}{\gamma' \tau^l p!} \stackrel{?}{=} \sum_{p=0}^{n-1} \frac{(n\Delta t)^p e^{-n\Delta t/\tau}}{\tau^p p!}. \tag{A.23}$$

At $\Delta t = 0$ Eq. (A.23) is fulfilled, if and only if

$$\mathcal{O}(\Delta t^0): \quad \frac{\gamma'}{\tau} \stackrel{?}{=} \sum_{p=0}^{n-1} \frac{\mathcal{T}'}{\mathcal{T}_0} \frac{k(2p)!}{n\, 2^{2p}(p!)^2}. \tag{A.24}$$

This is the zeroth-order condition (in Δt). For $n = 1$, Eq. (A.24) reads $k\tau\mathcal{T}' = \gamma'\mathcal{T}_0$ [in agreement with (10.16)], and is a *sufficient* condition. It can only be fulfilled for $k \geq 0$. Further, for $n > 1$, it yields a *necessary*, but not sufficient condition due to higher order contributions in Eq. (A.23). For example, the first order in Δt, which is relevant for all $n > 1$, yields the additional condition

$$\mathcal{O}(\Delta t^1): \quad \frac{\gamma'}{\tau}(n\Delta t) \stackrel{?}{=} \frac{\mathcal{T}'}{\mathcal{T}_0} \frac{k}{2n}(n\Delta t). \tag{A.25}$$

Combining this with (A.24) gives

$$\sum_{p=0}^{n-1} \frac{\mathcal{T}'(n\Delta t)}{\mathcal{T}_0} \frac{2(2p)!}{2^{2p}(p!)^2} \stackrel{?}{=} \frac{\mathcal{T}'(n\Delta t)}{\mathcal{T}_0}, \tag{A.26}$$

which can only be fulfilled if $\mathcal{T}' = 0$. However, at $\mathcal{T}' = 0$ condition (A.24) is violated. Hence, the FDR is broken for all choices $\mathcal{T}'$ when $n > 1$.

We have, in fact, also investigated a generalization of (10.11), where all temperatures are allowed to be different, i.e., X_j has temperature $\mathcal{T}_j$ (generalizing the derivations presented in Sect. 6.1.1 to this case is straightforward). Even in this case, the FDR cannot be fulfilled if $n > 1$.

References

1. Kloeden P, Platen E (1992) Numerical solution of stochastic differential equations. Springer, Berlin
2. Buckwar E (2000) Introduction to the numerical analysis of stochastic delay differential equations. J Comput Appl Math 125:297–307
3. (2020). Weblink: www.math.sci.hiroshima-u.ac.jp/~m-mat/MT/emt.html
4. Novikov EA (1965) Functionals and the random-force method in turbulence theory. Sov Phys JETP 20:1290–1294
5. Hall M, Reginatto M (2016) Ensembles on configuration space. Fundam Theor Phys 184
6. Geiss D, Kroy K, Holubec V (2019) Brownian molecules formed by delayed harmonic interactions. New J Phys 21:093014
7. Hänggi P, Thomas H (1982) Stochastic processes: time evolution, symmetries and linear response. Phys Rep 88:207–319
8. Munakata T, Iwama S, Kimizuka M (2009) Linear stochastic system with delay: energy balance and entropy production. Phys Rev E 79:031104

About the Author

Loos studied physics at Saarland University and Technical University of Berlin (TUB), and spent one semester at Duke University in North Carolina, USA. During her studies, she was granted a scholarship from the Studienstiftung des Deutschen Volkes, and further received a scholarship from TUB to study abroad. She wrote her doctoral thesis at TUB and is now a Postdoc at Leipzig University. Loos received several awards, including the Carl-Ramsauer Preis by the German Research Society for her dissertation, a prize by the German Research Society for her Master thesis, and a poster prize by Europhysics Letters.

Loos worked on numerous research projects in the fields of nonlinear dynamics and statistical physics, with a focus on non-Markovian systems, time-delayed feedback control, and stochastic thermodynamics.

Besides research activities, Loos engages in science communication and public outreach. She has organized a session of the Conference of Women in Physics 2019 in Berlin, and organized several scientific symposia and seminars at TUB and Leipzig University. She further participates in events for mentoring of high school students, and was a speaker at the Soapbox science event in 2019, a public outreach platform to increase the visibility of female scientists in society.

© The Editor(s) (if applicable) and The Author(s), under exclusive license to Springer Nature Switzerland AG 2021

S. A. M. Loos, *Stochastic Systems with Time Delay*, Springer Theses,
https://doi.org/10.1007/978-3-030-80771-9

About the Author